Contemporary Topics in Computer Graphics

Veysi İşler / Haşmet Gürçay / Hasan Kemal Süher /
Güven Çatak (eds.)

Contemporary Topics in Computer Graphics and Games

Selected Papers from the Eurasia Graphics Conference Series

PETER LANG

Bibliographic Information published by the Deutsche Nationalbibliothek

The Deutsche Nationalbibliothek lists this publication in the Deutsche Nationalbibliografie; detailed bibliographic data is available online at http://dnb.d-nb.de.

Library of Congress Cataloging-in-Publication Data

A CIP catalog record for this book has been applied for at the Library of Congress.

This publication is funded by Bahçeşehir University.

ISBN 978-3-631-80212-0 (Print)
E-ISBN 978-3-631-80522-0 (E-PDF)
E-ISBN 978-3-631-80523-7 (EPUB)
E-ISBN 978-3-631-80524-4 (MOBI)
DOI 10.3726/b16279

Peter Lang – Berlin · Bern · Bruxelles · New York · Oxford · Warszawa · Wien

This publication has been peer reviewed.

www.peterlang.com

Preface

This book includes peer-reviewed papers from EURASIA GRAPHICS conferences organized in years 2014, 2017 and 2018. EURASIA GRAPHICS Conference is a platform that brings together researchers, practitioners, students, and the industry to share the latest developments and research in the field of Computer Graphics, Animation and Gaming. Eurasia Graphics plays an active role in providing an environment where the results of the novel researches and experiences can be shared.

EURASIA GRAPHICS 2014, 2017 and 2018 conferences were held at Hacettepe University in Ankara, at Bahçeşehir University in İstanbul and at Hasan Kalyoncu University in Gaziantep, respectively. In those conferences, there were submissions from more than 10 different countries including Mexico, Germany, Ireland, USA, Brazil, UK, Iran, Azerbaycan and Pakistan.

The interdisciplinary character of the conference allowed people to present a wide range of approaches that helped stimulate intellectual discussions and the exchange of new ideas.

The goal of the editors of this book is to present interdisciplinary approach to computer graphics, animation, gaming and VR/AR/MR presented at these conferences. A total of 23 peer-reviewed full papers from various areas have been selected for this book.

The editors would like to thank the authors for their contributions to this book. We appreciate the organizers of EURASIA GRAPHICS in 2014, 2017 and 2018 from which we selected the articles. We also express our deep gratitude to Assoc. Prof. Dr. Barbaros Bostan and Dr. Güven Çatak from Bahcesehir University Game (BUG) Lab for their efforts in the process of publishing this book.

Table of Contents

Abbreviations

2D: Two Dimensional
3D: Three Dimensional
AHP: Analytical Hierarchy Process
AI: Artificial Intelligence
ANN: Artificial Neural Network
API: Application Programing Interface
ARCS: Attention, Relevance, Confidence, Satisfaction
AVE: Average Variance Extracted
BRDF: Bidirectional Reflectance Distribution Function
BSSRDF: Bidirectional Scattering Surface Reflectance Distribution Function
BVH: Bounding Volume Hierarchies
BVNDF: Bidirectional Visible Normal Distribution Function
CA: Cellular Automata
CAD: Computer Aided Design
CEGE: Core Elements of the Gaming Experience
CEGEQ: Core Elements of the Gaming Experience Questionnaire
CNN: Convolutional Neural Network
CPU: Central Processing Unit
CR: Compression Ratio
CT: Computed Tomography
CU: Compute Units
CUDA: Compute Unified Device Architecture
CUDPP: CUDA Data Parallel Primitives Library
D-G rho: Dillon-Goldstein rho
DLC: Downloadable Content
DoF: Degree-Of-Freedom, such as 2DoF or 6DoF
DV: Dependent Variable
EEG: Electroencephalogram
EFA: Exploratory Factor Analysis
EM: Expectation-Maximization
EU: Execution Units
FM: Fibromyalgia
FPS: First Person Shooter
FpS: Frames per Second
FRP: Fantasy Role Playing
FUGA: Fun of Gaming

GA: Genetic Algorithm
GAP: Game Approachability Principles
GCN: Graphics Core Next
GEnQ: Game Engagement Questionnaire
GExQ or GEQ: Game Experience Questionnaire
GMM: Gaussian Mixture Model
GPGPU: General-Purpose Graphics Processing Unit
GPPU: General Propose Processing Unit
GPU: Graphics Processing Unit
GSM: GPU Sharing Method
GUI: Graphical User Interface
HD: High Definition
HEEG: Heuristic Evaluation for Educational Games
HEP: Heuristics of Playability
HLSL: High Level Shader Language
HMD: Head-Mounted Display
HOG: Histogram of Oriented Gradient
HS: Holding State
HW: Hardware
ID: Index of Difficulty
IGE: Instructional Game Evaluation Framework
IR: Infrared
KMO: Kaiser-Meyer-Olkin
LF: Left Foot
LH: Left Hand
LSD: Least Significant Digit
M: Mean
MEC-SPQ: Measurements Effects Conditions Spacial Presence Questionnaire
MMORPG: Massive Multiplayer Online Role Playing Game
MPK: Multi OpenGL Multipipe SDK
MSD: Most Significant Digit
NASA-TLX: Nasa Task Load Index
NGH: Networked Game Heuristics
NMF: Non-negative Matrix Factorization
NPC: Non-Player Character
NSM: Network Sharing Method
OpenCL: Open Computing Language
PC: Personal Computer
PC: Player Character

PCA: Principal Components Analysis
PCG: Procedural Content Generation
PD: Proportional Derivative
PENS: Player Experience of Need Satisfaction
PHEG: Playability Heuristic Evaluation for Educational Computer Game
PIFF2: Presence-Involvement-Flow Framework
PLS-CFA: Partial Least Squares Confirmatory Factor Analysis
PRT: Precomputed Radiance Transfer
PSNR: Peak Signal-to-Noise Ratio
QEM: Quadric Error Metrics
QoL: Poor Quality Of Life
RAM: Random Access Memory
RANSAC: Random Sample Consensus
RF: Right Foot
RH: Right Hand
RMSE: Root-Mean-Square Error
ROI: Region of Interest
RPE: Rate of Perceived Exertion
RPG: Role Playing Game
SD: Standard Deviation
SDK: Software Development Kit
SEEM: Structured Expert Evaluation Method
SfM: Structure from Motion
SG: Spherical Gaussian
SH: Spherical Harmonics
SIMD: Single Instruction Multiple Data
SLBVH: Stackless Linear Bounding Volume Hierarchies
SMEQ: Subjective Mental Effort Questionnaire
SMX: Streaming Multiprocessor
SPES: Spatial Presence Experience Scale
SPGQ: Social Presence in Gaming Questionnaire
SPMD: Single Program Multiple Data
SRBF: Spherical Radial Basis Function
SS: Swinging State
SVD: Singular Value Decomposition
Tlf: Left foot swing wait duration
Tlh: Left hand swing wait duration
Trf: Right foot swing wait duration
Trh: Right hand swing wait duration

UMUX: Usability Metric for User Experience
VE: Virtual Environment
VR: Virtual Reality
VVS: Visualization Virtual Services
YOLO: You Only Look at Once

Games and Simulation

1. Controllers in VR Game User Experience: Perceived User Performance on a VR Puzzle Game

Mehmet Ilker Berkman
Visual Communication Design Department Bahcesehir University Istanbul, Turkey ilker.berkman@comm.bau.edu.tr

Barbaros Bostan
Game Design Department Bahcesehir University Istanbul, Turkey barbaros.bostan@comm.bau.edu.tr

Berk Yalcin
Game Design Department Bahcesehir University Istanbul, Turkey berkyalcin7@gmail.com

Abstract— This paper investigates the effects of controllers on virtual reality experiences and the feeling of presence associated with it. Virtual Reality (VR) technologies are in development since 1950s and they have been used for military and research purposes before. However, recent technological advances enabled VR to become a consumer product and more developers joined the market. Oculus Touch is one of the new devices developed for the new VR era and we focused on this input device in terms of its usability and possible contribution to a sense of presence. Oculus Touch and a classic gamepad have been compared in both usability and their effects on gameplay using a mixed methodology via questionnaires and semi-structured interviews. In conclusion, usability of Oculus Touch has been superior but this superiority does not affect the players' experience of the game in terms of presence, which also brings the question whether presence questionnaires designed in the 90s are still capable of measuring it in the new era or not.

Keywords— *virtual reality*; *controllers*; *presence*; *immersion*; *usability*

I. Introduction

After decades of research and development efforts, the HMDs (head-mounted displays) began to be mass-produced as a consumer-level hardware in March 28, 2016 by the company named Oculus VR, which is owned by Facebook. Although it was not the first attempt to make VR technology commercially available, it became a success unlike the precursors in 1990s. The 2016 was the year of VR, while the companies like Sony and HTC/Valve Corporation released their own HMDs that can be connected to personal computers and gaming consoles. By the meantime, VR had already become an experience that can be enjoyed via mobile devices such Google Cardboard released on March 2014 or Samsung Gear VR released on November 2015.

Among several reasons such as high-cost hardware and low-level graphic quality, which disappoints users, the 90s commercial HMDs also failed due to limited number of VR experiences offered to market. Learning from the past, Oculus followed a strategy of making the hardware available for the software developers as development kits before it is offered to end-users. Besides the hardware, the software development kit had been made available for free of charge, which was immediately integrated in popular game engines. As a result, a software habitat had been formed before the commercial HMDs were available.

On the other hand, early development kits did not contain any VR specific controllers. Some of the devices were shipped with 2-degree-of-freedom (2DoF) hand-held game controllers designed for game-consoles. As a result, the content created by the VR developers were mainly depended on the interactions using the HMD's positional tracking system. Additional interactions are based on keyboard and mouse inputs or 2DoF game controllers until October 2016, when Oculus released the 6DoF controllers called Touch. These tracked controllers, also called "wands", became a standard component of VR equipment, even for the mobile phone based HMDs since then. As HTC Vive was released with controllers, Sony's PlayStation VR supports PlayStation Move controllers. Samsung started to ship Gear VR with a controller and Google Cardboards successor; Daydream is also bundled with a 6DoF controller.

Overall intuitiveness that controller is perceived to have is the controller naturalness [1]. Tracked controllers naturally map to hand motion, can be visually co-located with the real hands, providing proprioceptive and passive touch cues [2] which leads to higher sense of presence and an enhanced user experience. On the other hand, many developers continued releasing applications that offer other means of user input, mostly the handheld controllers besides the tracked controllers. While one of their motivations in providing handheld controller options was "legacy support" for the owners of early development kits without 6DoF controllers, another reason is based on their targeted-users' habitations. Through years of use, gamers intuitively use game-controllers and became accustomed to these controls.

The purpose of this study to compare 6DoF tracked controllers and legacy controllers for virtual reality and provide information regarding important aspects of these controllers in virtual reality from users' perspective.

II. Related Work

Several studies were conducted to investigate both quantitative performance and subjective preferences of users on different types of input devices within different systems [3,4,5,6,7]. Sportillo et al. [8] compared the user performance and subjective impressions for a realistic steering wheel and a pair of 6DoF controllers on a driving simulation experienced with a HMD. Their results suggest that users performed better using the realistic steering wheel setup in the driving simulation, while the subjective measures on physical comfort, realism, ease of use and ease of adaptation were also slightly higher for the steering wheel setup. These findings were similar with previous research on non-VR driving simulations [9]. Lee et al. [10] intended to compare 6DoF controllers with data gloves in a HMD action - adventure game but they did not publish any findings on user performance. Lindsey [11] compared a gamepad, an optical hand-tracking interface and the touchpad on the HMD for performance, presence, preference and enjoyment on a selection task in VR. In her study, game controller was most preferred controller and she suggested that the result was due to prior high game experience of participants. There were not significant differences between controller's touchpad and game controller for performance measures, presence and enjoyment, but optical hand tracker yielded significantly poor scores, similar to the results provided in comparison of optical hand tracker to mouse and keyboard [12].

III. Methodology

A. Participants

3 male and 6 females, total of 36 participants, aged between 18 to 36 (M=25.33, SD=5.16) interacted with the stimuli using either a standard game controller or a pair of 6DoF wands. There were 18 participants on each group. Participants have been chosen from undergraduate and graduate game design students and VR enthusiasts. A modified version of Lifetime Television Exposure scale [13], which was adapted for gaming exposure, was applied to candidates. The question was modified as "During last 6 months, how often did you play a video game?" for "When you first wake/woke up in the morning?", "During Lunchtime?", "In the afternoon?", "During dinnertime?", "After dinner?", "Late at night, before going to bed?", During the day on Saturday?", "On Saturday nights?", "During the day on Sunday?", "On Sunday nights?", which can be answered within a 7 scale points from "Never" to "Almost Always". Based on the t-test results, there was not a significant difference between the weekend, weekdays and overall

TABLE I. LIFETIME TELEVISION EXPOSURE RESULTS

Controller		N	Mean	Std. Dev.	
Weekend Gameplay Frequency	6DoF Wand	18	4.875	1.15762	t(34)= -.4, p >.05
	Game Controller	18	5.069	1.70812	
Daily Gameplay Frequency	6DoF Wand	18	3.926	1.36389	t(34)= -.7, p >.05
	Game Controller	18	4.250	1.38532	
Overall Gameplay Frequency	6DoF Wand	18	4.306	1.16491	t(34)=-64, p >.05
	Game Controller	18	4.578	1.39476	

gameplay frequency scores of the groups, which can be seen as follows in Table 1, suggesting that the participants in either of the groups can be considered similar in terms of their gaming experience.

B. Stimuli

Participants have played the game "I Expect You to Die", which is an adventure game where users are required to solve puzzles by interacting with different objects in different ways. Users started with the tutorial and completed the first mission named "Friendly Skies". The tutorial informs the player about the game controls. In the "Friendly Skies" mission scenario, player spawns in car loaded in an airplane. The puzzles require interacting with the objects such as a car key to ignite the car, a screwdriver to open a lid on the car console, a knife to cut the wires to deactivate a bomb, shooting a handgun and pushing several buttons in the car. Interacting with distant objects requires an interaction design patterns called telekinesis [14].

C. Measures

To measure perceived workload, NASA-TLX (Task Load Index) [15,16] is employed participants were asked to rate six subscales; mental demand, physical demand, temporal demand, performance, effort and frustration level; within a 21 grade scale from low to high. After that, participants compare these dimensions pairwise for their importance while accomplishing given task. Two types of scores were obtained via NASA-TLX: a weighted score combined through a prescribed procedure to provide an overall index of the workload associated with a task or function or an unweighted score.

To measure perceived usability, we employed UMUX (Usability Metric for User Experience), which is a four-item scale [17] with a single dimension. It is verified to be sensitive to differences between similar purpose systems as it is intended to be while not being affected by interpersonal differences such as age, gender or native language [18].

As a measure of game experience in terms of immersion, flow, competence, positive affect, negative affect, tension and challenge, we employed Game Experience Questionnaire's (GExQ) 33-item core module [19]. Participants determine their level of agreement to each item within a 5-point scale.

Besides, we employed the 19-item Game Engagement Questionnaire (GEnQ) [20] as a 5-point measure with similar dimensions: absorption, flow, presence and immersion.

A single item scale, SMEQ (Subjective Mental Effort Questionnaire) [21] was used to assess the workload for four tasks in the "Friendly Skies" mission: Finding the key, defusing the bomb, unscrewing the panel and escaping out of plane. SMEQ is a visual analog scale, on which participants draw a line along. Around 5, the scale is labeled as "not at all hard to do" and the largest label read "tremendously hard to do" around 105, but the largest value on the scale is 150.

IV. Results

A series of independent samples t-tests did not reveal any significant difference on mean scores of any dimension of the Game Engagement Questionnaire, between the participants who used game controllers and the participants who used 6DoF wands. Mean scores are depicted as follows in Table 2, which are quite close for each group.

The comparison of Game Experience Questionnaire results also did not reveal a significant difference between the groups per dimension and for overall score. (Table 3)

Comparison of mean SMEQ results for 4 different tasks within the puzzle did not reveal a significant difference between the participant groups who used 6DoF wands and game controllers. For "Find the Key" task, both groups reported a similar mean score around 40, a little bit higher within the range of "fairly hard to do" on the scale. For the "Escape Task", both groups also had a similar mean score around 30 points on the scale, resembling to "a bit hard to do". The mean score of game controller users was higher in "Diffuse Bomb" while 6DoF wansd wand users yielded a higher score in "Screwdriver Task", but difference was not significant. Mean scores and t-test indicators are given below in Table 4.

When we compared weighted NASA-TLX dimension scores, we did not observe any significant effect of the controller used, as depicted below in Table 5. As well, there was not a significant difference on the overall NASA-TLX score

TABLE II. GAME ENGAGEMENT QUESTIONNAIRE RESULTS

Controller		N	Mean	Std. Deviation	
Presence	6DoF Wand	18	3.042	0.708	t(34)=.49, p >.05
	Game Controller	18	2.931	0.663	
Absorption	6DoF Wand	18	2.389	0.765	t(34)= -.181, p >.05
	Game Controller	18	2.433	0.707	
Flow	6DoF Wand	18	2.623	0.482	t(34)= .035, p >.05
	Game Controller	18	2.617	0.569	
Immersion	6DoF Wand	18	4.000	0.594	t(34)= .776, p >.05
	Game Controller	18	3.778	1.060	

TABLE III. GAME EXPERIENCE QUESTIONNAIRE RESULTS

Controller		N	Mean	Std. Deviation	
Overall GExQ Score	6DoF Wand	18	1.74	0.58	t(34)= -.127, p>.05
	Game Controller	18	1.76	0.72	
Positive Affect	6DoF Wand	18	4.07	0.61	t(34)= .638, p>.05
	Game Controller	18	3.91	0.83	
Competence	6DoF Wand	18	2.94	0.87	t(34)= -.688, p>.05
	Game Controller	18	3.16	0.97	
Immersion	6DoF Wand	18	3.69	0.73	t(34)= -.528, p>.05
	Game Controller	18	3.81	0.64	
Flow	6DoF Wand	18	3.00	0.82	t(34)= -.152, p>.05
	Game Controller	18	3.04	0.93	
Negative Affect	6DoF Wand	18	1.74	0.58	t(34)= -.127, p>.05
	Game Controller	18	1.76	0.72	
Challenge	6DoF Wand	18	2.98	0.72	t(34)= .235, p>.05
	Game Controller	18	2.92	0.70	
Tension Annoyance	6DoF Wand	18	1.85	0.79	t(34)= -1.456, p>.05
	Game Controller	18	2.26	0.89	

(t(34)= .741, p>.05) between the participants who used 6DoF Wand (M=53.75, SD=8.1) and the participants who used game controller (M=56.30, SD=12.1) The unweighted NASA-TLX score comparison resulted similarly to unweighted NASA-TLX score comparison, with an exception that Effort Workload was significantly higher for participant group who used Game Controller. Results are given below in Table 6.

TABLE IV. SMEQ RESULTS

Controller		N	Mean	Std. Deviation	
Find Key Task	6DoF Wand	18	40.01	22.16	t(34)= .971,
	Game Controller	18	39.72	24.40	p>.05
Diffuse Bomb Task	6DoF Wand	18	47.68	24.78	t(34)= .450,
	Game Controller	18	55.19	32.56	p>.05
Screwdriver Task	6DoF Wand	18	40.49	18.61	t(34)= .178,
	Game Controller	18	29.85	26.25	p>.05
Escape Task	6DoF Wand	18	32.50	22.06	t(34)= .690,
	Game Controller	18	29.35	24.28	p>.05

TABLE V. WEIGHTED NASA_TLX RESULTS

Controller		N	Mean	Std. Deviation	
Weighted Mental Workload	6DoF Wand	18	15.13	7.70	t(34)= .304,
	Game Controller	18	14.38	7.18	p>.05
Weighted Physical Workload	6DoF Wand	18	3.40	4.50	t(34)= -1.826,
	Game Controller	18	8.02	9.76	p>.05
Weighted Temporal Workload	6DoF Wand	18	10.58	7.52	t(34)= .810,
	Game Controller	18	8.77	5.70	p>.05
Weighted Performance Workload	6DoF Wand	18	8.19	7.23	t(34)= -.032,
	Game Controller	18	8.27	6.87	p>.05
Weighted Effort Workload	6DoF Wand	18	12.92	4.33	t(34)= 1.048,
	Game Controller	18	11.14	5.71	p>.05
Weighted Frustration Workload	6DoF Wand	18	3.54	4.26	t(34)= -1.020,
	Game Controller	18	5.71	7.96	p>.05

UMUX scale yielded significantly different scores ($t(34)=3.61$, $p<.05$). The 6DoF Wand users had a mean score of 67.8 (SD=25.5) while game controller users had a mean score of 37.7 (SD=24.5). When the mean results for each UMUX item were compared, it is also observed that there are significant mean differences between the groups, as given follows in Table 7.

TABLE VI. UNWEIGHTED NASA_TLX RESULTS

Controller		N	Mean	Std. Deviation	
Mental Demands	6DoF Wand	18	.24	.09	t(34)= -.612, p>.05
	Game Controller	18	.26	.09	
Physical Demands	6DoF Wand	18	.09	.09	t(34)= -1.223, p>.05
	Game Controller	18	.13	.13	
Temporal Demands	6DoF Wand	18	.19	.11	t(34)= .855, p>.05
	Game Controller	18	.16	.08	
Performance	6DoF Wand	18	.18	.09	t(34)= .018, p>.05
	Game Controller	18	.18	.09	
Effort	6DoF Wand	18	.23	.06	t(34)= 2.507, p<.05
	Game Controller	18	.18	.08	
Frustration	6DoF Wand	18	.08	.08	t(34)= -.892, p>.05
	Game Controller	18	.11	.12	

V. Discussion

Our results indicated the controller in a puzzle game that requires the player to reach and use several objects in the game world did not affect the measures of game engagement and game experience. This finding is consistent with Nacke [5] who did not observe any difference on Game Experience Questionnaire results comparing a single 6DoF wand and a game controller, although he observed a correlation with EEG data and Negative Affect dimension. As the items of Game Engagement Questionnaire and Game Experience Questionnaire query the participants with a focus on the gameplay regardless of the controller, we think that participants responded the items with a focus on their overall gaming experience. The scores of Game Engagement Questionnaire subdimensions were slightly higher for Presence and Immersion dimensions but identical for Absorption and Flow. Unlikely, Immersion was slightly higher in Game Experience Questionnaire for game controller, but Flow was also identical as well as Positive Affect, Negative Affect and Challenge. For Competence and Tension/ Annoyance dimensions, 6DoF wand users provided relatively lower scores. Thus, it is possible to say that naturalness of controller did not interfere with the player's engagement with the game, since the focus of the player is on finding the correct actions to solve the puzzle rather than interact with the objects in the environment. Besides, the experience of the game did not depend on the physical capabilities of the player reflected through the controller but merely on

TABLE VII. UMUX SCALES

Controller		N	Mean	Std. Deviation	
UMUX: This controller's capabilities meet my requirements.	6DoF Wand	18	5.39	1.46	t(34)= 3.989, p<.05
	Game Controller	18	3.44	1.46	
UMUX: This controller is a frustrating in virtual reality experience. (reversed)	6DoF Wand	18	5.11	1.49	t(34)= 2.501, p<.05
	Game Controller	18	3.61	2.06	
UMUX: This controller is easy to use in virtual reality.	6DoF Wand	18	5.33	1.61	t(34)= 5.151, p<.05
	Game Controller	18	2.67	1.50	
UMUX: I have to spend too much time correcting things with this controller during virtual reality experience. (reversed)	6DoF Wand	18	4.44	2.01	t(34)= 1.640, p<.05
	Game Controller	18	3.33	2.06	

the mental/cognitive capabilities of the player. This is verified by the unweighted NASA-TLX results, with the highest score on Mental Workload dimension score was the highest compared to others, while the Physical Workload score was the lowest. Although the participants evaluated their performance with the same score for both conditions, the effort score, which refers both physical and mental effort, was significantly higher for the 6DoF wand condition. However, it is not possible to explain whether this was due to participants' effort to learn the controls of the 6DoF wand, which they were not familiar before. Besides, we cannot explain the difference in SMEQ in Screwdriver Task, but it should be noted that the difference was not significant.

Previous studies [22,23] suggested that playing a game with more naturally-mapped controls affect the game experience, in terms of perceived realism, enjoyment or presence. However, we would not be able to verify these results in our study. However, it should be noted that the stimuli employed in the previous work was games that depend on the physical abilities of the user, such as a tennis game, a fighting or boxing game. As the stimulus in our study is a puzzle game, the effect of the game controller may not have been significant on the gaming experience.

On the other hand, the UMUX results were significantly different since the items of UMUX query the participants directly about the controller, rather than their subjective gaming experience or the workload of the tasks in the game. For the ease of use item in UMUX, "This controller is easy to use in virtual reality." was scored twice as much higher in 6Dof wand condition. The reversed item that indicates efficiency, which refers to the time spent on errors was %25 lower for the game controller condition. The item indicated the satisfaction reversely, mentioning the frustration and the item that indicates effectiveness as "meeting user's requirements" were also quite highly scored by 6DoF wand users. Although the study was conducted with "between-subjects" design, these sharp difference on UMUX strongly show that users have a higher perceived sense of usability for the 6DoF Wand, compared to game controller.

VI. Conclusion

Our results were not adequate to claim that 6DoF wand type controllers enhance the gaming experience and provide a more acceptable workload for the players in a virtual reality puzzle game which depends on interacting with the objects in the VE. On the other hand, the perceived usability of this device, which provided a more natural way of interaction was higher. However, our study reports the usability as a subjective measure. Further analysis is required in order to determine the performance metrics such as time spent in the VE for each task or number of failures per task. Besides, a qualitative analysis is required to understand user actions during gameplay. Furthermore, the effect of the controller on user experience in VR should be investigated for different types of applications, such as game that have higher physical demand and simulations that mimic use of vehicles and more complicated tools.

References

[1] P. Skalski, R. Tamborini, A. Shelton, M. Buncher & P. Lindmark. "Mapping the road to fun: Natural video game controllers, presence, and game enjoyment". New Media & Society, 13, 2011, 224–242.

[2] J. Jerald. The VR book: Human-centered design for virtual reality. Morgan & Claypool, 2015.

[3] C. Forlines, D. Wigdor, C. Shen & R. Balakrishnan. "Direct-touch vs. mouse input for tabletop displays". In Proceedings of the SIGCHI conference on Human factors in computing systems, 2007, pp. 647–656.

[4] L. Zaman, D. Natapov & R.J. Teather. "Touchscreens vs. traditional controllers in handheld gaming". In Proceedings of the international academic conference on the future of game design and technology, 2010, pp. 183–190.

[5] L.E. Nacke, "Wiimote vs. controller: electroencephalographic measurement of affective gameplay interaction". In Proceedings of the international academic conference on the future of game design and technology, 2010, pp. 159–166.

[6] K.M. Gerling, M. Klauser & J. Niesenhaus. "Measuring the impact of game controllers on player experience in FPS games". In Proceedings of the 15th International Academic MindTrek Conference: Envisioning Future Media Environments, 2011, pp. 83–86.

[7] A.M. Limperos, M.G. Schmierbach, A.D. Kegerise & F.E. Dardis. "Gaming across different consoles: exploring the influence of control scheme on game-player enjoyment". Cyberpsychology, Behavior, and Social Networking, 14(6), 2011, 345–350.

[8] D. Sportillo, A. Paljic, M. Boukhris, P. Fuchs, L. Ojeda & V. Roussarie. "An immersive Virtual Reality system for semi-autonomous driving simulation: a comparison between realistic and 6-DoF controller-based interaction". In Proceedings of the 9th International Conference on Computer and Automation Engineering, 2017, pp. 6–10.

[9] K.D. Williams. "The effects of dissociation, game controllers, and 3D versus 2D on presence and enjoyment". Computers in Human Behavior, 38, 2014, 142–150.

[10] S. Lee, K. Park, J. Lee & K. Kim. "User study of VR basic controller and data glove as hand gesture inputs in VR games". In Ubiquitous Virtual Reality (ISUVR), 2017 International Symposium, 2017, pp. 1–3.

[11] S. Lindsey. "Evaluation of Low Cost Controllers for Mobile Based Virtual Reality Headsets". Unpublished Master's Thesis. College of Aeronautics at Florida Institute of Technology. Melbourne, Florida, 2017.

[12] M.C.B. Seixas, J.C. Cardoso & M.T.G. Dias. "The Leap Motion movement for 2D pointing tasks: Characterisation and comparison to other devices". In Pervasive and Embedded Computing and Communication Systems (PECCS), 2015 International Conference, 2015, pp. 15–24.

[13] K. Riddle. "Remembering past media use: Toward the development of a lifetime television exposure scale". Communication Methods and Measures, 4 (3), 2010, pp. 241–255.

[14] R.M. Clifford, N.M.B. Tuanquin & R.W. Lindeman. (2017, March). "Jedi ForceExtension: Telekinesis as a Virtual Reality interaction metaphor". In 3D User Interfaces (3DUI), 2017 IEEE Symposium, 2017, pp. 239–240.

[15] S.G. Hart & L.E. Staveland, "Development of NASA-TLX (Task Load Index): Results of empirical and theoretical research". In Advances in psychology, Vol. 52, 1988, pp. 139–183.

[16] S.G. Hart. "NASA-task load index (NASA-TLX); 20 years later". In Proceedings of the human factors and ergonomics society annual meeting, Vol. 50, No. 9, 2006, pp. 904–908.

[17] K. Finstad. "The usability metric for user experience". Interacting with Computers, 22(5), 2010, 323–327.

[18] M.I. Berkman & D. Karahoca. "Re-assessing the usability metric for user experience (UMUX) scale". Journal of Usability Studies, 11(3), 2016, 89–109.

[19] W. IJsselsteijn, Y. De Kort, K. Poels, A. Jurgelionis & F. Bellotti. "Characterising and measuring user experiences in digital games". In International conference on advances in computer entertainment technology, Salzburg, Austria, 2007, Vol. 2, pp. 27.

[20] J.H. Brockmyer, C.M. Fox, K.A. Curtiss, E. McBroom, K.M. Burkhart, and J.N. Pidruzny. "The development of the Game Engagement Questionnaire: a measure of engagement in video game-playing," Journal of Experimental Social Psychology, vol. 45, no. 4, 2009, pp. 624–634.

[21] F.R.H. Zijlstra & L. van Doorn. "The construction of a scale to measure subjective effort". Technical Report, Delft University of Technology, Department of Philosophy and Social Sciences, 1985.

[22] R. McGloin, K.M. Farrar & M. Krcmar. "The impact of controller naturalness on spatial presence, gamer enjoyment, and perceived realism in a tennis simulation video game". Presence: Teleoperators and Virtual Environments, 20(4), 2011, 309–324.

[23] D.M. Shafer, C.P. Carbonara & L. Popova. "Controller required? The impact of natural mapping on interactivity, realism, presence, and enjoyment in motion-based video games". PRESENCE: Teleoperators and Virtual Environments, 23(3), 2014, 267–286.

2. A Taxonomy and Terminology Study on Embedded Narrative: A Case Study of Bloodborne

Güven Çatak
Game Design Department Bahcesehir University Istanbul, Turkey guven.catak@comm.bau.edu.tr

Barbaros Bostan
Game Design Department Bahcesehir University Istanbul, Turkey barbaros.bostan@comm.bau.edu.tr

Ali Burak Ankaralı
Game Design Department Bahcesehir University Istanbul, Turkey aliburakankarali@gmail.com

Abstract— This paper aims to term and to classify the embedded narrative elements in terms of game design, having regard first to their presentation to the player and then to their representation by the designer. The selected game for this study is Bloodborne, which is considered as one of the best examples of an encompassing narrative where smaller stories are embedded within a greater story and the game's mechanics are also designed in a similar vein to the nature of the story. The embedded narrative elements in Bloodborne are analyzed by close-playing technique with an aim to create a common terminology for game design and the twenty-six identified elements are classified by their temporal and spatial attributes. The Diegetic/Non-Diegetic dichotomy is used for spatial and the Sequential/Non-Sequential dichotomy is used for temporal classification, resulting in four types of narrative elements: Non-Diegetic/Non-Sequential, Non-Diegetic/Sequential, Diegetic/Non-Sequential, and Diegetic/Sequential.

Keywords— game stories; game narrative; embedded narrative; bloodborne

I. Introduction

It is very difficult to find a comprehensive definition of narrative but Onega and Landa defined narrative as the semiotic representation of a sequence of events, meaningfully connected in a temporal and causal way. [1]. Among the various other definitions in literature, some use the word 'event' and some prefer the word 'action', where an event implies transformation and an action involves agents [2] but all researchers agree that without these there is no narrative at all.

There is also ambiguity over related terms but Abbott [3] makes the clear distinction between narrative, story, and narrative discourse. Story is an event or a sequence of events. Narrative discourse is the story as represented. Narrative is the representation of events and consists of these two.

Although some video games do not have a story at all, many games have a story but the gaming experience cannot be reduced to a story and when games tell a story they tell it differently than other media. Compared with the conventional narrative, the distinctive feature of video games is interactivity but this creates a tension between the freedom provided to the player and the structure of the game story. The more freedom is given to the player, the more difficult it becomes for the designer to control the structure of the story. In this regard, emergent narratives where the story emerges from the gameplay are an exception but the focus of this article is on embedded narratives or environmental storytelling where the story elements are embedded within the game. Jenkins [4] defined two different techniques for an embedded narrative, the first is to have a rather unstructured narrative controlled and explored by the players and the second is to embed the pre-structured narrative within the mise-en-scene which waits to be discovered by the player.

Embedded narrative or environmental storytelling is also defined as the act of staging player-space with environmental properties that can be interpreted as a meaningful whole, furthering the narrative of the game [5] The virtual environment of a video game provides narrative context by communicating the history of what happened in the virtual world, who lives in it, the virtual society and communities in it, predictions about the future, the functional purpose and the mood of the game world. When these environmental elements are integrated in a meaningful way, the player interprets them as a whole and this strengthens the narrative quality of a game.

II. Bloodborne as an Embedded Narrative

Bloodborne takes place in the city of Yharnam, which is inflicted with a mysterious illness that has transformed most of the residents into bestial creatures. At the beginning of the game, the player is given little information on how and why these things are happening in the city. The only vital information is given with a note that says: "Seek Paleblood to transcend the hunt." No other information is given on what Paleblood is or why the hunt is taking place in the streets of Yharnam. The game narrative is obscure and the game has a hard to follow storyline. The story is more like a puzzle made up of different pieces, which are scattered through the various levels. Solving this puzzle requires attention to details such as the architecture, item descriptions, item details, little snippets of conversations and so on. It is the player's job to find, associate and interpret the scattered disparate elements as a meaningful whole. According to the classification of Jenkins [4], this is the type of narrative where the designers embed the pre-structured narrative within the mise-en-scene, awaiting for discovery. In this regard, drawing the distinction between the narrative and the story defined by Feldman et al. [6], Bloodborne can also be considered a game with an encompassing narrative with smaller stories embedded within it. Gathering these fragmented stories together is actually the game's central story and the player's main purpose. The story can also be considered a retrospective narrative, which is defined as the process of updating the representation of the past events and integrating the raw historical facts into the network of an intelligible historical sequence as the plot progresses [7]. The game's mechanics are also designed with respect to the embedded nature of the story, such as the Insight mechanic, which represents the depth of inhuman knowledge and increases with defeating bosses and other minor actions. Certain enemies, items, locations, vendors, and characters in the game are affected by the player's current Insight level.

In this game, when the player dies he finds himself in Hunter's Dream, also known as Dream Refuge, which acts as a central hub for accessing the various levels of the game. Written on page near one of the altars that can be found in the Hunter's Dream is a piece of knowledge about the hunt: "To escape this dreadful Hunter's Dream, halt the source of the spreading scourge of beasts, lest the night carry on forever." The story, transformed by the existence and the function of the Hunter's Dream, is also circular in nature. According to Richardson [8], this type of stories have a circular temporality that transforms the linear chronology of events, always returning to and departing from its point of origin. The ending of the game is also founded on this circular nature where the player can break the circle by accepting the offer of Gherman the First Hunter's offer to die or create a similar but new circle by defeating Gherman and taking his place in the Hunter's Dream. This NPC, when the player first arrives in the Hunter's Dream, is described by the Future Press game guide as: "Gehrman is a kindly old man whom you'll first meet upon awakening in the Hunter's Dream. He is the workshop's founder and the very first hunter, though due to his advanced age he now serves only as an advisor; as such, he has a wealth of experience and will provide you with his wisdom and guidance throughout your exploration of Yharnam." An ordinary character considered as friendly throughout the game is transformed into the final boss at the unexpected ending of the story. Langmead [9] also stressed the same kind of circularity for Bloodborne, with an emphasis on the idea of death. The beasts slayed by the player are reborn when the player dies or when the player leaves an area and returns to it. Thus, the beasts, similar to the player, are caught in an endless cycle of death and rebirth.

Rogers [10] discussed the general design of Bloodborne and suggested that the story of the game is the world and is narrated by the player. This asserts a general idea that captures the very essence of the narrative concept of Bloodborne: The game wants you to think about your place in its world, on both the macro and micro scale. Schniz [11] focused on the cryptic narrative qualities of the game, with special emphasis on the Reddits, wikis and other digital platforms that become hotspots of story analysis and information exchange. Ball [12] explained how the Bloodborne community tried to solve the story of the game as if solving a puzzle as:

> "Close readings of item descriptions, environmental designs, and locations of objects in the game world yield a confounding tangle of mythology for dedicated fans to unravel. A robust fan discourse engaged in these lore hunts has emerged in which fans, across various Internet platforms, collate in-game and intertextual resources to theorize Bloodborne's story These conversations often take the shape of speculative fiction; at times they generate conflicting territories of authorship and agency within the fan community."

III. Data and Methodology

The data employed in this paper is Bloodborne, the game itself, including The Old Hunters expansion DLC, on version 1.09. The game was played on PlayStation 4 platform without online features. Bloodborne is developed by Japanese video game development company FromSoftware and published by Sony Computer Entertainment on March 24, 2015. The expansion DLC is released on November 24, 2015. Bloodborne is an action role-playing video game with a third-person perspective. It has a bleak and gothic atmosphere in general. There are different enemies, bosses, and different levels but there is no difficulty setting in this game. The game can be replayed again after the final boss with the New Game Plus option. The game starts and ends with cutscenes. There is no save and load mechanism. Upon death, the player character is resurrected in a checkpoint location with the collectible Blood Echoes dropped at the location of the death. There are secret levels and passages, and there are also areas called Chalice Dungeons, some of which can be generated procedurally. There are three endings in the game and all of them were played. Narrative elements were noted during and after the play sessions.

These play sessions use the methodology known as close-playing process. According to Bizzocchi and Tanenbaum [13] this process requires the player must be a fresh participant of the game, ready to absorb all its nuances. But at the same time, as a researcher, the player should keep his distance from the experience, remembering and recording the details. The scholar constantly oscillates between these states. The same researchers proposed different strategies that make this dialectical process more effective. These are: the imagined naïve reader who explores a new experience, a performed player stereotype with a particular bias in order to discover certain aspects of the game, and as a researcher that employs focused readings with an analytical lens. The game is played with each approach respectively. First, it was finished for a couple of times, without giving much attention to detail, then every level is explored with an aim to get every achievement, and the final gameplay focused on details with taking notes and screenshots.

With an aim to create taxonomy of narrative elements, two dichotomies were used as classifying factors. These dichotomies refer to the spatial and temporal attributes of the narrative elements. These attributes are used for discovering the fundamental and distinguishable features of a narrative element. A Diegetic element implies that it exists within the game-world, but a Non-Diegetic element resides outside of the game-world but is still within the game, such as the user interface. A Sequential element consists of a continuous set of actions that occur in motion within a duration of time whereas a Non-Sequential element may be perceived in the game as a moment frozen in time, as if the player could stop the gameplay but still experience them.

IV. Findings and Discussion

The *Non-Diegetic/Non-Sequential* elements are: Character Nameplate, Character Statistic, Death Notification, Item Description, Item Icon, Item Location, Item Nameplate, Level Nameplate, Note Description, Object Nameplate and Travel Point Nameplate. The *Diegetic/Non-Sequential* elements are: Character Costume, Character Figure, Character Location, Environment and Landscape. The *Diegetic/Sequential* elements are: Ambient Sound, Character Abilities, Character Attitude, Character Awareness, Character Gesture, Character Metamorphosis, Combat, Cutscene, Dialogue and Scene. A narrative element that belongs to the *Non-Diegetic/Sequential* type could not be found in Bloodborne but theoretically it is possible to embed a narrative element in this way.

NON-DIEGETIC/NON-SEQUENTIAL ELEMENTS

Character Nameplate: Characters in a video game are similar to the characters in a book or actors in a film. In video games they can be the player character or characters, the friendly non-player characters (NPCs), the enemy mobs, or the enemy bosses. Nameplates are user interface elements that refer to a character or an object. A character nameplate is the name of a character represented outside of the game-world and on the graphical user interface as a text. For example, Ludwig, a boss in the game, has two phases during the combat. After the first phase, a cutscene is introduced and he starts using different combat tactics. But another subtle change occurs and his nameplate changes on the screen after the transition. Initially his name was "Ludwig the Accursed" but it becomes "Ludwig, the Holy Blade" in the next phase.

Character Statistics: Stats are data that represent character attributes in terms of game mechanics. They are generally of numerical value. For the most part they are purely an aspect of mechanical design. They are usually in accordance with the general narrative theme and sometimes the interaction between these statistics may provide information about the narrative as well. For example, characters in Bloodborne may have weaknesses in terms of gameplay. Creatures that are tagged as beast take extra damage from silver weapons. Creatures that resemble beasts but that are not completely so, do not take. There is even a unique type of enemy in the chalice dungeons of the game that does not completely resemble a beast but takes extra damage from silver. This statistic provides information about the nature and the type of the creature when the visual information is not there.

Death Notification: When the player character dies or when a boss is killed, a screen wide message appears to notify the player of their accomplishment or demise. For example, the standard message for the death of a boss is "PREY SLAUGHTERED" for the early and middle part of the game. The messages themselves may seem of little value at first but there are some nuances during the late game. For example, when certain enemies are killed, the displayed message is: "NIGHTMARE SLAIN". Most of these enemies are also identified as "the Great Ones". For bosses that are open to speculation about their kind, this variation provides the necessary information to the player.

Item Description: Items are the objects that players can interact with and most of them are represented in the user interface, such as an icon in the inventory screen or as a glowing object in the world waiting for the player to take. Items may have a name, an icon, a location, and a description, presented to the player within the user interface. Most of the textual information in the game comes from the item descriptions. Player can read the description of items they have in their inventory. Loading screens may also show random item descriptions, providing a non-linear narrative both spatially and temporally. For example, the description of one of the "One Third of Umbilical Cord" items is given below and it provides fragments of information about the game world.

> *"Every Great One loses its child, and then yearns for a surrogate. This Cord granted Mensis audience with Mergo, but resulted in the stillbirth of their brains."*

Item Icon: Each item has an icon on the left of its description and icons may provide visual cues. For example, while the Blood Vial item has red blood in its icon, Iosefka's Blood Vial is white. White blood's significance is also mentioned many times via textual elements during the gameplay, thus this color change reveals information about the type of blood Iosefka gives to the player character.

Item Location: Some items are scattered around the world, but some are dropped from killed monsters or are given by non-player characters. Item locations may have an important role and provide contextual information rather than its textual content or design. The placement of some items indicates a conscious effort in accordance with the narrative. For example, the Red Messenger Ribbon found near the Goliath Pig implies that the owner of the ribbon was killed on her way to the chapel.

Item Nameplate: The name of the item presented in the user interface whether above the item description or as pop-up text when picked up or shown in the inventory list. The item name itself may provide some information on the nature of the item. For example, Blood of Arianna tells that that blood belonged to Arianna.

Level Nameplate: The levels are isolated or interconnected zones of the game-worlds which usually have their own characteristics. At the bottom right of the screen, when the player enters a new level, the level name is displayed. For example, if the name of the level is displayed as Hemwick Charnel Lane, the player easily makes the connection between the level and the Witch of Hemwick.

Note Description: Notes are sparkling papers scattered around the world and when interacted they open a pop-up window. Both the sparkle and the window are part of the graphical user interface. For example, a note found early in the game has the text below, which provides information about the Byrgenwerth spider.

> *"The Byrgenwerth spider hides all manner of rituals, and keeps our lost master from us. A terrible shame. It makes my head shudder uncontrollably."*

Object Nameplate: Objects are similar to items but they are represented within the virtual world. Their names can be seen when interaction command appears on the screen. For example, a nameplate of a tombstone can be seen when interacted with.

Travel Point Nameplate: Travel points are like sublevels within a zone to travel to and from the main hub or between other points. For example, the nameplates "1st Floor Sickroom" or "Tomb of Odeon" provide information about the world and the lore otherwise not mentioned in other narrative element.

DIEGETIC/NON-SEQUENTIAL ELEMENTS

Character Costume: Bloodborne characters equip clothes or protective gear in the game. The design and the color of these clothes or gear provide subtle information on the nature of the character wearing them. The attire can also affect the gameplay, such as the encounter with Adella the nun. You need to be wearing Father Gascoigne or the White or Black Church attire sets to get her to talk to you.

Character Figure: This is the physical appearance of a character, including the face and the body. For example, the first hunter Gehrman creates the Doll and Lady Maria was Gehrman's student. The face of Lady Maria and the Doll seem very much the same, giving information about a relationship between the two.

Character Location: The location of a character may also contain contextual information and may change with player interaction. Location may also refer to their initial role within the mise-en-scene. For example, in the Nightmare of Mensis level, which is affiliated with the School of Mensis, the player can encounter a hostile hunter named Edgar, Choir Intelligencer. His presence there indicates that the Choir is spying on the School of Mensis.

Environment: This is the immediate surroundings of the player character. Like other elements, their content may become meaningful within the context of other narrative elements. Environment plays an important role in the visual storytelling aspect of Bloodborne and may contain information about the lore of the world. For example, there is a figure on some of the tombstones in Hunter's Dream that resembles the icon design of the Hunter Rune item. There are also some corpses hanged in the same fashion, one notably within the frame when the player first enters the Fishing Hamlet, which also resembles the hanged man of the tarot decks.

Landscape: Landscape is not within the immediate environment the player character inhabits at the moment. It can be seen from afar but cannot be reachable at that point. For example, the Moon changes after some key points in the game and with its change some other aspects of the game are also affected by it.

DIEGETIC/SEQUENTIAL ELEMENTS

Ambient Sound: Ambient sound is a sound embedded in the environment whose source cannot be seen. This happens for a period of time and is sequential. Sometimes it is also accompanied with visual information and is not entirely audial. For example, at certain places in the game, the player can hear a baby cry or church bells toll. These elements enhance the mood of the virtual world and may provide information about it.

Character Abilities: The special abilities and the spells character cast during the gameplay. For example, a Choir hunter casting the Augur of Ebrietas may indicate that the Choir has contact with Ebrietas.

Character Attitude: The manner characters interact with others may reveal interesting information on whether they are hostile, indifferent or friendly. For example, Mergo's Attendants that reside in the Mergo's Loft are not particularly aggressive and are usually not hostile unless they are provoked.

Character Awareness: Character awareness occurs when a character reacts to the changes in another character or the environment. This is very subtle and can often be missed. The game does not provide any clue on what may or may not happen. For example, if the player finds the Ring of Betrothal and equips it, it may be used to trigger a special dialogue with Annalise, the Queen of the Vilebloods. The player may propose marriage to the queen but she will reject the offer.

Character Gesture: The gestures are mainly used as an interaction technique between players. The multiplayer aspect of the game will not be discussed here but some non-player characters also react to gestures. For example, the Doll can mimic the gesture the player has done, or it is possible to contact the Brain of Mensis by the Make Contact gesture to drop a rune.

Character Metamorphosis: A common theme in literature, the transformation of a character is an effective tool both visually and textually, revealing info about the character. For example, the boss Ludwig mentioned above in the Character Nameplate section, looks and fights like a horse in his first phase, but after the transition he fights like a human, showing that after all that he has done and suffered, he still retains his human aspects.

Combat: Combat is the hostile interaction between characters that may include visual and audial cues. For example, enemies in Bloodborne may mumble a prayer or initiate a speech during a fight. Common mobs may also have voice lines to taunt the player.

Cutscene: There are some brief cutscenes in Bloodborne but contrary to other video games these cutscenes do not entirely reveal the story of the game. For example, the intro cinematic of the boss Father Gascoigne provides an enigmatic info about both the character and what may happen in the future with a few sentences: "...Beasts all over the shop... You'll be one of them, sooner or later..."

Dialogue: The dialogue is initiated when the player presses the corresponding button for the talk action. The dialogue in this game is different from what most of the role-playing games offer. The player or the hero is silent, but consequent dialogues, re-initiated by the player's interaction, contain information about what the player might have said. In other words, the lines of the player character are implied or assumed. For example, the player may speak with the characters behind the windows that have lit red lanterns. It is important to note that this cue regarding the lanterns is part of the environmental storytelling.

Scene: A scene resembles a cutscene but occurs without breaking the gameplay. It is a sequential event in the game world that happens with or without character interaction and may also be the consequence of player choices. It is more than a static dialogue and it does not resemble the narrative bits sprinkled over the combat sequences. For example, after giving the summons to the Alfred, the player can find him in queen's chamber where he has slain her. He talks about his accomplishments and then goes mad.

V. Conclusion

This paper introduces a new classification of embedded narrative elements in video games. In this classification there are two dichotomies, one is spatial, the other is temporal. Similar dichotomies are also used in text or media studies to describe other phenomena. Spatially, the narrative elements may be Diegetic or Non-Diegetic, that is to say within the narrative world of the game or not. Temporally, they may be Sequential or Non-Sequential, depending whether they occur in a period of time or not. Thus, the taxonomy comes with four different types of narrative elements. Though it should be noted that there are no Non-Diegetic/Sequential elements in Bloodborne.

The game is analyzed within the limits of the embedded narrative provided by the designers. In Bloodborne, players can summon each other's characters to get help defeating a boss or can invade one another's world for player versus player combat. Players can communicate through a set of gestures and they can also leave a note on the ground in their world. These notes can be seen and rated in other worlds. The notes, which can only contain words from predetermined sets of phrases, establishes it's own language within the game community. Players may also impersonate some non-playable characters by wearing their set of armors. It may be speculated then that this emergent narrative can be both Diegetic or Non-Diegetic and both Sequential and Non-Sequential. Though, listing the different types of emergent narrative elements is beyond the scope of this study. Future studies can also investigate emergent narratives with this taxonomy.

References

[1] Onega, and J. A. G. Landa, "Introduction." In Onega and Landa (eds.) Narratology: An Introduction, London: Longman, 1996, p. 3.

[2] M.-L. Ryan, "Toward a definition of narrative". In D. Herman (Ed.), The Cambridge companion to narrative, UK: Cambridge University Press, 2007, pp. 22–35.

[3] H.P. Abbott, The Cambridge introduction to narrative. Cambridge: Cambridge University Press, 2008.

[4] H. Jenkins, "Game design as narrative architecture". In N. Wardrip-Fruin & P. Harrigan (Eds), First Person: New media as story, performance, game. Cambridge: MIT Press, 2004.

[5] H. Smith and M. Worch. "What Happened Here? Environmental Storytelling". Talk. Game Developers Conference 2010, 2010. Retrieved from http://www.gdcvault.com/play/1012647/What-Happened-Here-Environmental

[6] M.S. Feldman, K. Skoldberg, R.N. Brown, and D. Horner. "Making Sense of Stories: A Rhetorical Approach to Narrative Analysis", Journal of Public Administration Research and Theory, 14(2), 2004, pp. 147–170.

[7] M.-L. Ryan, "Embedded Narratives and Tellability", Style, 20.3, 1986, 319–40.

[8] B. Richardson, "Narrative poetics and postmodern transgression: Theorizing the collapse of time, voice, and frame", Narrative, 8, 2000, 23–42.

[9] O. Langmead, "'Grant us eyes, grant us eyes! Plant eyes on our brains, to cleanse our beastly idiocy!': FromSoftware's Bloodborne, and the New Frontier of the Gothic", Press Start, Vol. 4, Num. 1, 2017, pp. 53–64.

[10] T. Rogers, "Bloodborne: You are the experience points", 2015, https://www.gamasutra.com/view/news/240839/Bloodborne_You_are_the_experience_points.php [accessed on 15 June 2018].

[11] F. Schniz, "Skeptical Hunter(s): A Critical Approach to the Cryptic Ludonarrative of Bloodborne and Its Player Community", In The Philosophy of Computer Games Conference, Malta, 2016.

[12] Ball, K. D. "Fan Labor, Speculative Fiction, and Video Game Lore in the Bloodborne Community." Transformative Works and Cultures, 25, 2017.

[13] Bizzocchi, J. and Tanenbaum, J. "Well read: applying close reading techniques to gameplay experiences". In Well played 3.0, Drew Davidson (Ed.). ETC Press, Pittsburgh, PA, USA, 2011, 262–290.

3. Design by Play: Utilizing Sims 4 in Preliminary Architectural Design

Güven Çatak
Faculty of Communication Department of Digital Game Design Bahcesehir University Istanbul, Turkey guven.catak@comm.bau.edu.tr

Çakır Aker
Faculty of Communication Department of Digital Game Design Bahcesehir University Istanbul, Turkey cakir.aker@comm.bau.edu.tr

Selin Sop
Institute of Social Sciences Game Design Graduate Program Bahcesehir University Istanbul, Turkey selinsop@gmail.com

Abstract— The aim of this study is to determine if the bestselling PC game The Sims 4 can be used as a tool for interior architectural design. The life simulation game comes with a highly detailed "build mode" where players can build & decorate their dream houses from scratch. Being fairly easy to learn and efficient to use, the build mode of the game might prove to be an alternative to popular interior design software such as AutoCAD, 3D Max and SketchUp; which are professional tools with high learning curves and long rendering times. To analyze the effectiveness of the game being used in the context of preliminary design, a draft design project using the game was tested. Moreover, to inspect if and/or how much the game makes a positive impact on the process from the architects' perspective, qualitative tests via observations and semi-structured interviews were conducted. As a result, it was identified that The Sims 4 build mode might provide the essential tools for a fast paced drafting stage. Moreover, even in its flawed state, Sims 4 build mode can be used for preliminary architectural design, however for limited purposes only.

Keywords— Game Design; Sims; Preliminary Design; Architectural Design Tool; Playful Experience

I. Introduction

Interior design and architecture, apart from being a professional field of study; has a lot of unprofessional enthusiasts. This fact is already being exploited by the gaming industry, and as a result there are many virtual home decoration games published in almost all platforms. These games provide little else than leisure.

However, the possibility of gaining practical, even professional benefit from such games still lies unexplored. Virtual environment building games hold the potential of not just providing casual entertainment to enthusiasts, but aiding professional interior architects in their line of work as well.

During the early stages of an interior architectural project, the architect produces many drafts to outline the work flow, and to estimate the final look of the project. At further stages of planning the project, there are lots of things to consider; such as constraints, proportions, measurements; hence the use of a professional tool is a must. On the other hand, during the preliminary project stage, which generally includes the process of sketching a rough draft for the purpose of showing the project idea to the client; the visualization might not need such realistic boundaries. The main point at this time is to understand the needs of the client and sketching it not for visualizing the project for yourself as the architect, but to get confirmation from the client. Several researchers [1, 2] have mentioned that designers can benefit from abstract external representations (i.e. sketches) particularly when they are in the early conceptual design stages. However, as Ibrahim and Rahimian mentioned, conventional analogue format of design ideation tools such as sketches, are yet to be replaced with appropriate digital formats [3].

With the advancement of technology and access to information; the modern and popular approach of all creative industries has shifted into the direction of getting the consumer/customer more involved. Interior architecture and design is no different. As Coleman states, "Design is becoming more open and democratic-architects and designers can no longer hope to avoid a genuine creative dialogue with powerful and articulate end users." []. However, the majority of existing geometric modelling software entail a high degree of specialization from the users to achieve the final forms that designer's expect. However, not all designers and/or enthusiasts can and need to reach this distinctive degree of skills [5]. As a result, these limitations may hamper the effectiveness of the design process and the collaboration that goes along with it [6].

In this study, we explored the possibility of using a game, 'Sims 4' as an alternative drafting tool for interior architectural design. It's possible to create realistic looking environments in Sims 4, and achieve visuals somewhat similar to preliminary project sketches drawn/modeled in a conventional design software. To determine if Sims 4 can indeed be used as a tool, we have conducted case studies with interior architects; and get them to execute the preliminary project step for a presumed interior architectural design contract, with another test subject assuming the role of "client" present. By doing so, we looked into if/how much the game makes a positive impact on the process from the architects' perspective. We restricted the case study for preliminary project design stage only;

as the tools within the game will not be sufficient to fulfill the requirements of further, more advanced planning.

II. Related Work

Many researchers have conducted studies regarding the use of sketches compared to design software. These studies involve the evaluation different media during a short-term design activity in a laboratory condition involving design students [7, 8, 9]. The number of these studies compared conventional computer aided design (CAD) tools against conventional sketching approach and concluded that manual sketching tools are superior compared to conventional CAD media during the conceptual architectural design process.

Radford [10], conducted a study on utilizing games for learning about architectural forms. The study aimed to inspect the relation between games and the act of play with developing student confidence and ability in spatial modelling, designing compositions and form creation. The study argued that learning about form in Architecture through playing games was beneficial to students.

Correlatively, utilization of games and playful interactions for design inspirations started to be seen often. The Swedish home furnishing company IKEA had made investments in both VR and Augmented Reality. The company first released a virtual kitchen decorating app for Steam platform on April 2016 [11]. The users were able to tour around a virtual kitchen, and furnish it with IKEA products. Afterwards, on September 2017 an augmented reality app was released for IOS and later Android. The app received mixed feedback, however it was still very innovative and ground breaking being the first of its kind.

In terms of the game at hand of this study, the Sims games has been the focus of many academic studies in the past. Most of these studies are either about the narrative aspect of the game, or about analyzing the game within sociological norms. One exemplary study regarding the use of the game in terms of design was conducted in Essex University (UK), using a modified version of Sims with their custom built simulator, in terms of analyzing the sustainability of intelligent buildings [12]. They were able to prove that the computer game software can be beneficial for creating such systems.

III. The Preliminary Stage and Current Software

Preliminary design is the process of visualizing the project in realistic, but flexible manner. Different than concept design, the scales and proportions of details are accurate; as well as the indication of main architectural elements (such as columns, walls, doors etc.), but it lacks the technical detail required for application, and it is highly stylized. Therefore, they mention that the CAD tools are not much effective and desired during the conceptual phase. Preliminary design schematics lets the architect/designer to communicate with their clients, as a mean of visual common language, as this is the only way for the client to comprehend to outcome of the process. An important thing to note, is that the architect avoids making solid decisions, as well as spending a lot of time shaping the details at this stage; as everything is subject to change with customer feedback. [13]. A study by Kwon et al. [6] attributes this inadequacy to the limitation of intuitive sketching capabilities of the conventional computer aided design (CAD) software.

There are several software being used during preliminary design stage in the architecture industry. AutoCad is one of the most CAD software used by the industry. It's known for its ability to produce extremely realistic looking visuals; as well as its high price, difficult learning curve and incompatibility with other software. Similarly, SketchUp software is being employed for creating preliminary design drafts. It was developed much later than AutoCad, and it's mostly preferred by the new generation of professionals. It is fairly easier to learn and use, has a vast asset library, and allows access to 3rd party plug-ins. Without additional plug-ins however, SketchUp 7 is not equipped to create high quality 3d visuals, and lacks controlled light sources; which is vital for realistic interior design renders. Although more advanced in terms of complexity, 3Ds Max is also widely used for producing 3D assets and visuals. It's more of a 3D modeling and animation software, so it has a lot of tools completely irrelevant to interior architecture design; therefore it has a complicated interface. It requires an in-depth knowledge of the program regarding which functions can be used for interior architectural design. In addition, it is slow and expensive like the other software. Similarly, Revit is a highly detailed software for architects, engineers, contractors and designers. It supports a modeling workflow that enables people from different disciplines to collaborate simultaneously. Revit is highly flexible, and widely used for its ability to create unified models that contain real-life data.

Because of various reasons including ease of use, decreased costs and the possibility of sharing options, The Sims 4 game has the potential to be used as an alternative drafting tool for interior architectural design. First of all, the game is equipped with a lot of building tools and assets to make this possible. Everything is designed from scratch in 3D, and designs can be exported and shared as high quality visuals that are pretty realistic and pleasant looking. Furthermore, The Sims 4 can run on low end machines, as it was specifically made to be played on laptops. This enables the architect to simply take their laptop, go to the project site and start sketching everything right on spot. And finally the most obvious advantage is the price of Sims 4, which is as of today around 100 times cheaper compared to conventional CAD prices.

IV. The Game

The Sims is a life simulation game first published by Maxis in 2000. The original creator of the game Will Wright; who had already developed numerous simulation games in the past; had the idea of making a virtual doll house after losing his own house in a firestorm. [14] The franchise continued to this day, and its counted amongst the most successful computer games ever made; having sold near 200 million copies worldwide. [15].

The Sims games has been the focus of many academic studies in the past. Most of these studies are either about the narrative aspect of the game, or about analyzing the game within sociological norms. There is a one study conducted in Essex University (UK), using a modified version of Sims with their custom built simulator, in terms of analyzing the sustainability of intelligent buildings [12]. They were able to prove that the computer game software can be beneficial for creating such systems.

Sims titles, staying loyal to the idea of creating a "doll house" to play with, always came with a separate in-game mode called the "Build Mode". Sims 4 is no

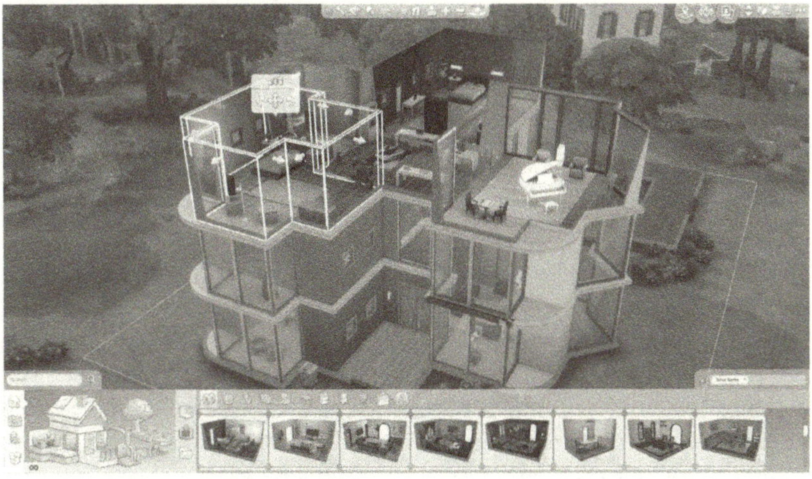

Figure 1: The Sims 4 build mode

different, with the exception of merging the "Buy Mode"; a separate mode present in previous 11 titles which was for furnishing and decorating purposes only, within the build mode itself. For a rather significant amount of Sims players, the build mode is the real game itself alone. [16] These players prefer to immerse themselves with the creative process of building / decorating lots rather than actually playing the life simulation aspect of the Sims games.

As seen above (fig.1), the build mode can be accessed by clicking the tools icon on the interface, located at the upper right corner of the game screen. This causes the game to go into pause mode, and the interface changes. There are a total of four main sections on the build mode interface, in which three of them are observed to be useful regarding preliminary interior design.

V. Methodology

To test the usability of Sims 4 build mode for preliminary design purposes, we conducted the following qualitative research methods;

The designer/architect was asked to work on a case study, using Sims 4 build mode to draft a project. The subjects were professional architects working in the industry. They had a time limitation of one and a half hour to work on the program. Both semi-structured interviews and questionnaire methods were used to collect data from participants, depending on their availability and the location of where the case studies were conducted. There were three different sets of questions, marked as Q1, Q2 and Q3 (Table 1). Q1 was conducted before each participant starts the case study, and it was aimed to help gather information about which architectural CAD software they use, the reasons behind their preference in using them, and strong/weak aspects of these programs from their personal point of view. Second part of the questionnaire, Q2, is to assess their opinion of the Sims 4 build mode and what they like/dislike about it. Finally, during Q3, each participant was asked their final verdict on rather or not they would want to use Sims 4 build mode for their preliminary design projects.

We have selected 9 participants who are all professionals with architectural education backgrounds, with varying years of experience in the industry with a minimum of 5 years of experience. For the tests, a MacBook Pro (Retina, 15-inch, Late 2013) 2,3GHz Intel Core i7 16GB Ram, with Nvidia gt750m graphics card was employed. To analyze the data collected, firstly a deductive approach was conducted to determine similarities, patterns and differences in participants' answers. Everything each participant have mentioned was transcripted individually. After this stage, an inductive approach was taken to determine relations

TABLE I. SET OF QUESTIONS USED DURING THE INTERVIEWS

Q1	*Q2*	*Q3*
Which CAD software(s) do you currently use?	How was your overall experience?	Would you use Sims 4 build mode?
Wjhat are the key factors for you to prefer using a CAD software?	What were the positive aspects of the Sims 4 build mode?	Would you use Sims 4 build mode if improvements were made?
What are the pros of the CAD software(s) you are currently using?	What were the weak aspects of the Sims 4 build mode?	Would you use a new CAD software with positive aspects of Sims 4?
What are the cons of the CAD software(s) you are currently using?	Do you recognize any properties of the Sims 4 build mode that you wish to see in other CAD software?	

and connections between the contexts of their answers. Each part of the questionnaire was visualized with mind maps, to indicate relations and connections, and how often each key concept was mentioned by the participants.

VI. Results & Discussion

9 out of 9 participants indicated that they use AutoCAD, making the most commonly used CAD software among our participants. It's followed by Revit (7/9), 3ds Max and SketchUp (both 4/9), and finally Unreal Engine and ArchiCAD (both 1/9). The key factors that designate our participants' choice of CAD software are; technical properties (6/9), popularity in the industry (also 6/9), susceptibility for revisions (5/9), interface (3/9) and learning curve (3/9). Other side factors that our participants value are interdisciplinary coordination (4 out of 9 people stated they use Revit especially for this reason) and realistic visuals (2 out of 9 people complimented 3ds Max for enabling this factor). The results confirm that neither of these CAD software can be considered flawless, and standalone sufficient to fulfill the architects' needs during preliminary design. Most common complaints were lack of visual reference of progress for AutoCAD (4/9), their dependence of 3rd party programs (3/9) for both AutoCAD and SketchUp, long render times (2/9) and inflexibility for making changes (2/9) on 3Ds Max.

In Q2, when asked our participants to comment on the positive and negative aspects of Sims 4 build mode. All case study subjects said they had an overall positive experience when they were trying out Sims 4 build mode. 2 out of 9 participants stated they were already familiar with it as they played the Sims 4 before. Two participants mentioned the software can be used by people from other lines of work. One participant stated that interior designers, not necessarily with architectural education background might benefit from the program a lot more than a regular architect would, as the software focuses on design and not technicalities. Real estate offices were also mentioned, as they could use the game to showcase replicas of the residences they have in their portfolios instead of making table-top models. Two participants stated that the game looks very convenient to make presentations of their design to the customers. Furthermore, individual participants complimented the useful shortcuts, ready-to-place assets for quick planning, and the existence of visual references to library assets which makes it extremely easy to find a specific object or tool.

As for the negative remarks the Sims 4 build mode received, there was one factor mentioned unanimously by all participants; the lack of proper measurement units present in the game. 9 out of 9 participants stated that the lack of measurement units made it extremely difficult for them to work with the program. In Sims 4 build mode, the environment is divided into grids. Although there is no official say regarding the exact size of these grids, looking at the size ratio of several objects give us an estimation of 1 grid being equal to 0,7*0,7 m2. This is in no way official or definite. As for heights, no means of measurement exist. Not even grids. The objects that require height simply come in three different sizes, and there is no way to measure anything except comparing it with the size of other objects. 9 out of 9 case study subjects claimed this was a huge downside of the program, and the main reason why it needs some update if it was to be recognized by the industry as a legitimate tool for creating preliminary design projects. Other negative comments made about the program were individual to each participant. One of them said the visuals exported from the program for presentation purposes were not hyper-realistic as they prefer. One participant had a little trouble with camera controls during their first 10 minutes using the software, although they also stated this was not a significant downside. The lack of an exact 2D top down view and the inability to export designs in formats compatible with other CAD software were also individually mentioned.

Finally, at the end of case study, the participants were asked to give a final verdict to determine if they believe Sims 4 build mode might be beneficial for preliminary architectural design. When they were asked if they would ever use Sims 4 build mode (in its current state, with its current flaws present), nobody gave a negative response. 2 out of 9 people straightforwardly replied "Yes.", while the others chose to limit their usage for more specific purposes; such as "Only for interior design." (3/9) or "Only for presentation." (2/9). To the question of rather or not they would choose to use an updated/modded version Sims 4 build mode; where their concerns regarding measurement units and lack of assets (and others) were addressed and corrected; 8 out of 9 participants replied "Yes." The remaining participant explained that she would not use the program personally, but would give the task of using it to one of her junior employees in her firm. To the last question of whether or not they would be interested in trying a new architectural CAD software containing the elements they liked about the Sims 4 build mode; again the answer was a unanimous "Yes." (9/9).

Looking at the results and interpreting them in detail, we can say with confidence that these professionals saw a potential in Sims 4 build mode, and confirmed that it can be beneficial for their work. Their feedback showed us that even in its flawed state, Sims 4 build mode can be used for preliminary architectural design, however for limited purposes only; and that its potential would increase even more if certain flaws were corrected. As a result of the qualitative tests, it was possible to make suggestions to increase the games potential for being preferred by architects as a recognized software for preliminary design.

The Sims community, starting from the very first game, had been extremely active about creating and adding custom content to the game. Apart from tools like Photoshop, Maya, 3Ds Max etc.; the custom content creators of the Sims community usually use these two tools for creating additional content; The Sims 4 Studio (a modding software developed by a player that enables making textures, recoloring, meshing etc.) and Blender (an open source software for creating 3d models). Both of these programs have detailed tutorials. Even if the architect is not interested in using such a program to create their own material, they can simply browse Sims custom content websites such as The Sims Resource and have access to extra textures, materials and object models that the library was originally lacking. It is possible to add limitless custom content to the game's built-in library.

Writing a simple script to add measurement units, by defining each grid as the equivalent of one meter square; and defining an additional vertical grid system to determine height in a similar fashion, would fix the problem the case study architects were addressing.

Lastly, there are already visual references regarding how much building something would cost in the game. This feature could be further enhanced by giving the architect the option to personally define the cost of each texture/material used; so while they are building within the game they will be able to determine the cost of the project more realistically.

VII. Conclusion

The main purpose of this study was to test to usability of Sims 4 build mode as an alternative to computer aided design software for making preliminary design schematics. We first went through the essential steps of architectural design, narrowing our focus on the preliminary stage as that was the process we wanted to base our study on. Briefly exploring three popular computer aided design software (AutoCAD, SketchUp and 3Ds Max), we summarized each software's advantages/disadvantages in terms of time, effort and resource. Later we looked into the Sims games franchise and especially Sims 4 and its build mode. We conducted nine case studies with nine professional architects, giving them the task of building an industrial themed penthouse within a time limit of one and a half hours using the Sims 4 build mode. Simultaneously, we conducted a semi structured interview with our participants, asking them questions before, during and after the case study.

After collecting and analyzing the responses from our participants, we saw a common pattern for both positive and negative criticism of the game. Regarding the supporting feedback from our case study participants, we finally came to the conclusion that it is possible to use the Sims 4 build mode as a beneficial tool for preliminary architectural design projects, though mostly for limited purposes in its current state. The game holds potential, yet also have a lot of room for improvement; as our participants agreed they would find more use for it if certain alterations were made. In the last part of our discussion, we talked about what these improvements might be, and how they might be made.

The Sims community, starting from the very first game, had been extremely active about creating and adding custom content to the game. Apart from tools like Photoshop, Maya, 3Ds Max; the custom content creators of the Sims community usually use these tools; Sims 4 Studio (a modding software developed by a player that enables making textures, recoloring, meshing etc.), Blender (an open source software for creating 3D models). Both of these programs have detailed tutorials. Even if the architect is not interested in using such a program to create their own material, they can simply browse Sims custom content websites such as The Sims Resource and have access to extra textures, materials and object models that the library was originally lacking. It is possible to add limitless custom content to the game's built-in library.

Writing a simple script to add measurement units, by defining each grid as the equivalent of one meter square; and defining an additional vertical grid system to determine height in a similar fashion, would fix the problem the case study architects were addressing. Adding visual representation of unit measurement to the game is a simple task. There is already visual references regarding how much building something would cost in the game. This feature could be further enhanced by giving the architect the option to personally define the cost of each texture/material used; so while they are building within the game they will be able to determine the cost of the project more realistically.

Developing an update for the game, with the content explained above; would make the process more beneficial as the participant architects also agreed during the semi structured interviews as well as might hold potential for further improvement of CAD software.

References

[1] M. Schütze, P. Sachse, A. Römer, Support value of sketching in the design process, 2003, pp. 89–97

[2] M. Suwa, T. Purcell, J. Gero, Macroscopic analysis of design processes based on a scheme for coding designers' cognitive actions, 1998, pp. 455–483

[3] Ibrahim & Rahimian, Comparison of CAD and manual sketching tools for teaching architectural design, 2010

[4] Coleman C. Interior Design: Handbook of Professional Practice, New York, McGraw-Hill, 2002.

[5] F. Levet, X. Granier, C. Schlick, 3D sketching with profile curves, LNCS, 4073, 2006, pp. 114–125

[6] J. Kwon, H. Choi, J. Lee, Y. Chai, Free-Hand Stroke Based NURBS Surface for Sketching and Deforming 3D Contents, PCM 2005.

[7] Z. Bilda, H. Demirkan, An insight on designers' sketching activities in traditional versus digital media, 2003, pp. 27–50

[8] K. Meniru, H. Rivard, C. Be´dard Specifications for computer-aided conceptual building design, 2003, pp. 51–71

[9] C. Stones, T. Cassidy, Comparing synthesis strategies of novice graphic designers using digital and traditional design tools, 2007, pp. 59–72

[10] Radford, A. Games and learning about form in architecture. Automation in Construction, 9(4), 2000, pp. 379–385.

[11] Dasey D. . IKEA Augmented Reality App: Try Before You Buy [online], 2017, https://highlights.ikea.com/2017/ikea-place/ [accessed April 30 2018]

[12] Davies M., Callaghan V., Shen L. Modelling Pervasive Environments Using Bespoke and Commercial Game-Based Simulators. In: Li K., Li X., Irwin G.W., He G. (eds) Life System Modeling and Simulation. LSMS 2007. Lecture Notes in Computer Science, vol 4689. Springer, Berlin, Heidelberg, 2007.

[13] Kilmer R., Kilmer W. O. Designing Interiors, New Jersey, John Wiley & Sons Inc. 2014, p 184–190.

[14] Taylor T., Will Wright: Inspired to Make The Sims After Losing a Home, Berkeleyside [online], http://www.berkeleyside.com/2011/10/17/will-wright-inspired-to-make-the-simsafter-losing-a-home. 2011. [accessed April 30 2018]

[15] Rhinewald S., McElrath-Hart, N., 2016 World Video Game Hall of Fame Inductees Announced, Museum of Play [online], http://www.museumofplay.org/press/releases/2016/05/2688-2016-world-videogame-hall-fame-inductees-announced, 2016. [accessed April 30 2018]

[16] LeTourneau T.,The Sims Builders, Games Radar [online], https://www.gamesradar.com/the-sims-builders/. 2006. [accessed April 30 2018]

4. Exploratory Research on the Gamification of Exercise for Fibromyalgia Using Virtual Reality

Katherine Hoolahan
Virtually Healthy LTD katie@virtuallyhealthy.co.uk

Abstract— Fibromyalgia (FM) causes debilitating pain and muscular stiffness, amongst other symptoms, which can be eased over time through regular exercise. As FM patients have an abnormal perception of pain, adherence to exercise programmes can be difficult due to the pain experienced whilst exercising. Virtual reality (VR) has been shown to distract users from physical and mental pain and therefore may be utilized to increase adherence to exercise. This project involved the creation of a VR game using the HTC Vive, designed to induce exercise at a moderate intensity and distract Fibromyalgia patients from pain experienced whilst exercising. This game was tested using 8 participants who completed a questionnaire regarding the enjoyability of the game as well as the intensity of exercise whilst playing the game. All except two participants felt as if they had exercised at a moderate intensity or higher whilst playing the game, with the remaining two participants just under the threshold for moderate exercise. All participants enjoyed the game and all except one would play it again and spend money on this or similar games. The participant who would not play this or similar games again fell into the 65+ age range, supporting research which states that over 65s have little interest in gamified VR. Half of the participants answered that their frequency of exercise (for at least 20 minutes) would increase if they owned this or similar games, with the remaining participants already exercising regularly.

Keywords—Fibromyalgia; Chronic Pain; Exercise; Mental Health; Virtual Reality; Motivation; Adherence to Exercise

I. Introduction

Fibromyalgia (FM) is the second most common condition that is seen by rheumatologists [1]. FM is a chronic condition which causes widespread, debilitating pain and muscular stiffness, along with various other symptoms and common comorbidities such as fatigue, insomnia and psychological issues [2]. Patients suffering with FM have an abnormal perception of pain, reacting strongly to stimuli that would not normally be thought to be painful. It is thought that the pain leads to anxiety disorders and fatigue leads to depression [3]. These symptoms associated with FM lead to a poor quality of life (QoL), with an average 'perceived present QoL' scored at 4.8 on a scale of 1-10 (1 being low and 10 being high) [4]. Although there is no standardized treatment for FM, treatment which consists of a combination of approaches such as psychological interventions, medication and exercise programmes are the most effective [5]. Psychological interventions generally include Cognitive Behavioural Therapy and medication may include painkillers, antidepressants and neuromodulating antiepileptics [2]. Drugs have recently been developed to specifically manage pain in FM patients, however these are thought to come with significant side effects and the efficacy of using them is questionable [5]. Exercise programmes for FM focus on increasing cardiovascular

fitness, muscular endurance and flexibility [6][7]. Increased physical activity has been shown to lead to a decrease in reported pain as well as an increase in the response of pain regulatory functions within the brain. However, exercise programmes created for patients with FM generally have low adherence rates due to the increased pain experienced whilst exercising, leading to poor physical health and a poor QoL [8].

Gamification refers to the production of game-like experiences which promote flow, mastery, achievement and intrinsic motivation with an aim to alter the user's behaviour and perceptions [9].

This study explores gamified virtual reality (VR) as a means of increasing motivation to exercise, with game design focused for FM patients. The game was designed to provide a distraction to the physical pain experienced by FM patients whilst exercising, and increase QoL through increasing cardiovascular fitness and flexibility (as well as strengthening when used with body weights). Although the game has been designed for FM patients, it could also be used by able-bodied players to increase motivation to exercise, or simply for entertainment.

II. Literature Review

VR has been shown to distract users from anxiety, depression and physical pain, possibly even changing "the way that the brain physically registers pain" rather than simply the way it perceives painful stimuli [10]. Due to these benefits, the use of VR in the gamification of exercise would be beneficial to affect the motivation of FM patients to exercise and improve QoL. Recent research has proven the effectiveness of VR for improving the self-management of diseases, physical activity, distracting from pain and discomfort, rehabilitating neuro-cognitive and motor performance and assisting in the treatment of psychological issues such as depression and anxiety [11]. It has been found that the more immersive the environment, the more distracted the user is from physical and mental pain

[12]. The near complete absorption into the environment whilst using a Head-Mounted Display (HMD), such as the HTC Vive or Oculus Rift, is best suited to help with more intense pain [11].

The most widely accepted model in explaining the mechanisms behind the pain relief induced by using VR is the 'gate theory' of attention, which states how VR absorbs and diverts attention away from pain, therefore reduces the perception of pain. Research into the cognitive-affective-motivational role of attention in pain processing supports the importance of gaming and play elements in VR interventions aimed at the reduction of pain [11].

Exercise interventions in FM patients have been shown to improve well-being through decreased fatigue, depression and pain as well as increased pressure that must be applied to a tender point before a pain response is triggered [8][13]. It is recommended that exercise should be performed at a moderate intensity (approximately 60 % of maximum heart rate) 2–3 times per week [13]. Setting realistic goals and expectations is important to help deal with the condition throughout daily life. A positive mindset of the patient is important for actively endeavouring to reduce symptoms through methods such as exercise [6]. The assistance with depression and anxiety that gamified VR presents (especially when combined with exercise – a treatment for depression) should be able to help with this [11].

Conflict was found between patients and healthcare professionals due to the limited resources available (such as time for consultations), and psychological support was not provided in a meaningful capacity. A gamified experience of exercising, allows patients to take a more pro-active approach which may alleviate conflict between FM patients and healthcare professionals caused by time restraints. The use of gamified VR exercise may also reduce the need for psychological support, which is being sought from healthcare professionals who feel they cannot provide the required support [5].

Gamification is thought to consist of three main parts – implemented motivational affordances, resulting psychological outcomes and further behavioural outcomes. Motivational affordances include aspects such as points, leader boards, achievements, levels, clear goals, stories, themes, feedback, progress and challenge. Gamification of an exercise programme encourages participation and positively influences attitude and behaviour towards exercise whilst allowing for easy goal setting and progression [14]. Goh & Razikin [15] found that gamification of exercise increases motivation to participate by shaping behaviour as well as improving the enjoyment of and general attitude towards exercise. Gamification has also been shown to decrease perceived pain [16]. These benefits are particularly important in FM patients due to the psychological and physical

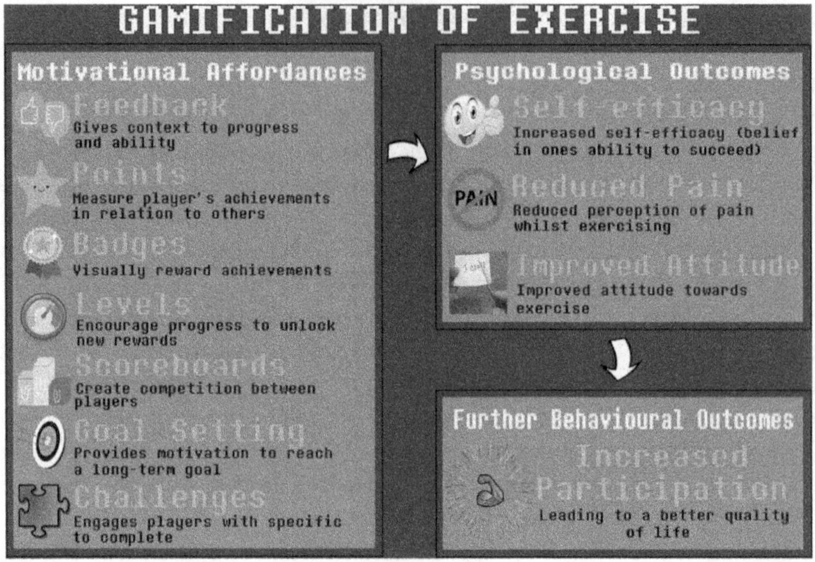

Figure 1: A breakdown of gamification in exercise

barriers to participation. Changing behaviour and perception of exercise may be the difference between inactivity, causing deterioration in QoL and an active lifestyle which vastly improves symptoms over time and increases QoL.

Although gamification mostly produces positive results, in some settings no effect is seen. This is thought to be down to a variety of factors such as the nature of the gamified experience, the motivation and preferences of the individual and design aspects [17]. Gamification may appeal to a large population of people, however there are populations and individuals that may not find these concepts as appealing. More consistently positive results have been found for individuals aged between 20–49 years, suggesting that this is the most effective age range for gamified VR. It is thought that people over the age of 65 have little interest in using VR [18]. These factors may affect the use of VR in therapy for some populations.

III. Methods

A. Game Development

Unity (5.6.0) [19] with SteamVR [20] was used to develop and build the game and a HTC Vive was used in the creation and playing of the game. Microsoft Visual Studio [21], Blender 2.78c [22] and Audacity 2.1.3 [23] were also used

Figure 2: Screenshots of the game being played

in the creation the game. The game was designed to induce moderate exercise which works the whole body – legs through bending to pick up a snowball, arms and chest through throwing (throwing with both arms is the most effective method) and the core through dodging snowballs.

B. Testing And Questionnaire

The inclusion criteria for participating in the testing of the game included; able-bodied or diagnosed with FM, physically fit enough to participate in moderate exercise and over the age of 18. Exclusion criteria included a diagnosis of any physical disability other than FM or chronic pain or any physical injury which may be affected by participation. Ethical approval was given for this study to be carried out. To ensure the safety of all participants, a risk assessment was also carried out and steps were taken to minimise the risks of participation. 8 participants were recruited who completed an informed consent form before participating.

The objectives of the game, the VR system and precautions to minimise the risks were explained to the participants before participation. Two 5-minute games were played – the first at an easy difficulty and the second at a slightly harder difficulty. After playing the game, participants completed a questionnaire containing 29 questions regarding the enjoyment, playability and effort put into the game. These were answered using a Likert scale with a section at the end for any comments. A table of showing the Rate of Perceived Exertion (RPE) scale [24] was also used to determine how hard the participant felt they exerted themselves whilst playing the game. The questionnaire was based off the 'need satisfaction' questionnaire [25] which evaluates exergames. Basic demographic data was collected to assess whether these factors may affect the answers given. In order to prevent bias, all answers were kept anonymous.

C. Data Analysis

Frequency tables, bar charts and pie charts were constructed to analyse the data as well as to create a visual representation of it. This data was analysed both qualitatively and quantitatively.

IV. Results & Discussion

The game was found to be fun, exciting, enjoyable and entertaining by all participants, as seen in figure 1.

Subjects 3 and 8 found the game to be boring; however, these participants also found the game to be fun, exciting, enjoyable and entertaining – which is contradictory. Subjects 3 and 8 also did not feel as if they had exercised at a moderate intensity, giving a rating of 11 on the RPE Scale (light exercise). This may imply that the game was too easy, which may be one reason as to why it was found to be boring and could be fixed by increasing the difficulty or adding in different game modes. The inadequate difficulty is supported by Subject 2's comment "if I was an actual gamer (regularly played console games) it would probably be too easy. Would be cool to have levels" as well as subject 4's comment "it would be more effective if there were more difficult levels". Subjects 3 and 8 also reported that the game was slightly uncomfortable to play. Subject 3 commented "maybe wireless goggles

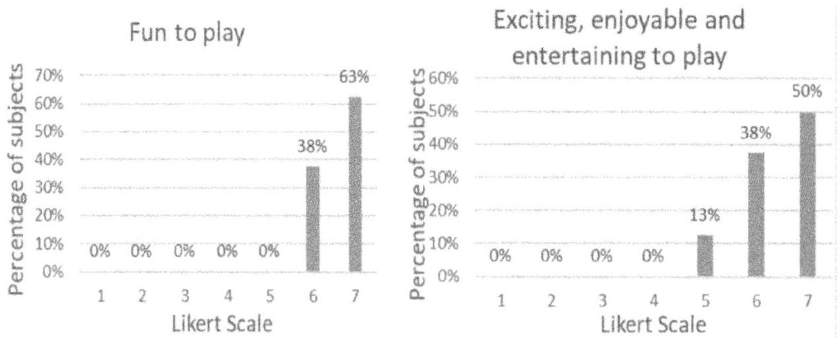

Figure 3. Bar chart showing questions regarding the enjoyability of the game. Likert scale: 1 = describes the game poorly, 7 = describes the game very well

would be better", suggesting that the trailing wire was the cause of the discomfort – the wire will likely become obsolete over time as the technology develops. However, this did not seem ruin the immersion as subject 3 still answered that they felt immersed in the game. Although subject 3 found the game to be boring and slightly uncomfortable to play, this participant would still play this game again and spend money on this or similar games. Subject 8, whilst finding the game to be fun, exciting, enjoyable and entertaining, also found it to be boring and slightly uncomfortable to play - this participant would not spend money on or play this or similar games again. As subject 8 is in the 65+ age range, the answers given support research that has shown that people aged 65 and over generally have little interested in VR and may not benefit as much from VR therapy [16]. Although subject 8 had not been diagnosed with FM, she believed herself to have FM, which may have contributed to the discomfort of playing this game. However, since no formal diagnosis of FM had been made for subject 8 and no comments were written, no conclusion can be made regarding the reason for discomfort.

All participants, with the exception of subject 8, indicated that they would play this or similar games once a week or more. Half of the participants answered that their frequency of exercise would increase upon owning this or similar games, this is shown in figure 2.

Subject 8 answered that frequency of exercise would significantly decrease upon owning this game. This is thought to be an anomaly - possibly due to not reading or understanding the question correctly. The remaining participants answered that they already exercised frequently – two of whom were already exercising at the maximum frequency given in the question, therefore frequency could not be increased. The final participant, subject 3, already exercised more than twice a week

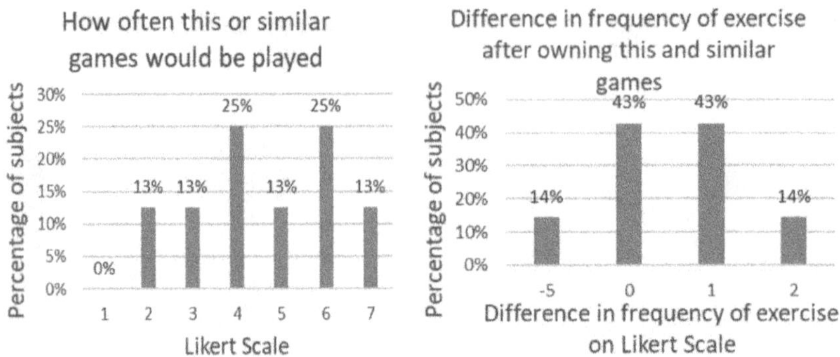

Figure 4. Bar charts showing questions regarding how often this or similar games would be played if owned. Likert scale: 1= never, 2 = occasionally, 3 = two or three times a week, 4 = once a week, 5 = twice a week, 6 = >twice a week, 7 = every day

and indicated that there would be no difference in frequency; which could be due to the large difference between exercising twice a week and exercising every day (the options given for 6 and 7 on the Likert scale). Therefore, if subject 3 would have exercised between one to three extra days extra per week, this may not have shown in the results of this question. This participant did however indicate they would play this or similar games more than twice a week, suggesting that their exercise frequency may increase. Subject 6, who already exercises daily, commented that the game was "surprisingly exhausting, however definitely a good workout and fun whilst being a challenge". Subject 1 also noted in the comments section "good way to distract someone, whilst getting exercise without thinking about it". This supports the gate theory of attention that the VR game acts as a distraction, taking away elements of attention that would otherwise be focused on pain or exhaustion from exercise. All participants also felt that they were immersed in the game. These points indicate that this game would likely distract FM patients from pain experienced whilst exercising – however further testing would be required to validate this assumption, and determine to what extent pain is eased.

All except 2 participants (subjects 3 and 8) felt that they had exercised at a moderate intensity or higher. Subject 7, who indicated that they exercised intensely (17 RPE) commented "I would like to start with 3 minutes work time up slowly". This indicates that a wider range of difficulties and different lengths of time for each game is required to meet the needs of different people, which could be easily implemented.

All participants indicated that they were motivated to get a high score and put a lot of effort into the game, as shown in figure 3, each getting a higher score

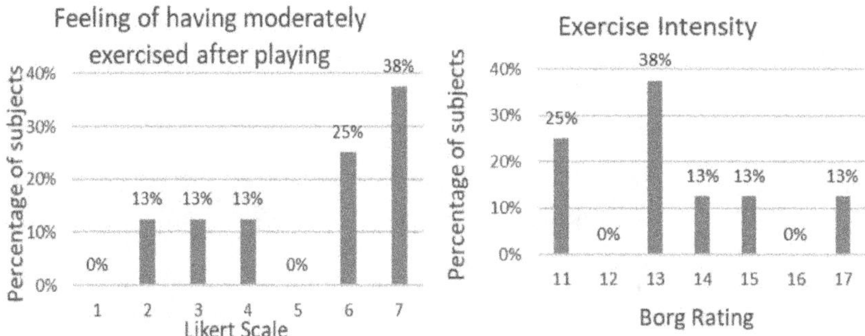

Figure 5. Bar charts showing the intensity of exercise whilst playing the game, using the RPE scale by Borg [24] & Likert scale: 1 = not true at all, 4 = somewhat true, 7 = very true

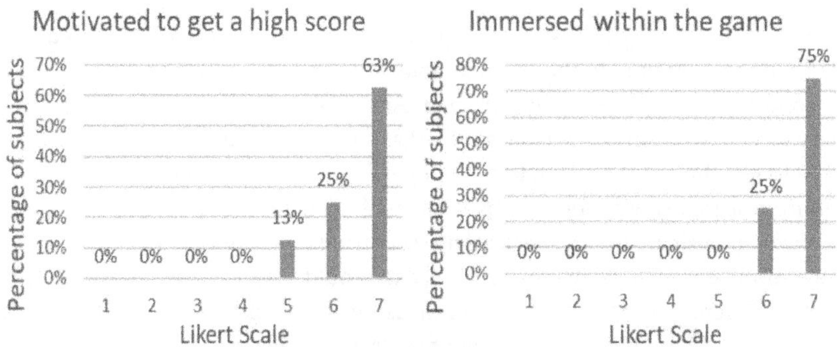

Figure 6. Bar charts showing questions regarding motivation and immersion within the game. Likert scale: 1 = not true at all, 4 = somewhat true, 7 = very true

on the second, herder, level than on the first. This shows that the motivational affordance of the scoring system had a positive effect on participants motivation and effort put into the game.

With the exception of subject 8 who did not answer the question, all participants answered that they would play this or similar games for 20 minutes or longer (up to 60 minutes) at a time. This meets and exceeds the 20 minutes of gameplay per session that this study was aiming for.

Overall, all participants rated the game as satisfactory to highly satisfactory. This, combined with the other results and comments, shows that the game was mostly successful at inducing moderate exercise in a fun and distracting manner,

Maximum time to spend playing this or similar games in one session (mins)

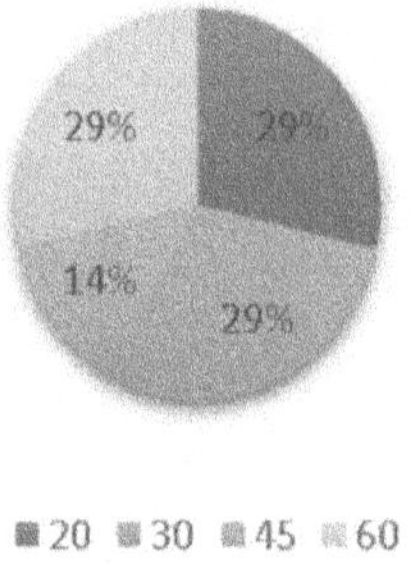

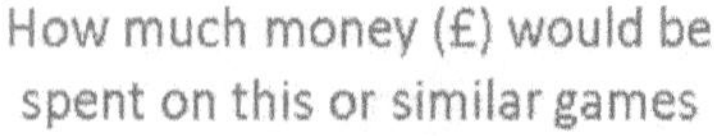

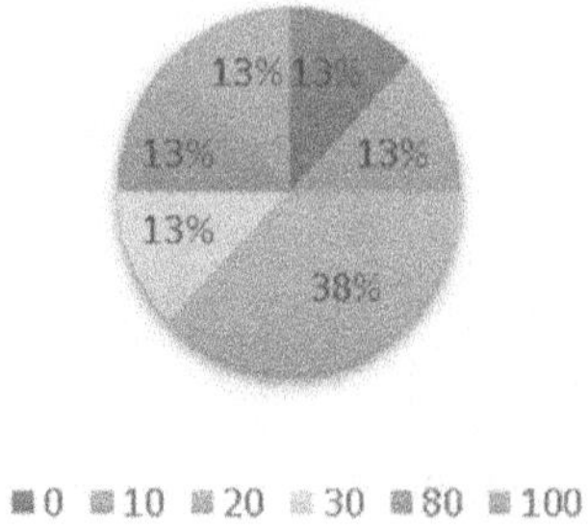

Figure 7. Pie charts showing the maximum amount of time per session and how much money would be spent on this game

which was satisfying to play. Although these results are not generalizable to a wider population due to the small sample size, this gives a good basis on which to conduct further research in this area. As this study only focused on preliminary research into the effectiveness of VR as a means of exercise, it was more feasible to test the game using able-bodied participants than participants with FM. Future research should focus on patients with FM and include a larger number of participants, including participants of various age ranges, to obtain more generalizable and reliable results. This data could also then be used to determine differences in the effectiveness of using VR to exercise between various populations. To obtain more accurate results, future research in this area should also test a wider variety of difficulties and length of games.

There are a variety of extras that could have been added into the game, however due to time constraints and ease of testing, the game was kept simple. Extras should be added in at a later date to increase variability, difficulty and induce different types of exercise. The implementation of an online scoreboard and platform to track progress would allow both participants and health professionals to monitor progress and set goals.

V. Conclusion

The game has been shown to be fun, exciting, entertaining and enjoyable whilst immersing participants within the game and providing a moderate intensity of exercise. Although two participants did not feel they had worked at a moderate intensity, their RPE was on the upper limit of light exercise, which may have been increased by raising the difficulty of the game, as testing was completed at a limited difficulty. Subject 8 (in the 65+ age range) found the game to be fun, exciting, enjoyable and entertaining, however would not choose to play this or similar games again. This supports previous research stating that people over 65 are uninterested in gamified VR - however as the younger population grows older alongside computer technology, this may change over time. The remaining participants would play again and spend money on this and similar games, with the majority answering that owning this and similar games would increase their frequency of exercise. With the exception of subject 8 who did not answer, all participants stated that they would play the game for 20 minutes or longer at a time - longer than was aimed for in this study. This suggests that the game would be effective on its own – however the option of playing other games as well would lower the risk of the one game becoming stale, i.e. a decrease in motivation to play that game over time. The scoring system provided motivation and incentive, with all participants stating that they put a lot of effort into the game and were motivated to obtain a high score. Online scoreboards would allow for competition between players as well as the option for allowing easy access for health professionals to monitor patients progress. From the results of this study, gamified VR is a viable means of exercise which is fun, exciting, enjoyable and entertaining and would likely increase motivation and frequency of exercise. This study provides preliminary research into the effectiveness of using VR for exercise, with a focus on FM. However further in-depth research is necessary for the benefits (and any disadvantages) of gamified VR to be fully understood.

References

[1] Clauw, D.J., Arnold, L.M. & McCarberg, B.H., 2011. The science of fibromyalgia. *Mayo Clinic Proceedings*, 86(9), 907–911.

[2] Bellato, E. *et al.*, 2012. Fibromyalgia syndrome: Etiology, pathogenesis, diagnosis, and treatment. *Pain Research and Treatment*, 2012, 1–17.

[3] Kurtze, N. & Svebak, S., 2001. Fatigue and patterns of pain in fibromyalgia: Correlations with anxiety, depression and co-morbidity in a female county sample. *Brit. J. Med. Psychol.*, 74(4), 523–537.

[4] Centers for Disease Control and Prevention, 2015. *Fibromyalgia.* Available at: www.staff.science.uu.nl/~ooste108/ExpC/website5/v1/arthritisbasicsfibromyalgia.html. Accessed [25/07/17].

[5] Briones-Vozmediano, E., Vives-Cases, C., Ronda-Pérez, E. & Gil-Gonzales, D., 2013. Patients' and professionals' views on managing Fibromyalgia. *Pain Research and Management*, 18(1), 19–24.

[6] Hawkins, R., 2013. Fibromyalgia: A clinical update. *J. Am. Osteopath. Assoc.*, 113(9), 680–689.

[7] Redondo, J.R. *et al.*, 2004. Long-term efficacy of therapy in patients with fibromyalgia: A physical exercise-based program and a cognitive-behavioral approach. *Arthritis Care & Research*, 51(2), 184–192.

[8] Busch, A.J. *et al.*, 2011. Exercise therapy for fibromyalgia. *Current Pain and Headache Reports*, 15(5), 358–367.

[9] Huotari, K. & Hamari, J., 2012. Defining gamification - A service marketing perspective. In *MindTrek*. Tampere: ACM, 17–22.

[10] Wiederhold, B.K., Riva, G. & Wiederhold M.D., eds. *ARCTT 2015: Virtual Reality in Healthcare*. 158–162.

[11] Trost, Z. *et al.*, 2015. The promise and challenge of virtual gaming technologies for chronic pain: the case of graded exposure for low back pain. *Pain Management*, 5(3), 197–206.

[12] Tong, X., Gromala, D., Amin, A. & Choo, A., 2015. The design of an immersive mobile virtual reality serious game in cardboard head-mounted display for pain management. In *Proc 5th Int. Symp. Pervas. Comput. Paradig. Mental Hlth*. Milan: Springer International Publishing, 284–293.

[13] Busch, A.J., Barber, K.A., Overend, T.J., Peloso, P.M. & Schachter, C.L., 2007. Exercise for treating fibromyalgia syndrome. In *Cochrane Database of Systematic Reviews*. Chichester, UK: John Wiley & Sons, Ltd.

[14] Hamari, J. & Koivisto, J., 2013. Social motivations to use gamification: an empirical study of gamifying exercise. In *Proc. 21st ECIS*, Utrecht, Netherlands, June 5–8.

[15] Goh, D. & Razikin, K., 2015. Is gamification effective in motivating exercise? *Human-Computer Interaction*, 608–617.

[16] Brauner, P., Valdez, A., Schroeder, U. & Ziefie, M., 2013. Increase physical fitness and create health awareness through exergames and gamification the role of individual factors, motivation and acceptance. In *Human Factors in Computing and Informatics*. Berlin: Springer-Verlag, 349–362.

[17] Hamari, J., Koivisto, J. & Sarsa, H., 2014. Does gamification work? — A literature review of empirical studies on gamification. In *47th HICSS*. Hawaii: IEEE, 3025–3034.

[18] Johnson, C., 2016. *The future of connected home health*, Great Chesterford: Plextek. 1–16.

[19] Unity (5.5.6), 2017. *Unity 3D*. Available at: https://unity3d.com. Accessed [20/03/17].

[20] Unity 3D, 2017. SteamVR Plugin. Available at: www.assetstore.unity3d.com/en/#!/content/32647. Accessed [20/03/17].

[21] Microsoft Visual Studio, 2017. Visual Studio. Available at: www.visualstudio.com. Accessed [07/06/17].

[22] Blender 2.78c, 2017. *Blender*. Available at: www.blender.org. Accessed [20/03/17].

[23] Audacity 2.1.3., 2017. *Audacity*. Available at: www.audacityteam.org/. Accessed [07/06/17].

[24] Borg, G.A.V., 1982. Psychophysical bases of perceived exertion. *Medicine and Science in Sports and Exercise*, 14(5), 377–382.

[25] Peng, W., Lin, J.H. Pfeiffer, K.A. & Winn, B., 2012. Need satisfaction supportive game features as motivational determinants: An experimental study of a self-determination theory guided exergame. *Media Psychol.*, 15, 175–196.

5. Deconstructing Game Stories with Propp's Morphology

Story Patterns in Role Playing Games and Cultural Differences

Barbaros Bostan
Game Design Department Bahcesehir University Istanbul, Turkey barbaros.bostan@comm.bau.edu.tr

Orcun Turan
Game Design Department Bahcesehir University Istanbul, Turkey orcun.turan@outlook.com

Abstract— This paper is concerned with the following: (1) the applicability of Propp's morphology to game stories; (2) the identification of the required changes in the original framework to analyze game stories on an act/mission level; (3) the discovery of the logical sequence of story functions that form story patterns; (4) and the identification of commonly repeated story patterns as well as the cultural differences between American and Japanese game stories. The scope of this study is limited to Role-Playing Games (RPG); believing that the narrative value of RPGs coupled with the freedom of choice they offer, make this genre more suitable for a structural study of game narrative. To focus on the cultural differences and to identify repeating story patterns, three RPGs from the east and three RPGs from the west are selected. Each game's overall story is broken down into acts and then each act is mapped to Proppian functions.

Keywords— game stories; game narrative; story analysis; narrative patterns; game stories and culture; Propp's morphology

I. Introduction

The structuralist study of the narrative [1,2,3] is concerned with the basic units of stories, such as actions, characters, and objects and influenced researchers from various fields. The most popular and well-known narrative study is 'Hero's Journey' [4] and the revision of it [5] but another influential study for the entertainment industry is Propp's Morphology of the Folktale [6]. In his study, Propp identified seven spheres of action and thirty-one functions that fit Russian folk tales. He is primarily concerned with an actor-action analysis but he also described how his framework might be used to generate tales. The entertainment industry, especially video games, is interested in the automatic construction of story plots and realized that the morphology may act as a blueprint for story generation systems.

Propp's morphology has been criticized for: having a linear sequence whereas computer game sequences are recursive and circular [7]; having a fixed sequence of events unrolling over time where the tale is destroyed if the sequence is broken [8]; being too constraining for an interactive drama where player actions are part of the resulting tale [9]; having severe limitations to story generation when used beyond its intended setting and the corpus where it arose from [10]; and being incapable of handling player choice or freedom [11]. In this context, researchers [12,13] also argued that since the conventional author and reader roles are not applicable to computer games, it is impossible to de-construct game narratives based on structural approaches. Therefore, it can be argued that the narratological theory of Propp may be more suitable for sequential stories or quest type storylines rather than interactive ones, though it has already been applied to a number of game studies that have a narrative focus. The framework so far has been used for developing an augmented-reality interactive storytelling system [14], generating story plots based on case-based reasoning [15,16], providing narrative macro structures and events for both the story and the discourse levels of a narrative generation system [17,18].

This article is concerned with the building blocks of a story from a structuralist approach and is interested in the applicability of Propp's model to computer game stories with special emphasis on RPGs. In this regard, rather than simply testing the suitability of the model for game stories, we aim to revise the original Proppian functions and identify new ones if necessary. Furthermore, we are also interested in finding the logical sequence of these story functions to define story patterns and determine the commonly repeated story patterns as well as the cultural differences between American and Japanese game stories.

II. Morphology of the Folktale as a Methodological Framework

Narrative analysis of games based on the popular story structure of Vogler [5] can be found for God of War [19] and Assassin's Creed [20] series, however, our primary aim is to question if Propp's structural narratology can be used to analyze the stories of computer games. More specifically, we are interested in finding repeating story patterns and identifying the cultural differences between game stories. The applicability of Propp's functions to game stories has already been assessed by Brusentsev, Hitchens and Richards [11] with three different games from three different genres (First Person Shooter, Action Adventure and RPG). The study found that Propp's functions can be mapped to the selected games at a higher level (overall story arc) and indicated that some additions are necessary to the original model to apply it to interactive stories. The researchers also stated that the selected RPG, Dragon Age Awakening, offers more freedom of choice and is more non-linear in terms of story progression. In this sense, the narrative value of RPGs coupled with the freedom of choice they offer directed our study towards this genre.

The scope of this study is limited to RPGs, believing that this genre is more suitable for a structural study of game narratives. Instead of selecting unrelated games, we chose a game series from the east and a game series from the west in order to facilitate the discovery of repeating story patterns. We selected three RPGs from the east (Final Fantasy X, Final Fantasy XII and Final Fantasy XIII) and three RPGs from the west (Mass Effect, Mass Effect 2 and Mass Effect 3) to focus on the cultural differences and to identify repeating story patterns. Massive multiplayer online RPGs, Final Fantasy XI and XIV, are omitted since studying massively multiplayer online environments is beyond the scope of this article. Each game's overall story or main quest is broken down into acts and then each act is mapped to Proppian functions. This approach takes the structural game story analysis of Brusentsev, Hitchens and Richards [11] to a lower level where each story act is analyzed individually.

Proppian approaches to analyze different folk tales and game narratives point to the fact that all the functions of the framework cannot be observed in every study. Wama and Nakatsu [21] found four storylines in Japanese folktales instead of the single storyline in Russian folktales and pointed out the need for a new morphology. When adapted to Sesotho Folktales, an average of seven functions are observed in all folktales [22]. A study that analyzed MMORPG quest structures in terms of Propp's narratology found 13 essential Proppian functions and added five more to the original framework [23]. At the quest level of narrative, Brusentsev and his colleagues [11] found partial mapping of Proppian functions to game segments. Thus, our first assumption is that changes in the original functions might be needed and the introduction of new functions might be necessary to cover all the acts of selected RPGs.

Differences in the American and Japanese stories have also been identified by various studies. Lanham and Shimura [24] examined folktales from two countries and identified differences in terms of apology, forgiveness and change of character. Similar differences can also be found in animated films for children where generic elements of 'western' animated examples are largely absent from the 'eastern' narratives [25]. A comparison of American and Japanese comics draws attention to the individualism vs. collectivism opposition and identified the independency vs. interdependency with other characters as a major difference [26]. In terms of RPGs, the Western hero of solitude is replaced by a 'strength through unity' notion in the East [27]. Thus, our second assumption is that cultural differences between the stories of American and Japanese RPGs will be revealed in the story structures and repeated story patterns.

For this study, both researchers played the games under examination and took their own notes during game sessions. At the end of each game session, an act of a game is transferred to an excel sheet with partial mappings to Propp's framework. After the transfer of six complete game stories, the researchers compared their sheets and decided on a unified structure. On the second phase of the analysis, functions that need major revisions were identified and new functions were proposed.

Seven Proppian functions were re-defined. The descriptions for these functions that arise from Russian folk tales have been replaced with a different set of descriptions. These are the struggle, the first function of the donor, the rescue, the transfiguration, the departure and the return. "Struggle" is defined as the battle between the hero and the villain but we divided it into two parts: the first one, "Confrontation", is the battle with the villain or the henchmen of him/her and the second one, "Struggle", is an obstacle to overcome provided by the villain. The first function of the donor is modified as "Outsider Help" since outsiders can directly help the player showing the right path, instead of giving an item. The original "Rescue" function represents a situation where the hero is saved from pursuit but in game stories there are many side characters that can be saved from various situations. The original transfiguration is the change in the hero but our new "Transfiguration" is the change in the villain, which usually results with new/more powers. Unlike folktales, most game stories do not end up at the starting place named as home. The player or the party leaves (departure) many locations; travels to new ones, and can also go back (return) to some of them. Thus, we combined these two functions as "Travel" representing each transition from a location to another. This function usually evokes another function to form a pair and if the player or the party does not go to that location the story does not progress. At the end of the coding, we had 31 original Proppian functions (see Table I), 6 modified and 15 new functions (see Table II).

To further explain the coding process, let us take a look at the deconstruction details of an example mission/act, "Stop the Collectors: Assemble a Team" from Mass Effect 2. The protagonist, Shepard, is charged with the duty to recruit a team (Difficult Task). The Illusive Man gives information to Shepard about potential recruits: Archangel, Convict, Professor and Warlord (Outsider Help).

TABLE I. PROPP'S FUNCTIONS AND EXPLANATIONS

Functions (Icon)	Explanation
1- Absentation (β)	One of the family members goes missing.
2- Interdiction (γ)	Hero is warned about an action.
3- Violation (δ)	The warning is ignored
4- Reconnaissance (ε)	Villain tries to learn/find something.
5- Delivery (ζ)	Villain is successful in his search.
6- Trickery (η)	Villain tries to deceive the victim.
7- Complicity (θ)	Victim helps villain unwittingly.
8- Villainy (A)	Villain does evil.
8a.- Lack (α)	The need of a family member is explained.
9- Mediation (B)	Hero learns about the lack.
10- Counteraction (C)	Hero chooses to go after the lack.
11- Departure (↑)	Hero leaves home.
12- First Function of the Donor (D)	Hero is prepared for the way to receive an item.
13- Hero's Reaction (E)	Hero succeeds or fails the preparation.
14- Acquisition (F)	Hero acquires a magical item.
15- Guidance (G)	Hero is sent after an object of search.
16- Struggle (H)	Hero and the villain battles.
17- Branding (J)	Hero is branded.
18- Victory (I)	Hero defeats the villain.
19- Resolution (K)	The initial lack is solved.
20- Return (↓)	Hero starts the journey back home.
21- Pursuit (Pr)	Hero is pursued by someone.
22- Rescue (Rs)	Hero is rescued from the pursuit.
23- Unrecognized Arrival (o)	Hero arrives home or in another country, unrecognized.
24- Unfounded Claims (L)	A false hero makes unfounded claims.
25- Difficult Task (M)	A task has been proposed to hero.
26- Solution (N)	The task is resolved.
27- Recognition (Q)	Hero is known by everyone.
28- Exposure (Ex)	False hero or villain is exposed to the world.
29- Transfiguration (T)	Hero is given a new appearance.
30- Punishment (U)	The villain is punished.
31- Wedding (W)	Hero is married and ascends to throne.

TABLE II. MODIFIED AND NEW FUNCTIONS WITH EXPLANATIONS

Video Game Functions	Explanation
32- Confrontation	Party faces with the henchmen or the villain.
33- Outsider Help	Party moves further with the help of a non-party character.
34- Rescue	Rescuing or helping someone in captivity.
35- Struggle	Party encounters an obstacle to overcome by the hand of the Villain.
36- Transfiguration	Villain changes and gains more power after defeat.
37- Travel	Movements of the party within the world.
38- Addition	A new member joins to the party.
39- Bargain	Party tries to make a deal with the villain.
40- Capture	A member or the whole party gets captured.
41- Comeback	A defeated enemy comes back.
42- Disband	Party members leave the party.
43- Escape	Party escapes from the villain's grasp.
44- False Victory	Defeating the villain hinders the party in some way.
45- Gain	Evens that help the party in their journey.
46- Gathering	Unification of separated party members.
47- Heroic Act	Heroism of the party members without a gain.
48- Loss	Events that hinder the party in their journey
49- Persuasion	Party tries to get help from non-party characters.
50- Planning	Party tries to find a way to exploit the villain's weaknesses.
51- Reveal	Explanation of the key points of the story to the player.
52- Sacrifice	The event of saving the party with self-sacrifice of a character.

Colored for story charts, orange as a modified, green as a new function

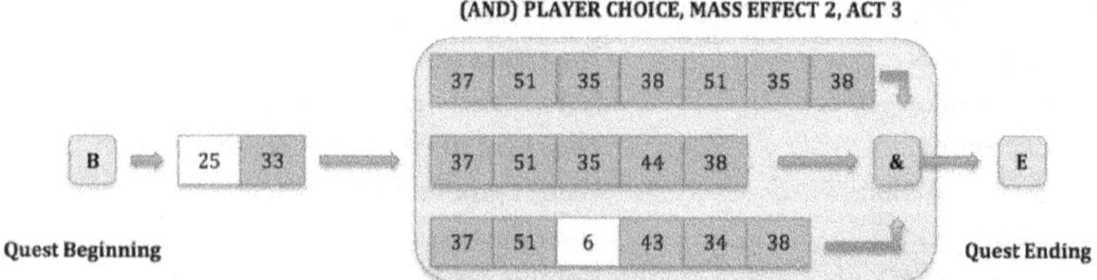

FIGURE I. Mass Effect, Act 3, Story Representation

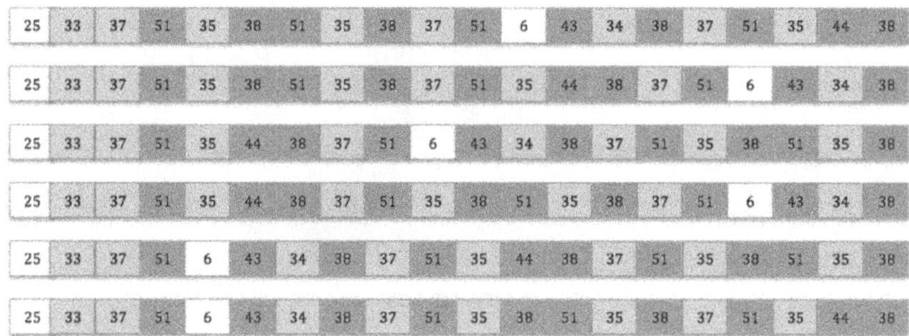

FIGURE II. Mass Effect, Act 3, Story Alternatives

At this point, the player has four Dossier missions with three choices: to travel to Omega for the Professor and the Archangel, to travel to Korlus for the Warlord, and to travel to Purgatory Prison for the Convict. If the party goes to Omega after Professor and Archangel (Travel), the player learns that the Professor is Mordin, who is trying to stop a plague created by the antagonists, Collectors (Reveal). Shepard helps Mordin to stop the plague (Struggle) and Mordin joins the crew after dealing with the plague (Addition). The player also learns that Archangel is Garrus and is fighting with three mercenary gangs (Reveal). Shepard starts fighting these gangs (Struggle). At the end, Garrus is found lying in a pool of his own blood but he survives and joins the crew (Addition). If the party goes to Korlus after the krogan warlord (Travel), they learn that the Warlord Okeer is trying to create a perfect krogan (Reveal). The mercenary group, Blue Suns, attacks the lab to kill Okeer and Shepard (Struggle). Shepard manages to stop the attack but Okeer dies (False Victory). Before he dies, Okeer leaves the perfect krogan soldier, Grunt, under the care of Shepard (Addition). If the party goes to Purgatory prison ship to get the biotic (Travel), they learn that Cerberus will purchase the biotic from the ship (Reveal). Warden attempts to capture Shepard for a bounty (Trickery) but the party escapes the trap (Escape), gets Jack (Rescue) and Jack joins the crew (Addition).

The functional sequence of this act is affected by a choice (see Fig. 1) and there are six different alternatives (see Figure 2). The choice is an AND choice where all the three alternatives have to be completed to progress in the story. The player can start with any Dossier mission and proceed to the next but at the end he/she has to complete three Dossier missions. This is more like an illusion of choice, rather than a choice affecting the story. There are also OR choices in the acts where the player has to make a real choice and complete one alternative to

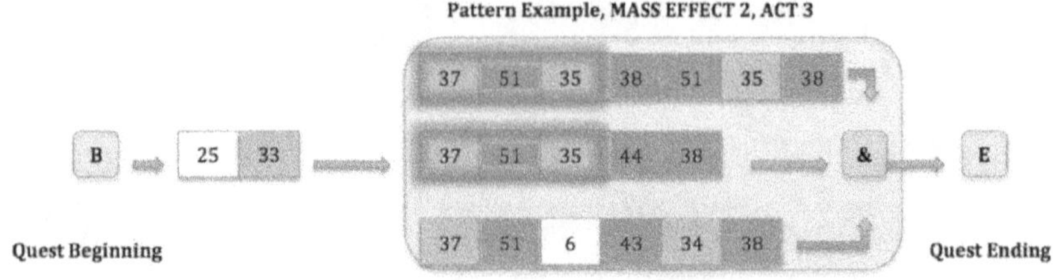

FIGURE III. Pattern Example, Mass Effect2, Act 3

progress in the story. This is more like a branching in the storyline and usually changes the story. Regardless of the nature of choice, what we get at the end is a sequence of story functions represented by numbers, such as the six alternatives given in Figure 2. Based on these linear number sequences, we define a story pattern consisting of n functions as a block of an act with consecutive n functions, such as the Travel-Reveal-Struggle three-function pattern represented by 37-51-35, encountered twice within the Act 3 of Mass Effect 2 (see Fig. 3).

The identification of all the possible patterns and the search for the recurring patterns in the story matrix (which represents all the acts of all the selected games) are handled by a pattern finder written in C#. The code finds all the possible patterns (n>=3) by searching through the matrix and saving a list of unique ones. Then it searches through the story matrix for the occurrence of these patterns based on our selection criteria (single game, series of games, all the games). At the end of the analysis, we focused on the similarities and differences between the Mass Effect and Final Fantasy series to identify possible cultural differences. When we take a look at the micro-patterns (n=3, n=4), the variety (number of different patterns) and the frequency (number of occurrences for all patterns) are higher in the Mass Effect series. As for the macro-patterns (n=5, n=6), these can only be found in the Mass Effect series and there are no macro-patterns in the Final Fantasy series (see Fig. 4 and 5).

The most frequently used 3-function pattern is the Reveal-Struggle-Reveal, which can be found twelve times in the Mass Effect series and seven times in the Final Fantasy series. Two other frequently utilized patterns are Travel-Reveal-Struggle and Difficult Task-Travel-Reveal. The most common 4-function pattern for both game series is the Travel-Reveal-Struggle-Reveal, which encompasses the 3-function Travel-Reveal-Struggle pattern. The occurrences of story patterns are not high enough to make a valid conclusion but three functions

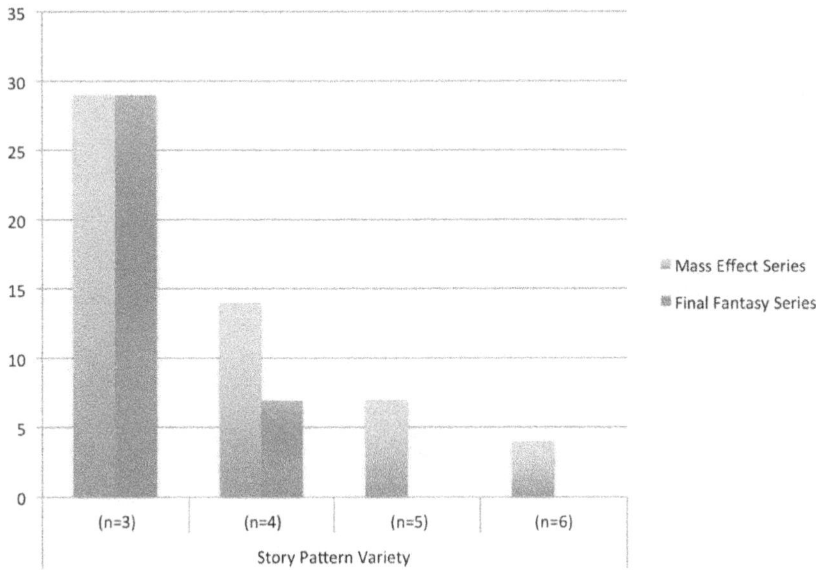

FIGURE IV. Story Pattern Variety, Mass Effect and Final Fantasy comparison

(Reveal, Struggle and Travel) are emerging as the dominant components of story patterns. Five and six-function patterns are only observed in the Mass Effect series.

III. Conclusion

The combination of narrativity and interactivity in computer games makes it difficult to adapt structural analyses of texts to computer games. Ryan [28] clearly made a distinction between the narrative game and the playable story. In a narrative game, the narrative meaning governs the player's actions and in a playable story, it is the player's actions that define the narrative meaning. Narrative games use a 'representological' concept of narrative where the player feels consciously present but physically absent in the expressed events and the playable stories use a 'presentological' concept of narrative where the player is physically anchored in the world of the story [29]. Ferri [30] also argued that video games are not literary or cinematographic "texts" as defined by structuralist semiotics but rather "interactive matrices", semiotic devices that produce many textual fragments. In this regard, the selected games for this study are playable stories where user choices affect the story and since the structure of game stories are different than

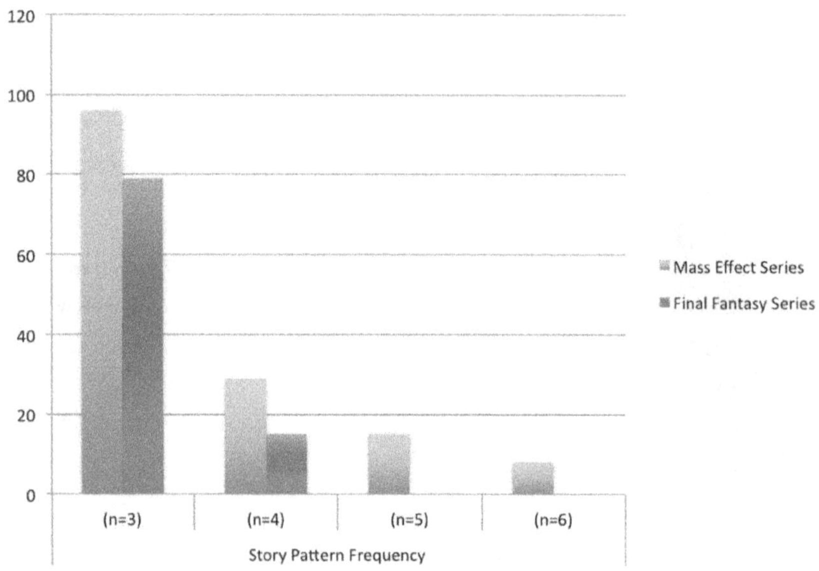

FIGURE V. Story Pattern Frequency, Mass Effect and Final Fantasy comparison

literary texts, it demonstrated that Propp's morphology needs revisions for mapping it to game stories.

On an act level narrative, as predicted, Propp's functions can only partially be mapped to game stories. Some of the functions need modifications and new functions are needed to cover the entire acts. The main difference between Russian folk tales and game stories is that the protagonist is not alone in his/her adventures. The addition of side characters further complicates the stories and the player's relationship with these characters is one of the main reasons for function modifications and inventions. The number of locations is also a major difference between Russian folk tales and game stories. Folk tales start with a departure where the hero leaves his/her home and at the end of the story the hero returns to the same location. The definition of 'home' for a game protagonist is usually blurry and the hero travels between various locations to complete the acts of a story. Finally, representing 'player choice' with Proppian structures is possible with AND/OR structures but it complicates the representation of story patterns.

The cultural differences between the American and Japanese stories can be observed from the variety and frequency of story patterns. American game stories use more patterns and the repetition of patterns is higher than the

Japanese counterparts. The non-existence of five and six-function patterns in Japanese game stories also needs attention. In line with this, we can imply that American games are using the same structures over and over in their stories whereas the Japanese games are using different story building blocks where the pattern repetition rates are lower. The identification of prominent n-function story patterns is not possible since the number of occurrence for each story pattern is usually very low but the Reveal, Struggle and Travel functions are the most popular individual patterns utilized in story patterns.

The applicability of Propp's model to computer game stories is addressed in this study with the revised and new functions, showing that an act level mapping is possible. Representing choice, finding the logical sequence of story functions and determining the commonly repeated story patterns are also accomplished within the scope of our study but it is not possible to make valid conclusions about the cultural differences between American and Japanese game stories. In this regard, the modified framework of Propp needs mapping to more Eastern and Western games in order to make a solid comparison. Differences can be observed in terms of pattern variety, pattern frequency and pattern structures but the generalizability of the results requires more research and analysis focusing on RPGs from both cultures.

References

[1] C. Lévi-Strauss, "The Structural Study of Myth," in Structural Anthropology, New York: Basic Books, 1963, pp. 206–231.

[2] R. Barthes, Elements of Semiology, trans. A. Lavers and C. Smith, New York: Hill and Wang, 1967.

[3] T. Todorov, Théorie de la littérature, Paris: Seuil, 1965.

[4] J. Campbell, The Hero With a Thousand Faces, Princeton, NJ: Princeton University Press, 1949.

[5] C. Vogler, The Writer's Journey: Mythic Structure for Storytellers and Screenwriters, M Wiese Productions, Studio City, CA, 1992.

[6] V. Propp, Morphology of the Folktale, University of Texas Press, 1968.

[7] D. Myers, "Computer Game Semiotics", Play & Culture, 4(4), 1991, pp. 334–345.

[8] J. Bruner, "What is a Narrative Fact?", Annals of the American Academy of Political and Social Science, 560, 1998, pp. 17–27.

[9] Z. Tomaszewski and K. Binsted, "The limitations of a Propp-based approach to interactive drama", paper presented at the Intelligent Narrative Technologies: Papers from the AAAI Fall Symposium, vol 7, 2007.

[10] P. Gervás, "Propp's morphology of the folk tale as a grammar for generation", in Workshop on Computational Models of Narrative, a satellite workshop of CogSci 2013: The 35th meeting of the Cognitive Science Society, Universität Hamburg, Hamburg, Germany, 2013.

[11] A. Brusentsev, M. Hitchens and D. Richards, "An investigation of Vladimir Propp's 31 functions and 8 broad character types and how they apply to the analysis of video games", in Proceedings of The 8th Australasian Conference on Interactive Entertainment: Playing the System (IE '12), 2012.

[12] G. Frasca, "Simulation versus Narrative: An Introduction to Ludology", in M. Wolf and B. Perron (eds.), The Video Game Theory Reader, Routledge, 2003, pp. 221–235.

[13] S. Louchart and R. Aylett, "Narrative theory and emergent interactive narrative", in Proceedings of the second international workship on Narrative and Interactive learning environments. Edinburgh, Scotland, 2002.

[14] G. Dieter and N. Braun, "A Morphological Approach to Interactive Storytelling", paper presented at the Living in Mixed Realities: Conference on Artistic, Cultural and Scientific Aspects of Experimental Media Spaces, 2001, 337–340.

[15] C.R. Fairclough, "Story Games and the OPIATE System Using Case-Based Planning for Structuring Plots with an Expert Story Director Agent and Enacting them in a Socially Simulated Game World", Doctoral Thesis, University of Dublin - Trinity College, 2004.

[16] P. Gervás, B. Díaz-Agudo, F. Peinado and R. Hervás, "Story plot generation based on CBR", Knowledge Based Systems, 18(4–5), 2005, 235–242.

[17] S. Imabuchi and T. Ogata, " A Story Generation System based on Propp Theory: As a Mechanism in an Integrated Narrative Generation System", in H. Isahara and K. Kanzaki (eds.) Advances in Natural Language Processing (pp. 312–321), Lecture Notes in Computer Science, vol 7614. Springer, Berlin, Heidelberg, 2012.

[18] S. Imabuchi, T. Akimoto, J. Ono and T. Ogata, T. "KOSERUBE: An application system with a Propp-based story grammar and other narrative generation techniques", in proceedings of the 6th International Conference on Soft Computing and Intelligent Systems & the 13th International Symposium on Advanced Intelligent Systems, 2012, 248–253.

[19] R. Cassar, "God of War: A Narrative Analysis", Eludamos. Journal for Computer Game Culture, 7 (1), 2013, pp. 81–99.

[20] A. Ngamgamol, "A Narrative Analysis of Assassin's Creed Series", Master Thesis, National Institute of Development Administration (NIDA), 2014.

[21] T. Wama and R. Nakatsu R. "Analysis and Generation of Japanese Folktales Based on Vladimir Propp's Methodology", in P. Ciancarini, R. Nakatsu, M. Rauterberg, M. Roccetti (eds) New Frontiers for Entertainment Computing. IFIP International Federation for Information Processing, vol 279. Springer, Boston, MA, 2008.

[22] P. Phindane, "The Structural Analysis of Sesotho Folktales: Propp's Approach", Southern African Journal of Folklore Studies, 24(2), 2014.

[23] K. Yong-Jae, "Structural Analysis of Game Quest-storytelling -Focusing on Applying Narrative Functions of Folk-tale by Propp", The Journal of the Korea Contents Association, 11(10), 2011, 69–76.

[24] B. Lanham and M. Shimura, "Folktales Commonly Told American and Japanese Children: Ethical Themes of Omission and Commission", The Journal of American Folklore, 80(315), 1967, 33–48.

[25] Y. Norman, "Miyazaki and the West: A Comparative Analysis of Narrative Structure in Animated Films for Children", İleti-ş-im: Galatasaray Üniversitesi İletişim Dergisi, 9 (9), 2014, 11–30.

[26] D. Matsuura, "Superman vs. goku: Different cultural values represented in superhero characters in american and japanese", Available from ProQuest Dissertations & Theses Global, (863229921), 2010, Retrieved from https://search.proquest.com/docview/863229921?accountid=15407

[27] K. G. Blasingim, "Hero Myths in Japanese Role-Playing Games", Master Thesis, Graduate College of Bowling Green State University, 2006.

[28] M. Ryan, "From narrative games to playable stories: Toward a poetics of interactive narrative", StoryWorlds: A Journal of Narrative Studies, 1, 2009, 43–49.

[29] T. Dubbelman, "Playing the hero: How games take the concept of storytelling from representation to presentation", *Journal of Media Practice*, 12(2), 2011, 157–171.

[30] G. Ferri, "Narrating machines and interactive matrices: a semiotic common ground for game studies", in proceedings of DiGRA 2007 Conference, 2007, 466–473.

6. Methodological Review of Playability Heuristics

Çakır Aker
Faculty of Communication Department of Digital Game Design Bahcesehir University Istanbul, Turkey cakir.aker@comm.bau.edu.tr

Kerem Rızvanoğlu
Faculty of Communication Galatasaray University Istanbul, Turkey krizvanoglu@gsu.edu.tr

Barbaros Bostan
Faculty of Communication Department of Digital Game Design Bahcesehir University Istanbul, Turkey barbaros.bostan@comm.bau.edu.tr

Abstract— This study reviews published scientific literature on the use of heuristics for evaluating player experience to (a) identify the potential contribution of the application of heuristics for assessing player experience, (b) present the status of playability heuristics evaluation procedures for evaluating player experience, and (c) define future research perspectives. By searching online bibliographic databases, 44 relevant articles were selected and included in the study. The aim, methodology, proposed heuristics and conclusions were studied separately for each article. The study indicated that a large variety of approaches on evaluating the experience of video games using heuristics were presented. Because of this, it is not possible to identify a generally accepted approach while studying evaluation of video games. This study intends to present and clarify a much-needed holistic point of view in terms of using heuristics for evaluating player experience because of the current dispersed state of the literature and interlaced heuristic evaluation approaches. The review study indicated that most of the articles presented new heuristics, either by iteratively improving the existing approaches or forming new ones. It is usually suggested that the presented sets of heuristics have been viable in general for assessing some aspects regarding the gaming experience. However, the heuristic approaches have neither proven to be including the experience in its entirety nor empirically tested adequately for validation to provide a possible de facto basis for further research. The implications of the articles were also studied for providing a common ground for future research in the field of heuristics evaluation of video games.

Keywords—playability; playability heuristics; heuristic evaluation; literature review; game experience

I. Introduction

Most of the usability evaluation techniques are not suitable for inspecting video games since the design considerations and aims for games are different from productivity applications. While productivity applications focus on solving problems and minimizing challenges, video games use challenges and problems to provide enjoyment. More so, video games do not have specific tasks to achieve a specific goal but depend on player choices and motivations. In order to satisfy user needs and enhance the overall experience for both the productivity and video game applications, there are several user evaluation techniques. One example is task-oriented user tests which the data gathered from a sample of users interacting with the application. Users are observed while following simple assignments representing a typical use of the application. The results are analyzed for indicating issues related to the user experience of the application. Differently, among those methods for evaluating an interactive system, heuristic evaluation offers the benefit of evaluation during the design process and do not require a task oriented inspection.

The overall importance of games, both in terms of industry and academia, has been growing rapidly but the evaluation of player experience has been complicated by the following problems: (1) The literature on player experience and playability is not vast; (2) there is no common definition for playability and there is no consensus on the heuristics for evaluating the playability in video games; and (3) researchers apply different approaches and therefore offer different heuristics for evaluation. In this review, our aim is to present both the goals and the procedures of using heuristics as a means of evaluating playability and player experience in video games. Likewise, we also aim to find out the future potential of various methods utilizing playability heuristics.

The game industry has been developing immensely although the methodology on evaluating player experience still lack a robust approach to evaluate the overall experience. Heuristic evaluation is an inspection technique that allows evaluators to examine an interface using statements of usability principles [1]. Korhonen mentions that heuristic evaluation is more effective for evaluating games compared to other methods because this approach does not require any tasks oriented tests to be conducted [3]. The studies on player experience using heuristics have the benefit of conducting a research with a rather cheap and fast manner. So far, researchers who studied the topic have not been able to present a holistic set of heuristics which could be considered as a common ground for evaluating player experience. Moreover, only a limited number of studies aimed to utilize existing approaches and even fewer attempted to combine different approaches and heuristics from previous works and validate them with empirical tests. Only a few of the researchers have tried to present new playability heuristics [7], [6], [4], [17], [11], [13] for games. Choosing heuristics which are applicable in a selected gaming platform is a challenging task and some of the researchers have proposed different heuristics for different platforms such as tabletop, computer, mobile, educational and web based social games [11], [13], [16], [20], [12], [22], [28], [30]. In addition, reviewed literature has presented that each heuristic method has pros and cons. This study aims to provide a comprehensive review on using heuristics for evaluating playability to improve the field of research regarding the evaluation of player experience.

II. Method

An extensive literature search was conducted with key-words of 'game heuristics, playability, playability heuristics, player experience, and game heuristics. A search in ACM Digital Library, IEEE, Springer, Taylor & Francis, Google Scholar and ISI databases was conducted. After this endeavor, a second examination of the abstracts led to an elimination of irrelevant journal articles and conference proceedings. The remaining 44 articles were examined in detail to find out the main contributions to the relevant research area. The findings of our literature study addressed some of the key differences for presenting heuristics for evaluating player experience. These key features were categorized in terms of what procedure the heuristics were based upon such as the choice of medium or sources for identification. The rest of the paper was structured as follows, first the categorization for procedures in the articles was explained. Four distinctive methods were identified; empirical evaluations, expert evaluations, inspections (literature & online game reviews) and evaluations using mixed methodology. Afterwards, studies in each category were summarized. Explanations of the articles utilizing empirical evaluation methods, expert evaluation methods, inspections and multi-modal methodology are presented followed by conclusions.

III. Procedure Types

In terms of the categorization of playability heuristics, several studies presented relevant articles in a chronological manner. Even though this structure for studying literature review has its benefits, such as indicating the iterative progress between testing methods and design methodology, it is not sufficient for examining the procedural differences between approaches. Moreover, our review focuses on structuring the literature in terms of methodology. Because of these reasons, categorizing differences in general between heuristic evaluation methods applied during tests and development of heuristic approaches hold a different perspective for analysis and potential for contribution. As a result, it is possible to conclude that there are four main approaches in the field. Among the eligible 44 articles, we identified 12 articles which would fit in the category of empirical evaluation, 17 articles for expert evaluation, seven articles for inspection and 8 articles for mixed-methodology.

Expert evaluations conducted by utilizing proposed heuristics is the most common approach in the literature, followed by empirical evaluations. Although many of the researchers claimed that the best approach for a valid and robust heuristic set is to combine different evaluation methods while testing them one of the least employed methodology is the mixed-method approach (Figure 1). Articles in each category is explained in a chronological fashion to provide an additional standpoint for indicating the extent of scientific progress.

A. Empirical Evaluation

The first observed approach is empirical evaluation method. This category includes studies conducted through user-testing methods such as surveys, interviews, focus groups and observations with a sample group of minimum 10 participants. According to our review, 12 of the relevant articles chose to evaluate either the heuristics or the games via user-tests rather than directly using the heuristic evaluation sets during the research.

Malone [4] proposed the first heuristics for encouraging the use of games in learning and teaching. He presented a set of heuristics for instructional games and suggested that there were three main heuristics for achieving entertainable interfaces. In the study, three empirical tests were employed to understand what gamers liked with a total number of 81 participants, all from elementary or secondary school students. The tests were conducted upon three games with eight versions each. As a result he proposed the three heuristics categories; challenge, fantasy and curiosity.

Fabricatore et al. [5] followed a different approach by first proposing heuristics and later evaluating it. The model was prepared in order to guide game designers for preparing better games. 53 participants, between the ages of 20 – 30, have attended the tests where they could make comments in every step. After play sessions, semi-structured in-depth interviews were realized. In the light of the findings derived from the play sessions, in-depth interviews and observation notes, the proposed heuristics were iteratively revised. 39 different games from different genres were tested during the study. At the end of each test, participants

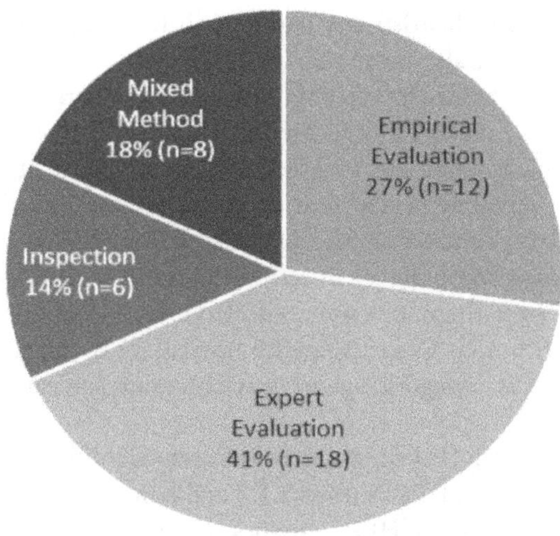

Fig. 1 . Distribution of Different Methods Utilizing Heuristics Evaluation for Playability and Player Experience

were interviewed for potential improvements of the heuristics. A hierarchical table was presented as determinants of the playability in games with several sub categories in each. The study indicated three main determinants; entity, scenario and goals. Item of entity consisted of four different sub categories; identity, energy, equipment and behavior. Scenario item consisted of view, spontaneous changes, transitions and interactions with entities. And the last item of hierarchy of goals consisted of complexity, linearity and interface. This study is regarded as one of the first qualitative playability evaluation models presented in the literature, therefore prepared a basis for the rest of the research field.

Inspired by heuristic evaluation approaches Röcker & Haar [7] investigated if the existing heuristics could be used for evaluating pervasive games [10]. Their study was based on Desurvire's HEP heuristics. They conducted the research via a focus group of 10 participants, between the ages of 32 - 38. The evaluated heuristics set was not shown to the participants during the tests for objectivity reasons. A smart home environment with pervasive computing capabilities were given as a scenario for the participants to make comments. After the interviews, participants were invited to join a focus group and asked to note their ideas to cards that were given to them. At the end, they were asked to do a card-sorting exercise to clarify the priorities and categories for each comment. Later, the researchers compared the heuristics with the proposed comments and indicated the need for additional heuristics to the set for it to be applicable for pervasive games. The researchers also noted that the peripherals could change the experience dramatically and this aspect would need separate heuristics.

Song & Lee [12] studied key factors of heuristics evaluation in games by taking the example of a well-known MMORPG game (World of Warcraft). They conducted both literature review and empirical research for their study and adopted post-surveys and a task oriented analysis. Participants were given specific tasks to follow and usability issues during the play time were noted. The results gathered from these tests were reflected to a new set of heuristics. They based their heuristics on Desurvire's HEP heuristics [7] and suggested 54 key factors under four key categories; game interface, gameplay, game narrative, and game mechanics.

Desurvire & Wiberg [19] conducted a research based upon Desurvire's previous approach of using HEP heuristics [7] and aimed not to just validate but improve it. During the study, HEP heuristics were modified for the game genres such as action, role playing game (RPG), action, adventure and first person shooter (FPS) and discussed with developers working at respected game development companies. After those discussions and refinement of heuristics, PLAY heuristics was proposed. Researchers also set nine categories for general principles of the heuristics: Game Play, Skill Development, Tutorial, Strategy & Challenge, Game/Story Immersion, Coolness, Usability/Game Mechanics and Controller/Keyboard. During the testing, three sets of surveys, depending on the game genre were prepared with a scale of points based on the score of the game that the game received from Metacritic website (www.metacritic.com). Participants were selected from attendees of an annual game conference and were chosen from people who played either the low rated games or high rated ones. In their study, researchers mostly aimed to explain how the PLAY heuristics were defined and how effective they were in a real-world application.

Tan et al. [22] presented a study conducted to analyze an educational game. In order to achieve this goal, it was aimed to develop a framework of heuristics which is called Instructional Game Evaluation Framework (IGE). The IGE framework had 42 heuristics and was based on "Events of Instruction" method [51], Keller's ARCS Model for Motivation [52], GameFlow model [9] and Nielsen's heuristics [1]. 12 primary school students participated in the tests while the research team of five attended as observers and supervisors. An instructional computer game was selected for testing the proposed heuristics. All the students attended pre-explanatory meetings and were given time to play the game without any restrictions. After the play sessions students were divided into three groups and attended focus group meetings. During those meetings, they were asked to comment on the heuristics proposed while the heuristics were simplified for them to understand their notions. The study indicated that including children at the early stages of formal evaluation was effective and valuable since there were revisions coming directly from the participants that effected the heuristics.

Zabion & Shirratuddin [30] conducted a study focusing on mobile based educational game by proposing a heuristics paradigm with four main modules: Game usability, mobility, gameplay, and learning content. The heuristics were based on Korhonen & Koivisto's playability heuristics modular approach [11] and proposed a module for learning content. In the first phase of the study, participants from primary school students were selected. They were asked to comment on the heuristics and fill surveys. At the second phase a new participant group from 80 exhibition attendees were recruited and asked to play a prototype game. Afterwards, they were asked to fill a Likert scale form representing the heuristics. At the end of the study, researchers revealed results regarding the games performance according to the heuristics they proposed.

Ülger [37] also proposed a modified version of heuristics, based on Nokia's Playability Heuristics for Mobile Games. Her study aimed to expand the existing heuristics set for new generation mobile devices and games by adding 4 different heuristics to the set; distribution of game items, user handedness, use of tilt sensors and haptic feedback. After the proposal, the heuristics were tested via Game Experience Questionnaire (GEQ) [55] and interviews. Four mobile games were tested. Two versions of the games were presented during the tests. By inspecting the relevant aspect of the game the heuristic at hand was analyzed. 10 participants were recruited for testing each of the games with a total number of 40.

Kornchulee Khanana & Effie Lai-Chong Law [38] conducted a study to use the Game Flow [9] heuristics on digital educational games. They tested four web-based computer games during their study with 100 primary school students. They also re-phrased the heuristics in a way that the students would understand easily so that the heuristics could be given in a survey format. As a result, they indicated the differences among games as well as future possibilities of using heuristics for educational aspects in games, yet they did not propose new heuristics.

Likewise, Rodio & Bastien [40] conducted a study to evaluate the PLAY heuristics from Desurvire & Wiberg [19]. In their research 120 amateur e-sport gamers were chosen as participants. A total of three games from different genres (Real Time strategy, MMORPG and FPS) were chosen during the tests. Each participant was asked to fill a five point Likert scale heuristics set which consisted of 47 heuristics. Furthermore, questions regarding the games were directed to the participants in order to receive comments on games they played during the tests. Results indicated that each genre had a different rating of importance in terms of given heuristics, yet the gameplay category carried a generic importance for all of the games.

Another heuristic set evaluation study was carried out by Marciano et al [41]. In their study, they aimed to propose a method for evaluating educational computer games. They used

Omar & Jaafar's Playability Heuristic Evaluation for Educational Computer Game (PHEG) [25] in order to analyze an educational computer game. The game developers themselves analyzed the tests and generated a survey format to address heuristics. This study also presented the use of an automated software being used for the application of heuristic evaluation.

Lastly, Guo & Goh [49] aimed to use heuristics evaluation instead of anecdotal research on educational games which is a semi-formal method that relies on anecdotal evidence such as user comments. In their study they applied HEP heuristics set to analyze a computer literature game. They presented a survey format of HEP heuristics with 43 items. 39 participants were selected among students. Participants were asked to fill out the survey within a five point Likert scale format after playing the game. Subsequently, the subjective comments of the participants were gathered through interviews. The study proposed suggestions to improve the HEP heuristics set and claimed that two new categories needed to be included; characters and pedagogical effects.

Cross references between heuristics were presented below (Table 1) with the notion of listing heuristics which were utilized in more than one research. This limitation of presented heuristics was necessary to avoid listing specific heuristics which involved specific areas of research such as educational context.

B. Expert Evaluation

The second approach in the review is conducting expert evaluations using the provided heuristics for evaluating player experience. It was reviewed that a sample group of minimum two participants have performed the evaluations. Even though Nielsen has stated that five experts are normally advised for conducting a heuristic evaluation [1], some of the studies have not followed this advice. According to our review, 14 of the relevant articles evaluated either the heuristics or the game via expert evaluation.

Federoff [6] did a research on existing game heuristics and collated them to analyze the 'fun' aspect of the games. Five people from a game development team were observed and interviewed to suggest a set of heuristics for evaluation of video games. Author analyzed the interviews and observation notes in order to

TABLE I. CROSS-REFERENCED HEURISTICS OF EMPIRICAL EVALUATION RESEARCH ON PLAYABILITY HEURISTICS

Summary of heuristics that are used at least more than one study										
Clear and varied outcomes	x									x
Variable difficulty level	x			x					x	
Embodiment of metaphors with physical or other systems that user understands	x		x							
Audio-visual supports the game	x			x		x				x
Support of a variety of game styles.	x			x	x			x		
Using humor appropriately	x			x						
Making effects of AI visible by ensuring they are consistent with the player's reasonable expectations		x								x
Game provides immediate feedback		x	x	x	x	x	x	x		x
Player can easily turn the game on/off, and be able to save in different states		x								x
The Player experiences the user interface as consistent but the game play is varied.		x		x	x	x		x		x
Interface/HUD as a part of the game.		x		x						x
Player has enough information to get started from the beginning		x		x						x
Context sensitive help		x	x	x						x
Meaningful sounds		x								x
Players do not need to use a manual to play game.		x		x						x
Non-intrusive interface		x								x
Make the menu layers well-organized and minimalist to the extent the menu options are intuitive		x	x		x			x		x
Quick involvement with tutorials and/or progressive or adjustable difficulty levels		x		x					x	x
Art should be recognizable to player, and speak to its function.		x								x
Always being able to identify score/status and goal		x	x	x	x	x		x	x	x
Standard conventions and natural mapping for controls			x	x		x				
Aesthetic and minimalist design			x				x			
Clear goals			x	x	x	x	x	x	x	

(*continued on next page*)

TABLE I. (continued)

Summary of heuristics that are used at least more than one study									
Appropriate rewards for effort and skill development		x	x	x	x		x		x
Challenge, strategy, and pace are in balance		x	x	x	x	x	x	x	x
Fun gaming, without repetitive or boring tasks			x		x				
Not being penalized repetitively for the same failure			x						x
Any fatigue or boredom was minimized by varying activities and pacing			x						x
Persistent game world			x						x
Application of the newly acquired knowledge / skill			x	x			x		x
Multiple ways to win.			x						x
Feeling in control			x	x	x	x	x	x	x
Empathy with the game character		x	x	x			x		x
Curiosity and exploration	x	x	x	x	x		x		x
Consistent learning curve with the industry			x						x
Screen layout is efficient, integrated, and visually pleasing			x	x	x				
Navigation is consistent, logical and minimalist.			x	x	x		x	x	
Player error is avoided.			x	x					
Player interruption is supported			x		x				
Total concentration				x		x	x		
Storyline relate to your life experiences and grabs interest				x			x		x
Visuals, animation and music able to capture interest				x			x		
Font types and sizes used allow easy reading				x			x	x	
Associations of new knowledge and skills with prior knowledge and skills				x			x		
Learning new concepts and skills		x		x			x		
Feeling of satisfaction and success after gameplay				x			x		
The game allows to do reflection on learning				x			x		

Previous Empirical Evaluation Research on Playability Heuristics	Guo & Goh (2016)	Marciano et al. (2014)	Rodio & Bastien (2013)	Khanana & Law (2013)	Ülger (2013)	Zabion & Shirratuddin (2010)	Tan et al. (2010)	Desurvire & Wiberg (2009)	Song & Lee (2007)	Röcker & Haar (2006)	Malone (1982)
The feedback and online help reinforce understanding			x				x				
Players want to play more of the game			x				x				
The player does not have to memorize things unnecessarily			x			x	x				
The words and phrases of the game is easy to understand			x				x				
Feeling confident playing the game			x				x				
Achieving the learning objectives			x			x	x				
Players easily get help during game play and find this “help” useful			x				x		x		
Warning messages and cues help make less mistakes			x				x				
The game helps to diagnose players own error			x				x				
Player can gauge the overall progress at each stage of the game			x				x				
The game rules assign a final score to the end of each session			x				x				
The game contains help						x			x		
The game story supports the gameplay and is meaningful	x					x		x	x		
Even if the game cannot be modeless, it should be perceived as modeless	x								x		
Player experiences fairness of outcomes	x								x		

form a list of heuristics. The data collected were compared to formal usability evaluation methods mostly with Nielsen's 10 usability heuristics [1]. As a result, Federoff presented a set of 10 heuristics for evaluating games yet the suggested heuristics lacked any validation.

Baauw, Bekker & Barengregt [8] conducted a study on the proposed Structured Expert Evaluation Method (SEEM) which was inspired from Norman's theory-of-action model [53] and Malone's concepts of fun [4]. SEEM model was presented to evaluate children's computer games. The aim of the study was mainly to validate the proposed model. They recruited 18 experts from the working area of children, usability and user testing. They also noted that the reason was to improve the SEEM method rather than analyzing the games. Four games were evaluated by the experts for approximately an hour for each game. Experts filled an interaction problem report sheet while conducting the tests. At the end, researchers claimed that the SEEM method was effective in general although they missed several problem categories such as goals, transition and physical action. They also mentioned that SEEM enabled mention some issues which were not revealed from the user-tests done before.

Sweetser & Wyeth [9], conducted a research on evaluating player enjoyment in video games. They proposed a novel set of heuristics, GameFlow model, for the evaluation of the games utilizing the term 'flow' [56] at its core. A holistic categorization aimed to evaluate and identify enjoyment in games. They suggested eight key elements including several heuristics in each of them. After suggesting the model, authors validated the model by evaluating two similar real-time strategy games via expert evaluation. As a result, the authors indicated that the model could be used as a guideline for an expert review or basis for other evaluations such as player-testing.

Korhonen & Koivisto [11] were first to publish playability heuristics for mobile games. They proposed a modular structure for their playability heuristics, which consisted of game usability, gameplay and mobility. Each had distinctive heuristics due to the category and the study was based on literature examination and mobile game reviews. They proposed 29 heuristics in total. Some of the categories and heuristics within those categories were developed from Nokia's Playability Heuristics for Mobile Games. There were two phases of the study. First part involved the use of the three categories of the heuristics with different mobile games. For the first version of the heuristic set, four experts analyzed five mobile games. The experts were either from the field of game design and development or productivity software fields. At the second phase, the set was iteratively improved and the experts conducted the test for the second time, but with different games. According to the results for the study, playability heuristics were effective for evaluating mobile games. Researchers also mentioned that the proposed heuristics could be used in other platforms and games because of its modular structure. Although the heuristics were not compared to previous work and lack empirical validation, playability heuristics and the novel modular structure was well received both in the academia and industry and became basis for other heuristic approaches [11]. Following their previous work, Korhonen & Koivisto [13] published a second paper on evaluating mobile multiplayer games. In their latter study they included another module for the multiplayer aspect of mobile games. They prepared the heuristics for the multiplayer category by examining three multiplayer mobile games and literature study.

Köffel & Haler [16] proposed heuristics for tabletop games. In order to define the heuristics, they incorporated literature reviews and comments from professionals. As a result they presented a modified set of heuristics with 11 items. 12 expert evaluators were asked to evaluate an augmented reality supported tabletop game. There were several sessions during the tests in which the experts were asked to define missing heuristics. In the end, they suggested an iteratively formed heuristic set but mentioned that the last version was not tested. Although this study applied iterative methodology for improving the heuristics, its findings could not be generalized since the focus of the study was on tabletop games.

Korhonen et al. [18] conducted a study for comparing two playability heuristic sets. In their paper, it was mentioned that the aim was to compare the sets of Korhonen & Koivisto and Desurvire's HEP approach [13], [7] since both of those heuristic sets were compatible. During the tests, eight experts were recruited. They were asked to play a mobile game and note the issues about the game in terms of playability. Later they were asked to compare the findings with the given heuristics. The study indicated that playability heuristics had to be improved to be applicable by game developers, in such that the items had to be less in number and more understandable in terms of terminology.

Pinelle et al. [20] proposed usability heuristics for networked multiplayer games. The study suggested a set of novel heuristics which they called Networked Game Heuristics (NGH). They adapted a previous methodology [17] which utilized online game reviews to define heuristics. To test the heuristics, 10 experts were asked to play two different games which had multiplayer capabilities via network. The experts were asked to fill out a Nielsen's Severity Scale [59]. Also the suggested heuristics were compared with the Groupware usability Heuristics [60] during the study. In the result section of their paper, researchers mention that Korhonen & Koivisto's [11] playability heuristics were viable in general. Additionally, they mentioned that the aim was to generate a set of heuristics specific to networked games. It was also indicated that the heuristic set was applicable in different platforms and genres, providing a generic property [20]. Other researchers criticized the article because the previous work which the heuristics were based on [17] was problematic due to the fact that the online game reviewers were not experts in terms of evaluation or game design. Because of this reason, it is possible to indicate that the suggested set might miss out several aspects of playability.

Koeffel et al. [27] conducted a study to inspect the use of heuristics to evaluate the overall user experience of video games and advanced interaction games (tabletop games). They presented a set of heuristics with three facets; gameplay, game story and virtual interface. The study aimed to develop a set of heuristics which could include more than one aspect of playability and player experience. Researchers based their set of heuristics on Pinelle's [17] and Sweetser & Wyeth's GameFlow approach [9]. By conducting an extensive research provided in the literature they put forward 29 items for their heuristics set. The authors claimed that the proposed set included heuristics about only the most important aspects of video games and assumed that it was necessary to investigate the usability/playability of a video game as well as the user experience/player experience to evaluate the overall quality of a game. To determine the effectiveness of the heuristic set, researchers compared the expert evaluation results to common game reviews. Five computer games were tested by two expert evaluators whom were experts in the field of usability and/or games during the tests. Experts were asked to play the games and evaluate them by using the given heuristics set while indicating results via Nielsen's Severity Scale [59]. The results of the tests (number of issues found through proposed heuristics) were later compared with online review scores. The results indicated that the heuristics were generic though lack the specificity for tabletop games. This study had the authenticity of comparing heuristic evaluation results with common reviews which was referred to Pinelle's approach.

Almeida et al. [23] conducted a heuristic evaluation of the web-based computer game 'FarmVille' by combining heuristics from Federoff [6], Desurvire [7] and Pinelle [17]. In their task oriented tests, they indicated 35 heuristics. Each given task during the gameplay was related with certain heuristics. Six evaluators were recruited to attend the tests to fill in the forms with yes or no answers. The study evaluated the game by only using heuristics and expert evaluations yet the participants lacked the expertise related to the field of gaming or playability or usability.

Suhonen & Vaataja [28] aimed to study the effect of using modular heuristics on health games. Five previous heuristic sets [7], [6], [61], [11], [13] were found to be eligible for being applied during the tests as the authors claimed that these sets complemented each other in terms of given heuristics. After inspection of the heuristic sets, Korhonen & Koivisto's playability heuristics [11] were found to be fit for the study. Also, the modular structure of the same study was adopted. Provided that the modular structure was perceived as useful and flexible and could be designed with consideration – given the example that the current heuristics modules could be improved and/or re-arranged. Therefore, to evaluate health games, researchers introduced two new modules to the set, namely for multimodality and persuasiveness. One computer game, one Nintendo Wii game and one mobile game were chosen for the tests. Two experts evaluated the games separately. Experts were asked to fill forms indicating the severity scale and frequency of issues. The results of the study indicated that adding separate modules according to the game genre could be efficient for evaluation purposes as well as being applicable with health games.

Omar & Jaafar [25] presented the Playability Heuristics for Educational Games (PHEG). They collated the first heuristics set by inspecting the literature for user experience, player experience and pedagogical use in games. Later, experts revised the suggested heuristics and filled a survey for evaluating the PHEG. Experts were also asked to prioritize given heuristic items. As a result of the tests, researchers presented a set with 43 items and five categories with indications of their priority. Researchers indicated that the PHEG set was specifically generated to be used for evaluating educational games hence improving the method by prioritizing the categories. However, their study was not without shortcomings since the heuristics were not examined or verified on an educational game, therefore the study did not involve the empirical validation of the PHEG heuristic set.

Ponnada & Kannan [33] researched how different mobile games created positive and immersive experiences for the players by using playability heuristics. They based their research on Korhonen & Koivisto's playability heuristic set [11]. Two expert evaluators were recruited for the examination of each mobile game. Four mobile racing games were chosen for the tests and the experts were asked to play them. After the gameplay, experts filled in the given heuristic forms with yes or no answers. No changes were made from the original heuristics and therefore the study had the value of being a direct implementation of the set. Researchers then compared the results with Android Market ratings and statistics. Researchers indicated that there were positive correlations between heuristic evaluations and statistics from the Android Market only for several games. Because of this reason, they indicated that a more advanced heuristics set had to be developed.

Hynninen [32] researched the differences between peripherals for first person shooter games using heuristic evaluation. Three games on iPod Touch platform was tested during the study. The author indicated that Pinelle's [17] heuristic approach was predicated. By reviewing the literature, a new heuristics set was proposed with the focus on first person shooter (FPS) games. Subsequently, the author tested the games using the heuristics to evaluate the iPod Touch games. The result indicated usability issues related to iPod Touch controls.

Carmody [34] followed a rather different approach. Instead of testing the heuristics, he applied a three session Delphi test process [57], [58] in order to generate collated heuristics. The aim of his study was to investigate which heuristics game designers were considering while developing serious games. Researchers interviewed game designers and generated a first draft for design challenges. After the categorization of the design challenges, they were linked with heuristics cited from the literature. Afterwards, researchers conducted a Delphi survey with 12 expert evaluators and analyzed the first 39 items which were proposed after the first draft. Being iteratively conducted by the tests, 19 items for the heuristic set were suggested as the final result, validated from the literature. In the paper, expert evaluators did not evaluate games but instead evaluated the proposed heuristics. At the end, the study presented a set of heuristics which could be a guideline for designers developing serious games.

Like Carmody's work, Mohammed & Jaafar [36], conducted a study on refining the previously explained Playability Heuristics Evaluation for Educational Computer Game (PHEG) heuristic set [25]. In the research, 15 expert evaluators were asked to evaluate the PHEG set. The study also aimed to prioritize the categories presented on the PHEG heuristic set. An Analytical Hierarchy Process (AHP) [62] was applied in order to achieve that goal. At the end, the researchers presented a version of the heuristics set indicating the order of importance for each category.

Wodike et al. [42] studied the efficiency of empowering teenagers as expert evaluators for analyzing video games in their paper. Based on Pinelle's [17] heuristic set, they recruited 20 male students as expert evaluators. A mobile game was evaluated during the tests and evaluators were asked to fill in a severity scale form. Even though the study provided results regarding the playability of the game, the researchers highlighted that empowering students as evaluators was non-effective for analyzing the game.

Barbosa et al. [47] conducted a research about heuristic evaluation of educational games, proposing a blend of items from HEP [7], PLAY [19], and GameFlow [9] heuristics. They suggested the set of Heuristic Evaluation for Educational Games (HEEG). The set was applied to five different educational games. Two researchers and one game developer were recruited as expert evaluators during the tests. At the end, the researchers suggested that the set could be a starting point for analyzing specific point of educational games.

Cross references between heuristics were presented below (Table 2) with the notion of listing heuristics which were utilized in more than one research. This limitation of presented heuristics was necessary to avoid listing specific heuristics which involved specific areas of research such as educational context.

TABLE II. CROSS-REFERENCED HEURISTICS OF EXPERT EVALUATION RESEARCH ON PLAYABILITY HEURISTICS

Summary of heuristics that are used at least more than one study																	
Clear and varied outcomes	x	x						x									
Variable difficulty level	x		x			x		x	x	x	x				x		
Embodiment of metaphors with physical or other systems that user understands		x							x								
Audio-visual supports the game		x		x		x		x	x	x		x					x
Support of a variety of game styles.			x	x		x						x					x
Making effects of AI visible by ensuring they are consistent with the player's reasonable expectations	x					x		x								x	
Game provides immediate feedback	x	x	x	x	x	x			x	x		x	x				x
Player can easily turn the game on/off, and be able to save in different states	x		x	x		x		x		x		x					
The Player experiences the user interface as consistent but the game play is varied.	x					x		x									
Interface/HUD as a part of the game.	x					x		x									x
Player has enough information to get started from the beginning						x		x									x
Context sensitive help						x			x	x				x			x
Meaningful sounds	x					x			x								
Players do not need to use a manual to play game.	x		x			x			x	x				x			x
Non-intrusive interface	x					x		x	x	x			x				
Make the menu layers well-organized and minimalist to the extent the menu options are intuitive	x					x		x	x	x							x
Quick involvement with tutorials and/or progressive or adjustable difficulty levels	x		x			x	x	x	x	x						x	x
Art should be recognizable to player, and speak to its function.	x					x		x						x			
Always being able to identify score/status and goal			x	x		x		x	x	x	x	x		x	x	x	x
Standard conventions and natural mapping for controls	x			x		x		x		x	x	x		x	x		x
Aesthetic and minimalist design									x		x				x		
Clear goals		x	x	x		x		x		x	x	x			x		x

(*continued on next page*)

TABLE II. (continued)

Summary of heuristics that are used at least more than one study																
Appropriate rewards for effort and skill development	x		x	x	x		x	x	x		x					
Challenge, strategy, and pace are in balance		x	x	x	x		x		x	x	x			x		x
Fun gaming, without repetitive or boring tasks				x	x				x		x					x
Persistent game world	x		x		x				x							
Multiple ways to win.	x		x		x			x	x				x			
Feeling in control		x	x	x	x		x	x	x		x					x
Curiosity and exploration		x			x											x
Screen layout is efficient, integrated, and visually pleasing				x	x				x		x					x
Navigation is consistent, logical and minimalist.				x	x				x	x	x	x		x		x
Player interruption is supported				x	x				x		x					
Total concentration			x						x							x
Storyline relate to your life experiences and grabs interest					x				x				x			
Visuals, animation and music able to capture interest	x	x	x					x	x				x			x
Font types and sizes used allow easy reading										x				x		
Learning new concepts and skills	x					x			x							
The player does not have to memorize things unnecessarily				x	x			x	x		x					
Warning messages and cues help make less mistakes	x							x	x							
The game contains help			x	x	x				x		x				x	
The game story supports the gameplay and is meaningful				x	x		x		x		x		x			
Even if the game cannot be modeless, it should be perceived as modeless	x				x				x							
Player experiences fairness of outcomes	x				x		x									
The game has unpredictable yet reasonable story elements.	x								x							
Multiple goals in each level	x							x								
Mechanics should feel natural and have correct weight and momentum	x								x							

Include a lot of interactive props for the player to interact with	x							x		x		
Giving hints (but not too many)	x							x		x		
One reward of playing should be the acquisition of skill	x							x				
Learning should be fun		x										x
Players should become less self-aware and less worried about everyday life or self		x						x				
Players should experience an altered sense of time		x						x				
Players should feel emotionally involved in the game		x			x		x	x				
Support competition and cooperation between players		x				x		x				
Games should support social interaction between players		x				x				x		
Games should support social communities inside and outside the game		x						x				
Device UI and game UI are used for their own purposes			x		x			x	x			
Visible indicators			x		x		x	x	x			
Player understands terminology			x		x				x			
Game controls are convenient and flexible			x		x				x		x	x
The player cannot make irreversible errors			x		x			x	x			
The game accommodates with the surroundings			x		x			x	x			
The first-time experience is encouraging			x		x		x		x			
Players can express themselves			x		x			x	x			
The game does not stagnate			x		x		x	x	x			
The game is consistent			x		x		x		x			
The game uses orthogonal unit differentiation			x		x			x	x			
The player does not lose any hard-won possessions			x		x				x			
The cognitive load of the player should not be overburdened				x								x
Challenge should be fun	x			x	x		x	x				

(*continued on next page*)

TABLE II. (continued)

Summary of heuristics that are used at least more than one study																	
The interpersonal communication and collaboration should be supported by the entirety of the game					x					x				x			
Simple session management							x			x							
Flexible matchmaking							x			x							
Appropriate communication tools							x			x							
Meaningful awareness information							x		x								
Identifiable avatars							x	x									
Manage bad behavior							x			x							
Easy to learn, hard to master	x					x		x		x							
The game should be replayable	x					x		x	x	x							
First action is obvious and gives immediate positive feedback						x		x		x							
The visual representation should allow an unobstructed view of the area that is tied to the location								x								x	
Allow customization options for controls								x	x	x			x			x	x
The game should allow customization for different aspects								x	x							x	
Players allowed to build content	x									x							
The game is paced to apply pressure but not frustrate the player	x					x				x							
Interesting and absorbing tutorial	x									x							
The design hides the effects of network (in online gaming)				x						x		x					
The game and play sessions can be started quickly				x						x		x					
Maximizes consistency and matches standards											x				x		
The interactivity of the game is suitable to learners level											x				x		
The integration of presentation means is well coordinated											x				x		
The uses of space, color and text are according to the principles of screen design											x				x		

Quality of user interface is acceptable		x	x
Provide specific and self- identified key for specific task (exit, glossary, main, objective)		x	x
Overall interface of the game is appealing		x	x
The activities are interesting and engaging		x	x
The design and the contents are reliable and proven.		x	x
Can be used as self- directed learning tools.		x	x
Support for self- learning skills.		x	x
Medium for learning by doing.		x	x
Considers the individual differences.		x	x
Performance should be an outcome-based.		x	x
Ability to work in their own pace		x	x
Reliable content with correct flow.		x	x
Clear and understandable structure of contents.		x	x
Supporting materials are sufficient and relevant	x	x	x
Materials are interesting and engaging.		x	x
Players able to understand the learning goal.		x	x
The content is chunk based on topic and subtopic		x	x
Major and minor topic is differentiate clearly		x	x
Usage of multimedia elements are acceptable		x	x
Combination of multimedia elements are adequate		x	x
The presentation of multimedia elements are well manage		x	x
Suitability of multimedia elements for specific use		x	x
Not too many multimedia element in one screen		x	x
The use of multimedia elements support meaningfully the text provided.		x	x
The quality of multimedia elements (text, image, animation, video and sound) used is acceptable.		x	x

(*continued on next page*)

TABLE II. (continued)

Summary of heuristics that are used at least more than one study																	
The uses of multimedia elements enhance the presentation of information.			x				x										
Games should provide a lot of stimuli from different sources	x			x											x		
Games should quickly grab the players' attention and maintain their focus throughout the game.	x														x		
Non-playable content can be skipped		x							x								
Previous Empirical Evaluation Research on Playability Heuristics	Barbosa, Rego, Medeiros (2015)	Wodike, Sim, Horton (2014)	Mohamed, Jaafar (2013)	Carmody (2012)	Hynninen (2012)	Ponnada & Kannan (2012)	Omar & Jaafar (2010)	Suhonen & Vaataja (2010)	Almeida, Mealha, Veloso (2017)	Koeffel et al. (2010)	Pinelle, Wong, Stach & Gutwin (2009)	Korhonen, Paavilainen & Saarenpää (2009)	Köffel & Haler (2008)	Korhonen & Koivisto (2006)	Sweetser & Wyeth (2005)	Baauw, Bekker & Barengregt (2005)	Federoff (2002)

C. Inspections

1) Literature Reviews

The fourth observed approach included articles which solely based on review of existing literature to achieve a more generic point-of-view towards heuristic evaluation.

Schaffer [14] proposed a white paper for evaluating usability in video games. The aim of the study was to suggest a guideline for evaluating video games via heuristics. It was indicated that with both the utilization of user-tests and expert evaluation methods, it would be possible to analyze the usability of games. With literature review and commendations from the developers, 21 heuristics were suggested with five categories: general, graphical user interface, gameplay, control mapping and level design. Highlighting the lack of empirical research on previous heuristics, the study also did not present test results.

Paavilainen [29] reviewed video game evaluation heuristics in the context of social games perspective. In the study, a diverse literature review was conducted and four heuristic sets were indicated as comparable among each other [6], [7], [11], and [20]. The focus of the study was social games; therefore a collation of items was prepared from the heuristics mentioned in the study. At the end of the study, the high number of heuristics were criticized and Korhonen & Koivisto's playability heuristics [11] was distinguished as the most effective. The author also indicated that user-testing methods combined with heuristics evaluation would provide the most effective analysis. However, the proposed collated set was not tested.

Jerzak & Rebelo [45] prepared a study for comparing existing heuristics evaluation methods for games with serious games on focus. They also aimed to represent the strengths and weaknesses of existing heuristics in their study. In their paper, they analyzed nine heuristic evaluation approaches. After the elimination of those heuristics, to reach a global view of the related works, authors chose to compare three different heuristic sets [6], [7], and [19]. They also defined the following three groups/categories for comparison; gameplay, learning & entertainment, usability & game mechanics. The rest of the procedure in the study involved literature inspection and effective aspects for each heuristic set was shown as a result.

2) Game Reviews

Another identified inspection method for developing heuristics is the collection of information from common (online) game reviews which have the potential of offering a much larger sample size.

Livingston et al. [26] presented a study on using critic reviews of games for refinement of heuristic evaluations. Pinelle's [17] heuristics were used in the study. Based on previous reviews, authors prioritized the problems which the critics indicated for the games. A modified and genre specific heuristic set was suggested in the study. The authors claimed that by inspecting online reviews, it was possible to prioritize heuristics in terms of severity. Authors also mentioned that even though the study could re-organize the heuristics, it did not encapsulate overall player experience.

Hara & Ovasaka [44] aimed to develop a heuristic set for action oriented games such as the games developed for Microsoft Xbox Kinect peripheral. The study inspected the reviews of 36 motion controlled games with a total number of 256 games. By the inspection of reviews of those games, authors developed new heuristics with 13 items. Although the authors mentioned that there were shortcomings of the use of subjective data gathered from online reviews, there was also the lack of testing the proposed heuristics.

Zhu et al. [50] utilized the notion of using online reviews to a different level by lexically analyzing 821,122 games with the help of a software. At the end of semi-automated inspections, the authors proposed a set of heuristics and claimed that the studies of Desurvire et al. [7], Federoff [6], Malone [4], Pinelle [17], [20] had deficiencies because of three basic reasons: use of small data sets, depending on qualitative data and not having been empirically testing, and lastly focusing on small number of games and therefore not being generic. As a result the authors presented 90 playability heuristics.

Cross references between heuristics were presented below (Table 3) with the notion of listing heuristics which were utilized in more than one research. This limitation of presented heuristics was necessary to avoid listing specific heuristics which involved specific areas of research such as educational context.

D. Mixed Method

The last and the fifth observed method involved mixed-method modality, combining empirical research, expert evaluation and/or inspection methods.

Desurvire et al. [7] proposed the Heuristics of Playability (HEP) framework and prepared a heuristics set of 43 items, based on literature and reviewed by several experts. The expert evaluator formed the HEP set while focusing on how each heuristic was indicating a playability issue. The HEP heuristics set consisted of four categories; gameplay, game story, mechanics and usability. This model was tested via a prototype game. During the study, the researchers conducted user-testing method for validating and comparing the results from

TABLE III. CROSS-REFERENCED HEURISTICS OF RESEARCH VIA INSPECTIONS ON PLAYABILITY HEURISTICS

Summary of heuristics that are used at least more than one study						
Support of a variety of game styles.					x	x
Making effects of AI visible by ensuring they are consistent with the player's reasonable expectations			x	x		x
Game provides immediate feedback			x		x	x
Context sensitive help			x			x
Meaningful sounds			x			x
Non-intrusive interface			x			x
Quick involvement with tutorials and/or progressive or adjustable difficulty levels		x	x	x		
Always being able to identify score/status and goal	x		x	x		
Standard conventions and natural mapping for controls	x		x		x	x
Clear goals	x		x			
Appropriate rewards for effort and skill development		x	x			
Challenge, strategy, and pace are in balance			x			x
Fun gaming, without repetitive or boring tasks	x		x		x	x
Persistent game world		x	x			
Feeling in control	x		x			
Storyline relate to your life experiences and grabs interest			x			x
The game story supports the gameplay and is meaningful			x			x
Visible indicators	x					x
Game controls are convenient and flexible			x	x		x
The player cannot make irreversible errors	x		x			
The game should be replayable			x			x
The game should allow customization for different aspects				x		x
Previous Research on Playability Heuristics via Inspections	Schaffer (2007)	Paavilainen (2010)	Jerzak & Rebelo (2014)	Livingston, et al. (2010)	Hara & Ovasaka (2014)	Zhu et al. (2017)

the heuristic evaluation. The heuristics were analyzed and evaluated through four participants in two-hour long sessions where they played the game and evaluated the aspects of the game using the given heuristic items. The evaluator logged the actions and observation notes during the testing period. The user-tests included think-aloud play sessions and satisfaction questionnaires as well as observation notes taken by the supervisors. At the end of the study, both the results from user-tests and HEP evaluations were compared and the overall findings indicated that HEP heuristics were much more effective for finding issues related to the playability of the game rather than user-tests. This study proposed a new set of heuristics which was then used by several researchers but it had problematic aspects such as the unclear wordings for heuristics statements.

Pinelle et al. [17] presented a study on evaluating early versions of the games via heuristics. They utilized 108 online game reviews to form 10 heuristics. Subsequently, 10 more heuristics were added to involve multiplayer aspects of the games. Proposed heuristics were prepared with reference to released game reviews. After gathering the reviews, authors presented 12 problem categories. Finally, preliminary tests were conducted with the suggested set of heuristics. Five participants were asked to test an under-developed computer game using the heuristics. The participants were asked to fill a report form and Nielsen's Severity Scale. Authors emphasized that the set could identify usability issues of the game and the study offered a novel approach for using online game reviews as basis for defining heuristics. Even though the heuristic set had insufficient preliminary evaluations, the study received criticism from other researchers because of the use of biased online reviews.

Febretti & Garzotto [21] conducted a research on long-term engagement effects of games and the relation of usability. The aim of their study was to determine the effects of the game interface and its relation to the long-term player experience. To achieve that goal, they based their approach on the comparison of usability and playability. By using inspection method, they blended and modified 22 heuristic items from the literature and presented the set in seven categories. Likewise, they followed a similar method for generating a usability focused heuristic set with five categories and 14 items. They applied both user-testing and expert evaluation methods in their study. They tested eight commercial long term games with eight groups of participants with a total number of 47 participants. The tests also involved 20 game design experts and inspectors of usability and playability. To evaluate the aspect of engagement, they conducted user-tests with and without supervisors. Ultimately, they investigated the correlations from both test results and claimed the study had the intrinsic value of focusing on engagement. They indicated that playability heuristics had higher correlation values rather than usability heuristics. The result of the study offered a methodological approach in general.

Papaloukas et al. [54] conducted a study with a multi-modal methodology, combining user-tests with expert evaluations. Since there were no adequate methods or methodologies for evaluating a game's usability, they proposed using a modified set of heuristics based on Nielsen's heuristics [1]. They conducted tests on 2 different games in different platforms (Nintendo Wii & web-based computer game). For the user-tests, 30 participants were selected for usability evaluations. A specialized software was used for gathering metrics including user logs, facial expressions and verbal reactions using a camera. Player actions were recorded and analyzed by three usability experts. On the other hand, experts played the games for a week and wrote down the heuristics they used to identify the problem. Authors resulted their study by indicating the importance of the combination of these two methods, noting that the final results were enriched with the data gathered during player observations.

Jegers [15] studied on defining the enjoyment in pervasive games. Three pervasive tabletop games were tested using the GameFlow model [9] in three phases. The first phase of the research involved user-testing with 58 participants. The second phase involved six expert evaluators testing the heuristics. Lastly, the third phase involved sessions with both groups conducting a playtest and a focus group study. The author presented 14 new heuristics to be added to GameFlow model.

Desurvire & Wixon [39] aimed to determine the effectiveness and advantages of using heuristics for evaluating video games in their study. The focus on the study was to identify differences between the findings provided by heuristics and informal usability inspections. In their research, both the PLAY [19] and Game Approachability Principles (GAP) heuristics [48] sets were analyzed. Two browser based computer games were evaluated by 22 experts from the fields of game development and game review in three sessions each. At first, evaluators were asked to perform informal evaluations without heuristics, later with using PLAY heuristics and lastly GAP heuristics. Experts were asked to mark their comments by coded representations. The overall results indicated that utilization of heuristics during the evaluations help not only spot problems and suggest solutions but also help participants recognize effective elements of the design and suggest improvements. The researchers suggested that both sets were not only sufficient for analyzing the games but also effective for generating suggestions related to the issues in the gameplay. The mean frequency of issues mentioned during the tests were higher for heuristic evaluation compared to informal evaluations. It was noted that using heuristics provided more issues and thus was a better choice for evaluation than previously conducted informal evaluations.

Desurvire & Wiberg [48] aimed to compare different evaluation methods to test GAP heuristic set. Also, they aimed to test this new set of heuristics on different gaming platforms. Researchers utilized usability and heuristics evaluation techniques to compare them. One researcher applied heuristics evaluation method utilizing heuristics gathered from playability and usability literature while the other applied user-tests. Four games were tested during the study. After the tests, researchers analyzed and compared the results from both methods. 32 participants attended to the empirical tests. In the result section, researchers claimed that GAP heuristics and user-tests supported each other while indicating the best approach for analyzing the overall experience in games was the use of both methods simultaneously. Desurvire noted that, like PLAY heuristics, GAP principles held a guiding purpose therefore not directly aimed to evaluate playability.

Hochleitner et al. [46] introduced a heuristic framework for evaluating user experience in games. The study aimed to improve previously presented heuristic approaches and correlated them with common game reviews. The study was complementary to Koeffel's research [27]. In order to measure the applicability of the heuristic framework, six games were tested. The online game review ratings were later compared with the results of heuristic evaluation. The proposition of the heuristics was based on the previous works of Malone [4], Federoff [6], Desurvire [7] Shaffer [14], Pinelle [17], Koeffel [27], and Korhonen & Koivisto [11]. However, the focal point of the suggested new heuristic set was Koeffel's [27] set with 29 items. At the end, a total set of 49 items was proposed. The games tested were selected due to online game review ratings and evaluated by three expert evaluators who had previous experience with the heuristics set. Consequently, it was stated that there was a correlation between average game review ratings and results obtained from the heuristics study.

Cross references between heuristics were presented below (Table 4) with the notion of listing heuristics which were utilized in more than one research. This limitation of presented heuristics was necessary to avoid listing specific heuristics which involved specific areas of research such as pervasive games context and provide a holistic point of view.

IV. Conclusion

By categorization of methodological differences between heuristic evaluation researches in this study, we aimed to present a novel perspective to the domain

TABLE IV. CROSS-REFERENCED HEURISTICS OF MIXED METHOD RESEARCH ON PLAYABILITY HEURISTICS

Summary of heuristics that are used at least more than one study								
Variable difficulty level	x				x	x	x	x
Audio-visual supports the game						x	x	x
Support of a variety of game styles.			x		x	x	x	
Using humor appropriately						x	x	
Making effects of AI visible by ensuring they are consistent with the player's reasonable expectations	x	x	x					x
Game provides immediate feedback	x				x	x	x	x
Player can easily turn the game on/off, and be able to save in different states	x				x			x
The Player experiences the user interface as consistent but the game play is varied.	x					x	x	
Interface/HUD as a part of the game.	x		x			x	x	x
Player has enough information to get started from the beginning	x					x	x	
Context sensitive help	x					x	x	x
Players do not need to use a manual to play game.	x				x	x	x	
Non-intrusive interface	x		x					
Make the menu layers well-organized and minimalist to the extent the menu options are intuitive	x							x
Quick involvement with tutorials and/or progressive or adjustable difficulty levels	x	x			x	x	x	x
Always being able to identify score/status and goal	x	x	x		x	x	x	
Standard conventions and natural mapping for controls				x		x	x	x
Clear goals	x		x	x	x	x	x	x
Appropriate rewards for effort and skill development	x		x		x	x	x	x
Challenge, strategy, and pace are in balance				x	x	x	x	x
Fun gaming, without repetitive or boring tasks						x	x	x
Persistent game world	x				x	x	x	x
Multiple ways to win.	x				x	x	x	

(*continued on next page*)

TABLE IV. (continued)

Summary of heuristics that are used at least more than one study								
Feeling in control	x				x	x	x	x
Application of the newly acquired knowledge / skill	x					x	x	
Visuals, animation and music able to capture interest				x		x	x	
Warning messages and cues help make less mistakes		x			x			
The game contains help	x							x
The game story supports the gameplay and is meaningful	x					x	x	x
Multiple goals in each level	x				x			
Players should feel emotionally involved in the game					x			x
The cognitive load of the player should not be overburdened	x			x				
Meaningful awareness information	x					x	x	
Easy to learn, hard to master						x	x	x
The players should not lose any hard won possessions.						x	x	
The game should be repayable	x							x
First action is obvious and gives immediate positive feedback		x	x	x				x
The game is paced to apply pressure but not frustrate the player	x							x
Curiosity and exploration						x	x	
If there is a game story, the player is eager to spend time thinking of the possible outcomes.	x					x	x	
Not being penalized repetitively for the same failure						x	x	
Game control should allow a smooth gaming experience without unnecessary pauses	x					x	x	
Provide consistency between the game elements and the overarching setting and story to suspend disbelief.	x							x
Empathy with the game character	x					x	x	
The game offers something different in terms of attracting and retaining the players' interest.	x		x			x	x	
Player error is avoided						x	x	

Previous Mixed Method Evaluation Research on Playability Heuristics	Desurvire, Caplan, Toth (2004)	Pinelle, Wong, Stach (2008)	Febretti & Garzotto (2009)	Papaloukas et al. (2009)	Jegers (2009)	Desurvire & Wixon (2013)	Desurvire & Wiberg (2015)	Hochleitner et al. (2015)
Consistent learning curve with the industry		x	x			x	x	
Games should respond to users' actions in a predictable manner		x						x
The game should provide views that allow the user to have a clear, unobstructed view of the area		x						x
Allow users to skip non-playable and frequently repeated content		x						x
Allow customization options for controls		x		x				
Game controls are convenient and flexible			x		x			
The game should provide different challenge levels for different players			x	x	x			x
Player interruption is supported						x	x	
Skills are useful						x	x	

TABLE V. COMMON PLAYABILITY HEURISTICS IN THE LITERATURE

Heuristics	Number of References
Support of a variety of game styles.	15
Making effects of AI visible by ensuring they are consistent with the player's reasonable expectations	13
Game provides immediate feedback	27
Context sensitive help	14
Non-intrusive interface	12
Quick involvement with tutorials and/or progressive or adjustable difficulty levels	22
Always being able to identify score/status and goal	29
Standard conventions and natural mapping for controls	21
Clear goals	26
Appropriate rewards for effort and skill development	21
Challenge, strategy, and pace are in balance	25
Fun gaming, without repetitive or boring tasks	14
Persistent game world	13
Feeling in control	23
The game story supports the gameplay and is meaningful	16

of playability evaluations. It was also aimed to present a holistic view to provide a guide for future research regarding methodological approaches for heuristic evaluation of games. The researches indicated that most of the authors suggested using more than one method in order to validate the proposed heuristics. It was also observed that literature review for defining the playability heuristics was the most common way to conduct studies.

The study represented that 15 heuristics were common for all of the methods (Table 5). Accordingly, the review indicated that studies combining heuristic sets were efficient but lacked validation. Multi-modality in research, such as using user testing in order to validate expert evaluations, yielded more heuristics in comparison with other methods. Further research might involve using collated playability heuristics via expert evaluations and empirical evaluations in order to validate the provided heuristics.

References

[1] J. Nielsen. "Usability inspection methods" In: *Conference companion on Human factors in computing systems*. ACM, 1994, pp. 413–414.

[2] T. Fullerton, C. Swain, S. Hoffman, Game design workshop: Designing, prototyping, & playtesting games. CRC Press, 2004.

[3] H. Korhonen, "Comparison of playtesting and expert review methods in mobile game evaluation," In: *Proceedings of the 3rd International Conference on Fun and Games*. ACM, 2010, pp. 18–27.

[4] T. W. Malone, "Heuristics for designing enjoyable user interfaces: Lessons from computer games" In: *Proceedings of the 1982 conference on Human factors in computing systems*. ACM, 1982. pp. 63–68.

[5] C. Fabricatore, M. Nussbaum, R. Rosas, "Playability in action videogames: A qualitative design model" In: *Human-Computer Interaction*, 2002, pp. 311–368.

[6] M. A. Federoff, "Heuristics and usability guidelines for the creation and evaluation of fun in video games," Doctoral Dissertation, Indiana University, 2002.

[7] H. Desurvire, M. Caplan, J. A. Toth, "Using heuristics to evaluate the playability of games" In: *CHI'04 extended abstracts on Human factors in computing systems*. ACM, 2004, pp. 1509–1512.

[8] E. Baauw, M. Bekker, W. Barendregt, "A structured expert evaluation method for the evaluation of children's computer games" In: *Human-Computer Interaction-INTERACT 2005*, 2005, pp. 457–469.

[9] P. Sweetser, P. Wyeth, "GameFlow: a model for evaluating player enjoyment in games," In: *Computers in Entertainment (CIE)*, 2005, v.3 n.3: 3–3.

[10] C. Röcker, M. Haar, "Exploring the usability of videogame heuristics for pervasive game development in smart home environments" In: *Proceedings of the Third International Workshop on Pervasive Gaming Applications–PerGames*. 2006, pp. 199–206.

[11] H. Korhonen, E. Ml. Koivisto, "Playability heuristics for mobile games," In: *Proceedings of the 8th conference on Human-computer interaction with mobile devices and services*. ACM, 2006, pp. 9–16.

[12] S. Song, J. Lee, Retracted: "Key factors of heuristic evaluation for game design: Towards massively multi-player online role-playing game," 2007, pp. 709–723.

[13] H. Korhonen, E. Ml. Koivisto, "Playability heuristics for mobile multi-player games," In: *Proceedings of the 2nd international conference on Digital interactive media in entertainment and arts*. ACM, 2007, pp. 28–35.

[14] N. Schaffer, "Heuristics for usability in games-white paper," 2007.

[15] K. Jegers, "Pervasive GameFlow: Identifying and exploring the mechanisms of player enjoyment in pervasive games," 2009, Doctoral Dissertation, Umeå Universitet, Inst för Informatik.

[16] C. Köffel, M. Haller, "Heuristics for the evaluation of tabletop games," In: Evaluating User Experiences in Games, Workshop at the 2008 Conference on Human Factors in Computing Systems Florence, Italy, 5–10 April 2008. CHI. Vol.8, 2008.

[17] D. Pinelle, N. Wong, T. Stach, "Heuristic evaluation for games: usability principles for video game design," In: *Proceedings of the SIGCHI Conference on Human Factors in Computing Systems*. ACM, 2008, pp. 1453–1462.

[18] H. Korhonen, J. Paavilainen, H. Saarenpää, "Expert review method in game evaluations: comparison of two playability heuristic sets," In: *Proceedings of the 13th international MindTrek conference: Everyday life in the ubiquitous era*. ACM, 2009, pp. 74–81.

[19] H. Desurvire, C. Wiberg, "Game usability heuristics (PLAY) for evaluating and designing better games: The next iteration," In: *International Conference on Online Communities and Social Computing*. Springer, Berlin, Heidelberg, 2009, pp. 557–566.

[20] D. Pinelle, N. Wong, T. Stach, C. Gutwin, "Usability heuristics for networked multiplayer games," In: *Proceedings of the ACM 2009 international conference on Supporting group work*. ACM, 2009, pp. 169–178.

[21] A. Febretti, F. Garzotto, "Usability, playability, and long-term engagement in computer games," In: *CHI'09 Extended Abstracts on Human Factors in Computing Systems*. ACM, 2009, pp. 4063–4068.

[22] J. L.Tan, D. H. Goh, R. P. Ang, V. S. Huan, "Usability and playability heuristics for evaluation of an instructional game," In: *E-Learn: World Conference on E-Learning in Corporate, Government, Healthcare, and Higher Education*. Association for the Advancement of Computing in Education (AACE), 2010, pp. 363–373.

[23] S. Almeida, O. Mealha, A. Veloso, "Heuristic Evaluation of "FarmVille". *VIDEOJOGOS 2010*, 2010.

[24] H. Mohamed, A. Jaafar, "Development and potential analysis of heuristic evaluation for educational computer game (PHEG)," In: *Computer Sciences and Convergence Information Technology (ICCIT), 2010 5th International Conference on*. IEEE, 2010, pp. 222–227.

[25] H. M. Omar, A. Jaafar, "Heuristics evaluation in computer games. In: International Conference on Information Retrieval and Knowledge Management: Exploring the Invisible World, CAMP'10, 2010.

[26] I. J. Livingston, R. L. Mandryk, K. G. Stanley, "Critic-proofing: how using critic reviews and game genres can refine heuristic evaluations," In: *Proceedings of the International Academic Conference on the Future of Game Design and Technology*. ACM, 2010, pp. 48–55.

[27] C. Koeffel, W. Hochleitner, J. Leitner, M. Haller, A. Geven, M. Tscheligi, "Using heuristics to evaluate the overall user experience of video games and advanced interaction games,". In: *Evaluating user experience in games.* Springer London, 2010, pp. 233–256.

[28] K. Suhonen, H. Väätäjä, "Assessing the applicability of modular playability heuristics for evaluating health-enhancing games," In: *Proceedings of the 14th International Academic MindTrek Conference: Envisioning Future Media Environments.* ACM, 2010, pp. 147–150.

[29] J. Paavilainen, "Critical review on video game evaluation heuristics: social games perspective," In: *Proceedings of the International Academic Conference on the Future of Game Design and Technology.* ACM, 2010, pp. 56–65.

[30] S. B. Zaibon, N. Shiratuddin, "Heuristics evaluation strategy for mobile game-based learning," In: *Wireless, Mobile and Ubiquitous Technologies in Education (WMUTE), 2010 6th IEEE International Conference on.* IEEE, 2010, pp. 127–131.

[31] Y. H. Liao, C. Shen, "Heuristic evaluation of digital game based learning: a case study," In: *Digital Game and Intelligent Toy Enhanced Learning (DIGITEL), 2012 IEEE Fourth International Conference on.* IEEE, 2012, pp. 192–196.

[32] T. Hynninen, "First-person shooter controls on touchscreen devices: A heuristic evaluation of three games on the iPod touch" M.Sc. Thesis, Department of Computer Sciences, University of Tampere, Tampere, Finland, 2012.

[33] A. Ponnada, A. Kannan, "Evaluation of mobile games using playability heuristics," In: *Proceedings of the International Conference on Advances in Computing, Communications and Informatics.* ACM, 2012, pp. 244–247.

[34] K. W. Carmody, "Exploring serious game design heuristics: a delphi study," Doctoral dissertation, Northeastern University, 2012.

[35] J. L. G. Sánchez, F. L. G. Vela, F. M. Simarro, N. Padilla-Zea, "Playability: analysing user experience in video games," *Behaviour & Information Technology*, 2012, v31,n.10: pp. 1033–1054.

[36] H. Mohamed, A. Jaafar, "Prioritization Of Playability Heuristic Evaluation For Educational Computer Games (Pheg) Technique Using Analytic Hierarchy Process," In: *Proceedings of the International Symposium on the Analytic Hierarchy Process 2013*, 2013.

[37] G. Ulger, "Playability heuristics for mobile games using touchscreen displays," M.Sc. dissertation, Middle East Technical University, 2013.

[38] K. Khanana, E. L. C. Law, “Designing children’s digital games on nutrition with playability heuristics,” In: *CHI’13 Extended Abstracts on Human Factors in Computing Systems*. ACM, 2013, pp. 1071–1076.

[39] H. Desurvire, D. Wixon, “Game principles: choice, change & creativity: making better games,” In: *CHI’13 Extended Abstracts on Human Factors in Computing Systems*. ACM, 2013, pp. 1065–1070.

[40] F. Rodio, J. M. Bastien, “Heuristics for Video Games Evaluation: How Players Rate Their Relevance for Different Game Genres According to Their Experience,” In: *Proceedings of the 25th Conference on l’Interaction Homme-Machine*. ACM, 2013, pp. 89.

[41] J. N. Marciano, L. C. Miranda, E. E. C. Miranda, “Evaluating multiple aspects of educational computer games: literature review and case study,” *International Journal of Computer Games Technology*, 2014, v.2014, n.14.

[42] O. A. Wodike, G. Sim, M. Horton, “Empowering teenagers to perform a heuristic evaluation of a game,” In: *Proceedings of the 28th International BCS Human Computer Interaction Conference on HCI 2014-Sand, Sea and Sky-Holiday HCI*. BCS, 2014, pp. 353–358.

[43] S. Soomro, W. F. W. Ahmad, S. Sulaiman, “Evaluation of mobile games with playability heuristic evaluation system,” In: *Computer and Information Sciences (ICCOINS), 2014 International Conference on*. IEEE, 2014, pp. 1–6.

[44] M. Hara, S. Ovaska, “Heuristics for motion-based control in games,” In: *Proceedings of the 8th Nordic Conference on Human-Computer Interaction: Fun, Fast, Foundational*. ACM, 2014, pp. 697–706.

[45] N. Jerzak, F. Rebelo, “Serious games and heuristic evaluation–the cross-comparison of existing heuristic evaluation methods for games,” In: *International Conference of Design, User Experience, and Usability*. Springer, Cham, 2014, pp. 453–464.

[46] C. Hochleitner, W. Hochleitner, C. Graf, M. Tscheligi, “A heuristic framework for evaluating user experience in games. In: *Game User Experience Evaluation*. Springer International Publishing, 2015, pp. 187–206.

[47] M. B. Barbosa, A. B. Rêgo, I. De Medeiros, “HEEG: Heuristic Evaluation for Educational Games,” In: *Proceedings of SBGames 2015*, 2015, pp. 224–227.

[48] H. Desurvire, C. Wiberg, “User Experience Design for Inexperienced Gamers: GAP—Game Approachability Principles,” In: *Game User Experience Evaluation*. Springer International Publishing, 2015, p. 169–186.

[49] Y. R. Guo, D. H. Goh, “Heuristic evaluation of an information literacy game,” In: *Digital Libraries: Knowledge, Information, and Data in an Open Access Society*. Springer International Publishing, 2016, pp. 188–199.

[50] M. Zhu, F. Zhao, X. Fang, C. Moser, "Developing Playability Heuristics Based on Nouns and Adjectives from Online Game Reviews," *International Journal of Human–Computer Interaction*, 2017, v.33,n.3, pp. 241–253.

[51] R. M. Gagne, L. J. Briggs, *Principles of instructional design*. Holt, Rinehart & Winston, 1974.

[52] J. M. Keller, "Using The ARCS motivational process in computer-based instruction and distance education," *New directions for teaching and learning*, 1999, v.1999,n.78, pp. 37–47.

[53] D. A. Norman, *The design of everyday things*. London, MIT Press, 1998.

[54] S. Papaloukas, K. Patriarcheas, M. Xenos, "Usability assessment heuristics in new genre videogames," In: *Informatics, 2009. PCI'09. 13th Panhellenic Conference on*. IEEE, 2009, pp. 202–206.

[55] W. IJsselsteijn, Y. A. W. De Kort, K. Poels, "The game experience questionnaire," *Manuscript in preparation*, 2008

[56] M. Csikszentmihalyi, *The psychology of optimal experience*. New York, 1990.

[57] G. Skulmoski, F. Hartman, J. Krahn, "The Delphi method for graduate research," *Journal of Information Technology Education: Research*, 2007, v.6,n.1, pp. 1–21.

[58] K. Cuhls, "Delphi method," Fraunhofer Institute for Systems and Innovation Research. Germany, 2003.

[59] J. Nielsen, "Severity ratings for usability problems," *Papers and Essays*, 1995, v.54, pp. 1–2.

[60] K. Baker, S. Greenberg, C. Gutwin, "Empirical development of a heuristic evaluation methodology for shared workspace groupware," In: *Proceedings of the 2002 ACM conference on Computer supported cooperative work*. ACM, 2002, pp. 96–105.

[61] F Garzotto, "Investigating the educational effectiveness of multiplayer online games for children," In: *Proceedings of the 6th international conference on Interaction design and children*. ACM, 2007, pp. 29–36.

[62] T. L. Saaty, "How to make a decision: the analytic hierarchy process," *European journal of operational research*, 1990, v.48,n.1, pp. 9–26.

7. Explorations in Game Experience: A Case Study of 'Horizon Zero Dawn'

Barbaros Bostan
Game Design Department Bahcesehir University Istanbul, Turkey barbaros.bostan@comm.bau.edu.tr

Mehmet İlker Berkman
Communication Design Bahcesehir University Istanbul, Turkey ilker.berkman@comm.bau.edu.tr

Abstract— The aim of this study is to find the relationship between the specific gameplay elements of a chosen game and the gameplay experience it provides which will be measured by a selected survey from the literature, the Game Experience Questionnaire (GEQ). Before illustrating the relation between the game elements and the player experience measured via GEQ, a confirmatory factor analysis was conducted to validate the factor structure of GEQ and GEQ is also evaluated for its unidimensionality, internal reliability, convergent validity and discriminant validity. The selected game for the study is Horizon Zero Dawn which is an open-world action role-playing video game developed by Guerrilla Games. Selling more than 2.6 million units worldwide in two weeks, the game has become the best-selling new first party IP launch on the PlayStation 4. Our study aims to answer the following questions: (1) Can we measure a successful and popular game's experience with GEQ? (2) What are the gameplay elements that make this game so popular? (3) Are there any correlations with the subscales of GEQ and the gameplay elements in question?

Keywords— game experience; gameplay elements; game experience questionnaire; GEQ; role playing games

I. Introduction

The concept of user experience is the subjective relationship between the user and the application [1] and player experience in computer games is usually associated with four terms: Engagement, Immersion, Flow and Presence. Engagement is the state of involvement with a video game and is closely related with how enjoyable the experience is. Immersion is a quantifiable description of a technology and includes the extent to which the computer displays are extensive, surrounding, inclusive, vivid and matching [2]. Flow is the enjoyable state experienced when engaged in a challenging activity [3] and presence is the sense of being inside a virtual world [4]. Discussing the relationship between these terms and the game experience is beyond the scope of this article but it may be argued that these terms are the outcome of extreme gaming experiences rather than the core elements of the interactive gameplay process itself [5].

Core elements of the interactive gameplay can be identified as mechanics, dynamics and aesthetics, based on the MDA framework [6]. The experience of gaming is a personal experience and is also affected by the genre or the game in question because each genre, even each game in a specific genre, uses different mechanics. Even if the mechanics seem similar, how these mechanics act on player inputs and with each other over time, i.e. dynamics of the game, changes. Aesthetics, the emotional responses evoked in the player also differ, even between the quests or levels of a single game. The aim of this study is to find the relationship between the specific gameplay elements of a chosen game and the gameplay experience it provides which will be measured by a selected survey from the literature. The gameplay elements in question can be conceived as the mechanics, dynamics and aesthetics of the MDA framework [6].

II. Related Studies

Modeling player experience is a diverse field of research. Wiemeyer, Nacke, Moser and Mueller [7] analyzed the following general and domain-specific psychological models of player experience: the Player Experience of Need Satisfaction (PENS) [8] based on Self-Determination Theory [9]; Keller's [10, 11] Attention, Relevance, Confidence, Satisfaction (ARCS); Sweetser and Wyeth's [12] GameFlow based on Flow theory [3]; the Presence-Involvement-Flow Framework (PIFF2) [13]; the Fun of Gaming (FUGA) [14]; Core Elements of the Gaming Experience (CEGE) [5]; and the Play Heuristics of Desurvire and Wiberg [15]. The researchers stated that the models either apply general concepts from other fields to gaming or address the process of player experience from different perspectives (players, developers, researchers). Some of the elements defined by these models are hardly separable and there are complex interactions between them. At the end, when they combine the results, the following (social-)psychological elements of player experience are reported: competence; autonomy and control; immersion, (spatial and social) presence, flow and GameFlow; involvement and (enduring) engagement; social relatedness and social interaction; challenge; tension; curiosity; fantasy; positive and negative emotions; intrinsic goals; feedback and evaluation.

Assessment of the player experience with surveys is also a research area with different approaches. One of the earliest experience scale is the Presence Questionnaire of Witmer and Singer [16] commonly used in VR (virtual reality) and VEs (virtual environments). The MEC spatial presence questionnaire (MEC-SPQ) [17], the Spatial Presence Experience Scale (SPES) [18] and the Social Presence in Gaming Questionnaire (SPGQ) [19] are three presence related questionnaires used to assess player experience. In terms of domain specific surveys three of them deserve special attention. Core Elements of the Gaming Experience Questionnaire, CEGEQ, consists of 38 items for 10 dimensions that measure enjoyment, frustration, CEGE, puppetry, video-game, control, facilitators, ownership, gameplay, and environment [5]. The Cronbach alpha for the whole questionnaire is .794. The Game Engagement Questionnaire (GEnQ) consists of 19 items that measure absorption, flow, presence, and immersion [20]. The Cronbach alpha for the whole questionnaire is .85. The Game Experience Questionnaire (GEQ) consists of 36 items representing 7 latent variables that measure competence, immersion, flow, tension, challenge, positive and negative affect [21]. The subscales of GEQ have an average Cronbach's alpha score of .81.

Considering the diversity of game experience surveys mentioned above, our survey selection criteria are: (1) choosing a domain specific survey rather than a general one applied to gaming and (2) selecting a survey which is tested with sufficient number of participants that play different types of games. Among the three domain specific surveys, CEGEQ [5] was tested with 15 participants who played Tetris. GEnQ [20] is tested twice, first with 153 junior high school students and then with 107 undergraduate students who played a first-person shooter named 'S.T.A.L.K.E.R.: Shadow of Chernobyl'. The typical game playing time per week based on six predetermined time categories were taken but which genres or games they usually played were not published. GEQ [21] is tested with 380 participants with diverse gaming backgrounds: First Person Shooter games (22 %), Role Playing games (14 %), Sport games (13 %), Puzzle/board/card games (11 %), Action adventure games (10 %), Strategy games (9 %), and other genres (e.g., simulation games, fight games, children's games, music games) (11 %).

Based on our selection criteria, GEQ seemed the most suitable choice. It is designed to assess the games in terms of playing experience. While it also have a Social Presence Module and a Post-Game module, the 36-item core covers a broader range of dimensions that are associated with gaming experience, compared to other questionnaires. The original development efforts depend on more participants that play games within a wider variety of genres. On the other hand, the psychometric qualities of GEQ were not completely established [22, 23], especially related with its factor structure that represents its construct validity. Brühlmann and Schmid [24] explored the factor structure of GEQ using principal axis exploratory factor analysis with oblimin rotation. Their results did not match the factor structure suggested in scoring guidelines published by Ijselstein et al. [21] and the factors evaluated through two different games were not similar. However, a confirmatory approach has not been employed yet, to validate the factor structure of GEQ. For this reason, we decided to run a confirmatory factor analysis as a part of this study before illustrating the relation between the game elements and the player experience measured via GEQ.

III. Methodology

The selected game for the study is Horizon Zero Dawn which is an open-world action role-playing video game developed by Guerrilla Games and published by Sony Interactive Entertainment for the PlayStation 4 on 28 February 2017 in North America, on 1 March in Europe and 2 March in Asia. Selling more than 2.6 million units worldwide in two weeks, including units sold at retail and digitally through the PlayStation Store, the game has become the best-selling new first party IP launch on the PlayStation 4. The game has a Metascore of 89 based on 115 critics and a user score of 8.3 based on 4584 ratings[1]. Game critics referred the gameplay design of Horizon Zero Dawn as the best of everything with the minimum number of flaws. Choosing a commercial game with such positive reviews for a player experience study has the risk of having extreme negative skewness within the collected data. However, our study aims to answer the following questions: (1) Can we measure such a successful title's game experience with GEQ? (2) What are the gameplay elements that make this game so popular? (3) Are there any correlations with the subscales of GEQ and the gameplay elements in question?

After an analysis of game reviews for Horizon Zero Dawn, 36 gameplay elements drew our attention. Seven of them are narrative elements: quality of the main quest, length of the main quest, quality of side quests, number of side quests, antagonists of the story, side characters of the story, and lore/fiction of the world. Four elements focus on the protagonist of the game, Aloy: Aloy's physical appearance, Aloy's dialogue options, Aloy's voice acting, and Aloy's facial expressions and animations. Six elements are related with the customization power or the interactivity range provided by the game: armor variety, weapon variety, machine variety, trap variety, ammo variety, and skill variety. There are 3 audio-visual elements: environment (visual) design, machine (visual) design and game soundtrack. 10 are core gameplay elements: enemy AI, crafting system,

1 Metacritic website, accessed on the 5th of June 2017, http://www.metacritic.com/game/playstation-4/horizon-zero-dawn

game controls, combat mechanics, machine overriding, game UI, save system, travel options, inventory system, and game feedback. Besides the side quests, the game offers several activities and collectibles, which are summarized by 6 elements: Bandit Camps, Hunting Grounds, Tallnecks, Cauldrons, Corrupted Zones and Collectibles. Participants were asked to rate their satisfaction levels about these 36 gameplay elements on a five-point Likert scale with a rating ranging from 1 for 'not at all satisfied' to 5 for 'extremely satisfied'. Participants were then asked to answer the GEQ questionnaire, followed by demographic questions.

Three hundred and thirty-seven participants (44 female; 276 male; 17 without gender declaration) took part in an online study, which was announced on social media platforms and several forums of the selected game. Ages ranged from 13 to 66 (M = 29.05, SD = 8.651). In terms of gaming habits, 6 participants without any declaration, the remaining 331 participants are classified into five groups: 0–1 h/week (.6 %), 2–5 h/week (12.2 %), 6–10 h/week (31.5 %), 11–15 h/week (26.1 %) and more than 15 h/week (27.9 %), indicating that the majority of the participants are "frequent" game players.

IV. Results

A. Psychometric Evaluation of GEQ

We evaluated each GEQ item for their skewness and identified that data on 14 of 36 items were skewed, when the skewness statistic divided by its standard error and the kurtosis statistic divided by its standard error are greater than z = ±3.29, suggesting that the distribution is severely skewed. [25]. Visual inspection of histograms also supports this finding. For this reason, we decided to embrace the partial least squares confirmatory factor analysis (PLS-CFA) approach instead of a covariance-based method, as PLS-CFA is not sensitive to skewed data [26, 27]. The skewness is due to the positive bias of participants, who have voluntarily evaluated what they have been playing. Besides, the sensory-imagery immersion construct of GEQ drives independent items that make it a formative variable, which can be inspected via PLS based methods. Using PLS-CFA, we evaluated GEQ for its unidimensionality, internal reliability, convergent validity and discriminant validity using construct development and validation guidelines of Lewis et al. [28] and Straub et al. [29], based on the predetermined structure of 7 latent variables described by Ijselstein et al. [21].

As an indicator of unidimensionality of a latent variable, each item should load with a high coefficient on one factor only, and this factor should consistently be the same for all items that are supposed to measure it, with an eigenvalue exceeding 1. A predefined latent variable with items that load on two factors exceeding eigenvalue 1 suggests that it is not, unidimensional. Thus, those items may refer to a latent variable in addition to the intended construct. The items intended for competence, immersion, flow, tension, positive and negative affect lead to a single construct with an eigenvalue exceeding 1, but the items intended for challenge construct revealed two factor with eigenvalue greater than 1. The first factor has an eigenvalue 1.94 while the second has 1.13. The results imply that items intended for challenge construct are also driven by some other latent variable. Eigenvalues of all latent variables can be followed on Table II.

The item CHA 23 has a higher load of .623 on a secondary dimension rather than its intended dimension, exceeding .6. Items CHA 32 and CHA 26 also have a medium load exceeding .4 on the same secondary factor, while the items CHA 11 and CHA 26 also have medium level load on a second factor. So, the resulting factor structure is violating the unidimensionality of the latent variable Challenge. These problematic loads are given on Table I.

These results imply that some variables assigned to constructs can be related with some other dimension that is considered or not considered within the scale.

The reliability indicator for each latent variable is Cronbach's alpha value, as well as Dillon-Goldstein (D-G) rho value. D-G rho is a better indicator of reliability since it does not assume that each manifest variable is equally important in defining the latent variable, while Cronbach's alpha makes this assumption [26]. Latent variable Negative Effect and Sensory-Imagery Immersion have Cronbach alpha values of .59, which are quite close to .6 that determine a reliable construct, while others exceed this limit, as given below. All D-G rho values are above the acceptable limit of .7, indicating that predetermined latent variables of GEQ have internal reliability. Reliability indicators can be followed on Table II.

The average variance extracted (AVE) index should exceed .5 to verify convergent validity, suggesting that a construct's items should assemble at a higher degree compared to items measuring other latent variables. The Flow and Tension-Attention constructs met this criterion with a value of .63, while others did not. This result suggests that most of the constructs of GEQ are not able to explain more than half of the variance of their indicators [27]. In other words, the low AVE's indicate that items do not represent their latent variables clearly. Results can be followed on Table III.

If the shared variance between two constructs is lower than the AVE for each individual construct, then each construct has different measures from each other, indicating the discriminant validity of the scale [30], which is called

TABLE I. VARIABLES / FACTORS CORRELATIONS OF GEQ

CHALLENGE					
	F1	F2	F3	F4	F5
CHA_11	0.663	-0.316	**0.568**	-0.249	-0.275
CHA_23	0.573	**0.623**	0.197	-0.143	**0.473**
CHA_32	0.603	**0.564**	-0.188	0.302	**-0.437**
CHA_33	0.658	-0.278	**-0.558**	-0.421	0.027
CHA_26	0.613	**-0.497**	-0.015	0.557	0.256
COMPETENCE					
	F1	F2	F3	F4	F5
COM_02	0.761	-0.072	-0.324	-0.045	0.555
COM_10	0.555	-0.524	0.632	-0.135	0.019
COM_15	0.733	0.109	-0.262	-0.500	-0.364
COM_21	0.509	0.745	0.415	0.097	0.063
COM_17	0.726	-0.156	-0.170	0.587	-0.274
POSITIVE AFFECT					
	F1	F2	F3	F4	F5
PE_01	0.705	-0.390	-0.291	0.515	-0.013
PE_04	0.643	0.424	0.562	0.216	-0.209
PE_14	0.773	-0.284	-0.111	-0.408	-0.378
PE_06	0.799	-0.164	0.228	-0.198	0.493
PE_20	0.509	0.696	-0.497	-0.056	0.082
NEGATIVE AFFECT					
	F1	F2	F3	F4	
NE_07	0.586	**-0.611**	-0.505	-0.171	
NE_08	0.681	-0.201	0.653	-0.266	
NE_09	0.803	0.102	0.009	0.587	
NE_16	0.600	**0.688**	-0.259	-0.317	
SENSORY IMAGERY IMMERSION					
	F1	F2	F3	F4	F5
SII_03	0.461	**-0.646**	0.533	0.160	-0.244
SII_12	0.438	**0.720**	0.472	-0.021	-0.260
SII_18	0.710	-0.130	-0.394	-0.515	-0.239
SII_19	0.674	0.098	-0.492	0.536	-0.086
SII_30	0.788	0.011	0.202	-0.076	0.576
FLOW					
	F1	F2	F3	F4	F5
FLO_05	0.571	**0.697**	-0.415	0.126	-0.006
FLO_25	0.773	-0.187	0.208	0.500	-0.271

(continued on next page)

TABLE I. (continued)

CHALLENGE					
FLO_28	0.699	0.358	0.554	-0.243	0.134
FLO_13	0.767	-0.267	-0.215	-0.446	-0.308
FLO_31	0.775	-0.385	-0.189	0.068	0.459
TENSION / ATTENTION					
	F1	F2	F3		
TA_22	0.846	-0.023	0.533		
TA_24	0.764	0.584	-0.274		
TA_29	0.773	-0.552	-0.312		

Fornell-Larcker criterion. When we inspected the shared variances between constructs and AVE's, we observed that shared variance of Positive Effects and Sensory-Imagery Immersion is .42, which is higher than the .38 AVE value for Sensory-Imagery Immersion, as seen on Table III. This finding violates the Fornell-Larcker criterion, providing evidence that disproves discriminant validity of the scale.

Another detailed indicator of the discriminant validity of a scale is that each item has the highest loading on their designated latent variable [27]. When we inspected the cross-loadings of GEQ items, we observed that each item has the highest load on its designated construct. However, some items that strongly load on Positive Effects have also moderately loaded on Sensory-Imagery Immersion.

Besides the EFA factor loads indicating unidimensionality, we also utilized cross-loadings to identify discriminant validity. As suggested by Chin [27], we used the squares of cross-loadings because this representation provides "more intuitive interpretation since it represents the percentage overlap between an item and any construct". As it can be investigated on Table IV, Items have loaded strongly on their intended construct at the first hand, providing evidence for unidimensionality of GEQ.

To investigate formative variables, we examined each formative indicator's weight (relative importance) and loading (absolute importance), using bootstrapping with a sample of 5000 at the 5 % significance level [32]. For formative variables, the recommended standardized path coefficients should be greater than .100 [31] or .200 [26]. Sensory – Imagery Immersion is the only latent variable considered as a formative construct. We followed Lohmöller's recommendation with a more liberal approach since many of our variables did not meet the criteria of Chin, as seen follows.

TABLE II. RELIABILITY INDICATORS AND EIGENVALUES OF GEQ DIMENSIONS

Latent variable	Cronbach's alpha	D.G. rho (PCA)	Eigenvalues
CHALLENGE	0.605	0.760	1.942
			1.131
			0.709
			0.662
			0.557
COMPETENCE	0.676	0.795	2.211
			0.871
			0.773
			0.625
			0.520
POSITIVE AFFECT	0.723	0.819	2.406
			0.924
			0.712
			0.521
			0.437
NEGATIVE AFFECT	0.589	0.765	1.812
			0.896
			0.748
			0.545
SENSORY IMAGERY IMMERSION	0.599	0.758	1.984
			0.962
			0.945
			0.585
			0.524
FLOW	0.766	0.843	2.601
			0.869
			0.604
			0.529
			0.397
TENSION / ATTENTION	0.708	0.837	1.897
			0.646
			0.456

TABLE III. CONVERGENT AND DISCRIMINANT VALIDITY INDICATORS OF GEQ

	CHA	COM	PE	NE	SII	FLO	TA	(AVE)
CHA		0.002	0.016	0.040	0.023	0.107	0.087	0.375
COM	0.002		0.258	0.014	0.217	0.140	0.032	0.436
PE	0.016	0.258		0.119	**0.422**	0.327	0.105	0.481
NE	0.040	0.014	0.119		0.081	0.064	0.282	0.450
SII	0.023	0.217	**0.422**	0.081		0.321	0.065	0.379
FLO	0.107	0.140	0.327	0.064	0.321		0.022	0.516
TA	0.087	0.032	0.105	0.282	0.065	0.022		0.630
(AVE)	0.375	0.436	0.481	0.450	0.379	0.516	0.630	

Based on the whole psychometric quality indicators, it is possible to say that GEQ is valid and reliable tool to investigate the player experience. Although our data revealed some flaws of unidimensionality in GEQ, it should be considered that the data set evaluated in this study is collected from the players of only one game.

B. Factor Analysis of Gameplay Elements

An exploratory approach is required to identify the constructs since our Gameplay Elements are developed only for evaluation of the selected game. The items were generated based on the elements within the game, rather than aiming to measure a set of theoretical constructs.

Prior to the factor analysis of gameplay elements, the individual responses to the items were screened to determine if there was substantial skewness or kurtosis, as well as inspected for outliers. The data were negatively skewed and this is not surprising for a satisfaction scale measured with a game with very positive reviews but concern arises when the skewness statistic divided by its standard error and the kurtosis statistic divided by its standard error are greater than $z = \pm 3.29$ [25]. As suggested by Tabachnick and Fidell [25] and Howell [33], Logarithmic (Log 10) transformation is applied to Gameplay Elements. The transformation was suggested for substantially negative skewness and is represented by NEWX = LG10(K – X) where K is a constant from which each score is subtracted so that the smallest score is 1; usually equal to the largest score + 1. As a result of this transformation method, item scores are reversed as 0 for "Satisfactory" and 1 for "Unsatisfactory".

A principal components analysis (PCA) was run on 36 transformed gameplay elements that measure satisfaction levels on 337 participants. The suitability of

TABLE IV. ITEM CROSS-LOADINGS OF GEQ ON LATENT VARIABLES

	CHA	COM	PE	NE	SII	FLO	TA
CHA_11	**0.254**	0.010	0.000	0.009	0.001	0.018	0.012
CHA_23	**0.368**	0.002	0.004	0.185	0.008	0.000	0.246
CHA_32	**0.392**	0.000	0.000	0.051	0.000	0.022	0.070
CHA_33	**0.383**	0.001	0.003	0.002	0.012	0.074	0.009
CHA_26	**0.480**	0.015	0.084	0.013	0.092	0.145	0.001
COM_02	0.001	**0.592**	0.148	0.016	0.124	0.056	0.021
COM_10	0.006	**0.305**	0.075	0.000	0.052	0.031	0.009
COM_15	0.006	**0.470**	0.103	0.011	0.059	0.062	0.012
COM_21	0.001	**0.173**	0.043	0.002	0.022	0.018	0.000
COM_17	0.005	**0.641**	0.187	0.012	0.203	0.135	0.029
PE_01	0.001	0.186	**0.475**	0.043	0.167	0.107	0.034
PE_04	0.016	0.066	**0.409**	0.047	0.162	0.122	0.054
PE_14	0.021	0.233	**0.621**	0.049	0.290	0.260	0.055
PE_06	0.004	0.111	**0.645**	0.131	0.280	0.218	0.084
PE_20	0.004	0.048	**0.253**	0.029	0.113	0.079	0.029
NE_07	0.045	0.002	0.013	**0.267**	0.000	0.000	0.114
NE_08	0.027	0.001	0.051	**0.438**	0.055	0.039	0.043
NE_09	0.040	0.026	0.092	**0.723**	0.071	0.039	0.258
NE_16	0.002	0.001	0.061	**0.372**	0.033	0.061	0.115
SII_03	0.013	0.020	0.079	0.005	**0.154**	0.060	0.006
SII_12	0.000	0.014	0.059	0.032	**0.164**	0.063	0.030
SII_18	0.011	0.173	0.154	0.025	**0.492**	0.185	0.025
SII_19	0.010	0.063	0.127	0.023	**0.287**	0.086	0.014
SII_30	0.019	0.150	0.381	0.073	**0.800**	0.230	0.054
FLO_05	0.025	0.043	0.223	0.074	0.172	**0.408**	0.032
FLO_25	0.089	0.078	0.113	0.030	0.131	**0.550**	0.001
FLO_28	0.032	0.105	0.211	0.034	0.247	**0.543**	0.030
FLO_13	0.079	0.081	0.179	0.021	0.132	**0.550**	0.004
FLO_31	0.078	0.058	0.110	0.015	0.137	**0.530**	0.003
TA_22	0.052	0.042	0.097	0.210	0.093	0.038	**0.776**
TA_24	0.038	0.005	0.076	0.192	0.025	0.011	**0.576**
TA_29	0.090	0.018	0.027	0.131	0.010	0.000	**0.537**

TABLE V. FORMATIVE VARIBLE WEIGHTS

Items	Weight
SII_03	0.112
SII_12	0.153
SII_18	0.372
SII_19	0.139
SII_30	0.625

PCA was assessed prior to analysis. Inspection of the correlation matrix showed that all variables had at least one correlation coefficient greater than 0.3. The overall Kaiser-Meyer-Olkin (KMO) measure was 0.917 which is classified as 'marvelous' according to Kaiser [34]. Bartlett's test of sphericity was statistically significant ($p < .0005$), indicating that the data was likely factorizable.

The initial PCA with oblimin rotation revealed nine components that had eigenvalues greater than one and but the 8th component had an eigenvalue of 1.085 and the 9th component had an eigenvalue of 1.007. To remove these two components with eigenvalues barely above 1, the extraction criteria were set to a fixed number of seven factors and the remaining components explained a cumulative variance of 38.975 %. The interpretation of the data was consistent with the gameplay attributes the questionnaire was designed to measure with strong loadings of core gameplay elements on Component 1, story protagonist items on Component 2, audio-visual items on Component 3, narrative items on Component 4, customization items on Component 5, a group of mixed items on Component 6, two prominent core gameplay elements, "game controls" and "combat mechanics", on Component 7. Component loadings are presented in Table VI. Component 1 is named as "Core Gameplay Elements", Component 2 is named as "Avatar Elements", Component 3 is named as "Audio-Visual Elements", Component 4 is named as "Narrative Elements", Component 5 is named as "Customization Elements", Component 6 is named as "Secondary Gameplay Elements", and Component 7 is named as "Playability Elements".

Six of the proposed core gameplay elements are grouped together in Component 1 with the addition of an activity element, "Bandit Camps" and the "Collectibles" element. "Skill Variety" and "Trap Variety", which are proposed customization elements, fall under Component 1 but they also have high loadings on Component 5 which is composed of the remaining customization elements. Four elements that focus on the protagonist of the game, Aloy, are grouped together in Component 2 with the addition of a narrative element, "Quality of

TABLE VI. ROTATED PCA SOLUTION

Structure Matrix

	Factor						
	1	**2**	**3**	**4**	**5**	**6**	**7**
Inventory system	,658	,240	,369	,303	-,411	,286	-,240
Crafting system	,625	,255	,314	,321	-,400	,425	-,271
Activities - Bandit Camps	,597	,320	,093	,348	-,363	,416	-,204
Skill variety	,547	,336	,254	,396	-,456	,339	-,202
Enemy AI	,511	,343	,246	,342	-,419	,237	-,251
Game UI	,497	,306	,313	,277	-,327	,292	-,388
Trap variety	,491	,159	,276	,380	-,487	,329	-,184
Game feedback	,484	,396	,392	,202	-,389	,386	-,231
Collectibles	,402	,237	,347	,366	-,386	,331	-,031
Save system	,395	,135	,361	,339	-,318	,374	-,253
Aloy`s dialogue options	,471	,624	,162	,185	-,412	,587	-,214
Aloy`s voice acting	,215	,590	,285	,233	-,231	,157	-,173
Aloy`s facial expressions and animations	,124	,585	,186	,205	-,132	,152	-,174
Aloy`s physical appearance	,266	,524	,314	,228	-,303	,345	-,109
Quality of the main quest	,153	,418	,366	,386	-,268	,276	-,281
Machine (visual) design	,225	,233	,602	,207	-,273	,229	-,256
Game soundtrack and sound effects	,293	,362	,554	,126	-,272	,299	-,204
Activities - Tallnecks	,303	,393	,409	,257	-,376	,327	-,049
Environment (visual) design	,123	,178	,328	,149	-,130	,144	-,244
Number of side quests	,347	,189	,164	,623	-,393	,284	-,207
Quality of side quests	,402	,352	,071	,575	-,417	,329	-,087
Length of the main quest	,128	,295	,234	,512	-,321	,132	-,220
Side characters of the story	,343	,493	,295	,501	-,315	,320	-,338
Activities - Cauldrons	,396	,184	,316	,404	-,260	,179	-,316
Activities - Hunting Grounds	,335	,156	,180	,381	-,240	,212	-,183
Weapon variety	,275	,135	,149	,300	-,710	,215	-,218
Armor variety	,453	,253	,203	,278	-,664	,444	-,087
Ammo variety	,324	,281	,341	,328	-,601	,272	-,242
Machine variety	,249	,216	,366	,471	-,564	,155	-,267
Machine overriding	,419	,333	,378	,403	-,473	,372	-,241
Travel options	,315	,214	,332	,251	-,336	,688	-,175
Antagonists of the story	,371	,436	,191	,416	-,330	,520	-,366

(continued on next page)

TABLE VI. (continued)

Structure Matrix							
Activities - Corrupted Zones	,491	,180	,238	,461	-,390	,511	-,241
Lore/fiction of the world	,079	,367	,276	,143	-,235	,374	-,308
Game controls	,352	,193	,290	,282	-,339	,319	-,635
Combat mechanics	,299	,248	,367	,320	-,452	,124	-,566

the Main Quest". The three proposed audio-visual elements are grouped together in Component 3 with the addition of an activity, "Tallnecks". Four of the proposed narrative elements are grouped together in Component 4 with the addition of two activities, "Cauldrons" and "Hunting Grounds". Four of the proposed customization elements are grouped together in Component 5 with the addition of "Machine Overriding". Component 6 is a mix of different variables, possibly highlighting the problematic elements in the questionnaire. Component 7 is comprised of two elements, "Game Controls" and "Combat Mechanics", which distinguish themselves from other core gameplay elements in a significant way. The overall factor structure is consistent with the proposed gameplay elements structure with the exception of activities. Although the developers named these elements as activities, they are not regarded as a group of variables and they fall under different components in the PCA.

Reliability of the Gameplay Elements items and explored constructs are evaluated based on their Cronbach's alpha value. A Cronbach's alpha of .923 yielded for the whole 36 items. This result suggests that there is redundancy within the items.

For the latent variables explored through the factor analysis, Cronbach's alpha values are as follows. Reliability of the "Avatar", "Secondary Gameplay" and "Playability" dimensions are below the threshold of .7, but Nunnaly [35] suggests that a Cronbach's alpha value above .6 is reliable for a freshly developed instrument. Although the items of "Audio Visual Elements" construct does not met this criterion, we decided to investigate these constructs' relation with GEQ constructs, whether to see it is possible measure the quality of a game through the player's evaluation of its gameplay elements.

C. GEQ and Gameplay Elements

We employed a Pearson's correlation to assess the concurrent validity of our approach based on assessment of Gameplay Play Elements with the GEQ based evaluation of player engagement. Results, given in Table VII shows that correlations

TABLE VII. RELIABILITY INDICATORS FOR GAMEPLAY ELEMENTS DIMENSIONS

Factor	Cronbach's Alpha
Core Gameplay Elements	.833
Avatar Elements	.66
Audio Visual Elements	.547
Narrative Elements	.705
Customization Elements	.739
Secondary Gameplay Elements	.644
Playability Elements	.611

between Gameplay Elements dimensions and GEQ dimensions are significant in most cases at .01 level marked with an "*". However, these correlations are low or medium.

"Negative Effects" dimension of GEQ correlates with all Gameplay Elements significantly, but the level of the correlations is quite low. "Positive Effects" dimension of GEQ correlates with "Playability Elements", "Core Gameplay Elements", "Secondary Gameplay Elements" at a medium level, while a little bit lower with others. "Competence" measures in GEQ are significantly related with all Gameplay Elements dimensions, however with a low level. "Challenge" does not significantly correlate with "Playability Elements", "Audio-Visual Elements" and "Avatar Elements", but correlate with others at a low level. Immersion measures of GEQ has the strongest correlation with Gameplay Elements dimensions, compared to other dimensions of GEQ. However, these correlations are mostly at a medium level. Tension/Attention has low correlations with Gameplay Elements, except the "Playability Elements". These correlations are positive, suggesting that satisfaction from dimensions of Gameplay Elements reduces the irritation, annoyance or frustration of user. Flow correlates with "Core Gameplay Elements" at a medium level, and its correlation with other dimensions are slightly below medium, except "Playability Elements".

V. Discussion

Results given above suggest that it is possible to uncover the underlying structure of gameplay elements with an EFA but the correlations between the gameplay elements and latent variables of GEQ are still low. This outcome is not surprising because it may not be possible to associate a single variable of GEQ, in example

TABLE VIII. CORRELATIONS OF GEQ AND GAMEPLAY ELEMENTS

	GEQ Negative Effects	GEQ Positive Effects	GEQ Competence	GEQ Challenge	GEQ Immersion	GEQ Tension Attention	GEQ Flow
Core Gameplay Elements	.167*	-.352*	-.241*	-.148*	-.346*	.137*	-.317*
Avatar Elements	.119*	-.307*	-.224*	-.062	-.358*	.127*	-.265*
Audio Visual Elements	.153*	-.282*	-.211*	-.074	-.319*	.137*	-.241*
Narrative Elements	.156*	-.281*	-.208*	-.106*	-.245*	.109*	-.210*
Customization Elements	.130*	-.252*	-.196*	-.130*	-.252*	.103*	-.229*
Secondary Gameplay Elements	.166*	-.337*	-.283*	-.094*	-.328*	.167*	-.283*
Playability Elements	.194*	-.350*	-.221*	-.039	-.286*	.073	-.173*

Flow, with a single gameplay factor such as the "Core Gameplay Elements". Flow is defined as an optimal state of consciousness where we feel our best and perform our best, so it is more meaningful to associate it with a group of gameplay elements, rather than trying to explain it with a single factor.

Our item set for evaluating the Gameplay Elements does not have an acceptable level of reliability on its dimensions. Some of the constructs emerged through EFA were not clear, while some items have loaded on more than one construct. This may be depending on the limitations of the study or measurement tools employed within the study. Also, the methods used for normalizing the skewed data influences results.

Our study could not conclusively determine that a game evaluation method involving the players can be based on assessment of gameplay elements. On the other hand, our results provided some evidence that players can assess game play elements which are also correlated with their experience with the game. Future work relying on gameplay elements oriented evaluation approach may benefit our factor analysis results to develop a better set of items.

As a measurement tool employed in this study, GEQ was not clearly evaluated for its psychometric quality before. However, our study contributed to the development of GEQ, providing evidence for its reliability and validity, while exposing some flaws in its unidimensionality. Developers of GEQ or any future scale for evaluation of gaming experience may consider our findings for their work. Future studies may also focus on other variables that affects gaming experience, in addition to gameplay elements.

References

[1] P. Wright and J. McCarthy, Technology as experience, MIT Press, 2004.

[2] M. Slater, M. Usoh and A. Steed, "Taking Steps: The Influence of a Walking Metaphor on Presence in Virtual Reality", ACM Transactions on Computer-Human Interaction (TOCHI) 2(3) September, 1995, 201–219.

[3] M. Csikszentmihalyi, Flow: The psychology of optimal experience, New York: Harper & Row, 1990.

[4] M. Slater and S. Wilbur, "A framework for immersive virtual environments (FIVE): Speculations on the role of presence in virtual environments", Presence: Teleoperators and Virtual Environments, 6 (6), 1997, pp. 603–616.

[5] E. H. Calvillo-Gámez, P. Cairns, A. L. Cox, "Assessing the core elements of the gaming experience", in R. Bernhaupt (ed), Evaluating user experience in games, Springer, London, UK, 2010, pp 47–71.

[6] R. Hunicke, M. LeBlanc and R. Zubek, "MDA: A Formal Approach to Game Design and Game Research", Proceedings of AAAI04 WS on Challenges in Game AI, 2004, pp. 1–5.

[7] J. Wiemeyer, L. Nacke, C. Moser and F. `Floyd' Mueller, "Player Experience", in Serious Games: Foundations, Concepts and Practice, Springer International Publishing, 2016, pp. 243–271.

[8] R. M. Ryan, C. S. Rigby, and A. Przybylski, "The Motivational Pull of Video Games: A Self-Determination Theory Approach", Motivation and Emotion 30(4), 2006, pp. 344–360. http://doi.org/10.1007/s11031-006-9051-8.

[9] R. M. Ryan and E. L. Deci, "Self-determination theory and the facilitation of intrinsic motivation, social development, and well-being", Am Psychol 55(1), 2000, pp. 68–78.

[10] J. M. Keller, "Development and use of the ARCS model of instructional design", JID 10(3), 1987, pp. 2–10.

[11] J. M. Keller, Motivational design for learning and performance: the ARCS model approach, Springer Science & Business Media, New York, 2009.

[12] P. Sweetser and P. Wyeth, "Game flow: a model for evaluating player enjoyment in games", ACM CIE 3(3), 2005.

[13] J. Takatalo, J. Hakkinen, J. Kaistinen and G. Nyman, "User experience in digital games: differences between laboratory and home", Simulation and Gaming 42(5), 2011, pp. 656–673.

[14] K. Poels, Y. A. W. de Kort, W. A. IJsselsteijn, "FUGA—The fun of gaming: measuring the human experience of media enjoyment", Deliverable D3.3: game experience questionnaire, TU Eindhoven, Eindhoven, The Netherlands, 2008.

[15] H. Desurvire and C. Wiberg, "Game usability heuristics (PLAY) for evaluating and designing better games: the next iteration", in A. A. Ozok and P. Zaphiris (eds), Online communities and social computing, LNCS 5621, Springer, Berlin, 2009, pp. 557–566.

[16] B. G. Witmer and M. J. Singer, " Measuring presence in virtual environments: a presence questionnaire", Presence: Teleoperators and Virtual Environments, 7(3), 1998, pp. 225–240.

[17] P. Vorderer, W. Wirth, F. R. Gouveia, F. Biocca, T. Saari, F. Jäncke, S. Böcking, H. Schramm, A. Gysberg, T. Hartmann, C. Klimmt, J. Laarn, N. Ravaja, A. Sacau, T. Baumgartner, P. Jäncke, "MEC spatial presence questionnaire (MEC-SPQ): short documentation and instructions for application", Report to the European community, project presence: MEC (IST-2001-37661), 2004.

[18] T. Hartmann, W. Wirth, H. Schramm, C. Klimmt, P. Vorderer, A. Gysbers, S. Böcking, N. Ravaja, J. Laarni, T. Saari, F. Gouveia, A. M. Sacau, "The spatial presence experience scale (SPES)", JMP, 2015, doi:10.1027/1864-1105/a000137.

[19] Y. A. de Kort, W. A. IJsselsteijn, K. Poels, "Digital games as social presence technology: Development of the social presence in gaming questionnaire (SPGQ)", in Proceedings of PRESENCE, 2007.

[20] J. H. Brockmyer, C. M. Fox, K. A. Curtiss, E. McBroom, K. M. Burkhart, J. N. Pidruzny, "The development of the game engagement questionnaire: a measure of engagement in video game-playing", Journal of Experimental Social Psychology, 45, 2009, pp. 624–634.

[21] IJsselsteijn, W., van den Hoogen, W.H.M., Klimmt, C., de Kort, Y.A.W., Lindley, C., Mathiak, K., Poels, K., Ravaja, N., Turpeinen, M. and Vorderer, P. "Measuring the experience of digital game enjoyment", In Proc. of Measuring Behavior, Maastricht Netherlands, 2008, pp. 88–89.

[22] Johnson, D., Watling, C., Gardner, J., and Nacke, L. E. "The edge of glory: The relationship between metacritic scores and player experience", In Proc. of the First ACM SIGCHI Annual Symposium on Computer-human Interaction in Play, CHI PLAY '14, ACM (New York, NY, USA, 2014), 141–150.

[23] Norman, K. L. "GEQ(game engagement/experience questionnaire): a review of two papers", Interacting with Computers 25, 4, 2013, pp. 278–283.

[24] Brühlmann, F. and Schmid, G. "How to Measure the Game Experience? Analysis of the Factor Structure of Two Questionnaires", CHI'15 Extended Abstracts, April 18–23, 2015, Seoul, Republic of Korea. ACM 978-1-4503-3146-3/15/04.

[25] Tabachnick, B.G. and Fidell, L.S. Using multivariate statistics (5th ed.). Boston: Allyn and Bacon, 2007.

[26] Chin, W.W. "The partial least squares approach to structural equation modeling", In Marcoulides G.A. (Ed.), Modern methods for business research (pp. 295–336). Mahwah, NJ: Lawrence Erlbaum Associates, 1998.

[27] Chin, W.W. " How to write up and report PLS analyses", In Vinzi, V.E., Chin, W.W., Henseler, J. and Wang, H. (Eds.), Handbook of partial least squares (pp. 655–690). Berlin Heidelberg: Springer, 2010.

[28] Lewis, B.R., Templeton, G.F. and Byrd, T.A. "A methodology for construct development in MIS research", European Journal of Information Systems, 14 (4), 2005, pp. 388–400.

[29] Straub, D., Boudreau, M., & Gefen, D. "Validation guidelines for IS positivist research", Communications of the AIS, 13, 2004, pp. 380–427.

[30] Fornell, C. and Larcker, D. F. 1981. Evaluating structural equation models with unobservable variables and measurement error. Journal of Marketing Research, 18, 1981.

[31] Lohmöller, J.-B. Latent variable path modeling with partial least squares. Heidelberg, Germany: Physica, 1989.

[32] Hair, J. F., Jr., Ringle, C. M., & Sarstedt, M. "PLS-SEM: Indeed a silver bullet", Journal of Marketing Theory and Practice, 19(2), 139–151, 2011.

[33] Howell, D.C. Statistical methods for psychology (6th ed.). Belmont, CA: Thomson Wadsworth, 2007.

[34] Kaiser, H.F. "An index of factorial simplicity", Psychometrika, 39, 1974, pp. 31–36.

[35] Nunnally, J. and Bernstein, H. Psychometric Theory. McGraw Hill. 1978

8. Hybrid Narrative Generation with Deferred Planning

Sercan Altun
Bahcesehir University Game Design Department
Istanbul, Turkey altun.sercan@gmail.com

Abstract— Narrative generation systems tackle the problem of creating a meaningful sequence of events to provide computer generated storytelling. Narrative planners generate cohesive story plans that comply to the author's specifications ahead of time and let story agents enact this plan. In contrast, emergent narrative systems do not use story plans. Instead, these systems utilize intelligent actor behaviors generating series of events through simulation, which can be perceived as a narrative. This paper presents Deferred Planning as a hybrid approach which bridges the gap between narrative planning and emergent narratives. Deferred Planning starts with a partially bound plan where some elements of the story are not determined ahead of time. As agents interact in the story world these elements are determined at runtime and the partial story plan is modified based on new happenings in the story world.

Index Terms—deferred planning; narrative generation; narrative planning; emergent narrative; interactive drama; procedural content generation

I. Introduction

In the current state of research, narrative generation systems tend to follow one of two major approaches in their design. On one side, narrative planners favor creating story plans ahead of time to create cohesive experiences. These story plans can be verified for their quality prior to their usage. By contrast, emergent narratives are defined as bottom-up systems that build narratives with events created in real-time through the interactions of intelligent story agents [1].

While the narrative generation research has a long history that stretches back to 70's [2], these techniques are rarely used in the production of commercial entertainment products. There are a couple of narrative generation systems released as games that are aimed at general public [3] [4]. By contrast, other seminal works are either accessible on research websites that are mainly visited by researchers, or otherwise not publicly accessible [5] [6] [7]. This brings up the question whether non-researcher authors can use these systems in the first place [8]. While some may argue that these works are still made with entertainment purposes in mind, they are clearly made for research first and won't be entertaining anyone anytime soon. In fact, real world applications of narrative generation systems that can be experienced by non-researchers are focused on training [9] and education [10]. This tendency can be attributed to the nature of research funding and it may be argued that such strides in the industry are not meant to be made by the researchers themselves but by the industry professionals. However, if we consider video game industry's quickness to adapt new research in other fields, the low adoption rate of narrative generation systems in the entertainment industry signals that current approaches are not enticing the video game professionals.

From the perspective of video game professionals working on commercial entertainment products, video games are produced for profit. With each product, game publishers seek to craft quality experiences for their customers. By the same token, a video games commercial success is dependent on how well this experience is crafted. Consequently, for a narrative generation system to be used in a game production pipeline, it is important for the developer to have authorial control over the system to shape the experience. At the same time, as opposed to other forms of media, video games offer agency to players [11]. Although authorial control is an important aspect of the video game production, leaving an open narrative space for players to explore and shape is also a pursuit of many game developers.

Deferred planning is situated in a different position in narrative generation systems as a hybrid approach. While the main algorithm works from a top-down perspective similar to narrative planners, it leaves details of the story plan partially resolved. A simulation component handles tasks from the story plan via story agents. As these agents take actions in the game world alongside with the player new events and existents emerge in the fabula. The previously planned partial story progresses, as the details of the story are fleshed out with these emergent changes in the fabula. Consequently, the final story incorporates impactful actions of the player. If story plan is interrupted, the deferred planner favors diverging from the initial plan rather than converging back to the similar outcome. As a result, deferred planning offers an authoring process similar to narrative planners while supporting emergent behavior.

II. Related Works

Deferred planning uses story plans to direct story agents into pursuing goals that would move the narrative forward. A modified partial-ordered planning algorithm is used to create these plans. Partial-ordered planners have been used in narrative generation systems before [12] [13]. However, in previous research, these planners produce complete plans where every story action, actor, and object is predetermined and verified. Whereas deferred planner does not establish all of these parameters initially, but make assignments and update the plan at runtime.

A recent research on delayed role assignment presented a method similar to deferred planning that creates partially bound story plans and assigns roles later [14]. This method generated complete plans first, then used a post-processing mechanism to extract roles out of this complete plan. As a result, this post-process produced a partially-bound plan to be instantiated later on. In contrast, deferred planning produces partially bound plans directly and does not define roles explicitly. More importantly, the core difference between these methods design is that delayed roles focus on the soundness of the just-in-time instantiated story plan and maintains the initial structure, whereas deferred planning accepts disruptions of stemming from an emergent environment and modifies the story structure to fit the new fabula along with new assignments.

Lifting limits over disruptions on story plans can improve player's narrative agency. However, when unconstrained, player actions can cause conflicts with narrative requirements of generated stories. This conflict between maintaining narrative constraints and providing narrative agency without constraints is called narrative paradox [15] [8]. Emergent Narratives eliminate strict narrative constraints by moving away from working with story plans. Emergent Narratives as a bottom-up approach favor giving both players and story agents agency in a behavior rich simulation. It is argued that narrative may emerge from these actions. However, emergent narratives do not guarantee resulting stories would be interesting [1]. The current research on emergent narratives takes inspiration from improvisational theater techniques to reconcile with the lack of story plans [15] [16] [17]. In emergent narratives, story agents are not mere characters in a story, but actors who can use empathy to know how dramatic an action is [18] and take note in the current dramatic pace of the story to pick actions that fit that pace [19]. As a result, emergent narratives can result in diverse outcomes without impairing players agency. On the other hand, not having direct control over narrative direction makes them harder to author. In order to relay themes, messages and outcomes deemed appropriate by the author, mechanisms of the simulation and individual agent behaviors has to be tweaked to make sure the emergent events would statistically lead to the desired experience [20].

III. Implementation

Deferred Planning is an alternative approach that operates at a middle point between plan-based systems & emergent narrative. The deferred planner uses

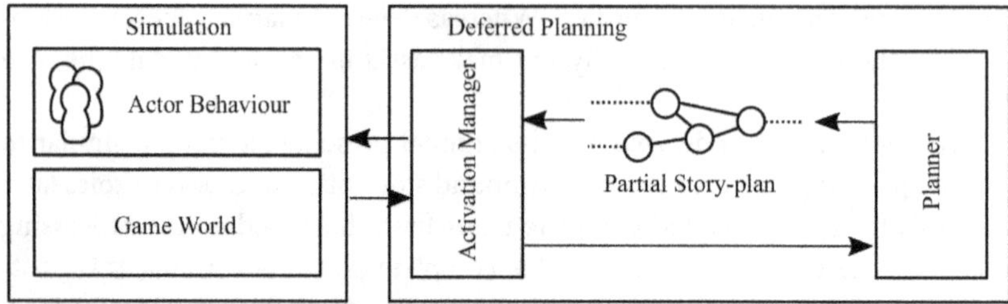

Fig. 1. Overview of process flow in Deferred Planning

partially bound plans to direct simulated agents while enabling new fabula events to emerge from interactions between these agents and the player. The deferred planner was implemented from scratch in a game engine as a partial ordered planner [21] customized to support partially bound plans and other functionalities necessary for the Deferred Planning process.

The design of the system focused on use cases where the narrative generator would produce new story plans one after another continuously, in an open-ended narrative where player actions can change the outcome of the story without restrictions. While this kind of process is best suited for openworld sandbox games, a game developer may easily modify it to impose design specific limits afterward. As such, the design goals are set to cover a set of requirements to improve the flexibility of story plans; reducing limits on players agency, allowing the simulation to interrupt story plan whenever necessary, providing open-ended stories without damaging narrative coherence.

The system itself only covers planning aspect of the processes, not the simulation. Deferred Planning is designed to work with any type of simulation driven system, including an emergent narrative system. The game developer may use any meta-narrative methods as part of its character simulation or opt for a simpler behavioral AI. Deferred planner only requires an interface to assign tasks to the story agents and monitor changes in the game world. The simulation can reject these tasks anytime (for instance if it is conflicting with characters internal mental state) and the planner will modify the plan. For the conducted feasibility study, story agents were given simple reasoning capabilities like moving to the appropriate location, autonomously taking actions assigned, and following a working schedule.

The planning domain consists of a set of story nodes that have requirements to fulfill before execution and resulting effects. While the notation is similar to other domain independent planners, the story nodes can both be described as atomic nodes or conceptual nodes. Atomic nodes denote a single character action without any sub steps. Whereas, conceptual nodes may include multiple characters, sub steps and extra contextual requirements that are needed for narrative cohesion. For example, in a corporate espionage domain, a "copy" story node is an atomic node that assigns a task to copy a file from a server. Similarly, "steal" story node is a conceptual node that assigns the exact same task. However, unlike "copy", "steal" node has extra conceptual requirements where the character has to be a hacker and previously hired by someone else. A contextual story node can also contain multiple tasks that are assigned to more than one agent. For instance, in "exchange" story node, the hacker and the employer both have tasks to exchange a stolen file.

The deferred planner itself does not validate or plan for how these individual tasks are going to be fulfilled. This responsibility is offloaded to the simulation itself, where the task can be processed by another planner operating in a task domain or a simple behavioral AI. Thus with this notation, tasks and story nodes are decoupled in terms of requirements and effects in the planning phase. Deferred planner only cares about the abstract story structure and leaves the execution details to simulation. As a matter of fact, failure to accomplish a given task is part of an ongoing narrative generation process. If a hacker is tasked to steal a file, the simulated agent then runs its behavior to infiltrate the server. If the hacker gets caught then the story node activation fails and the planner modifies the plan. This decoupling leaves room for the simulation to facilitate interactions within the game world in its own terms rather than directly enacting what the planner dictates.

A. Planning Phase

The planning phase in deferred planning is very similar to other partial ordered planning systems. The deferred planner was The planner is supplied with an initial world state, set of story nodes to accomplish and planning domain that includes all possible story nodes available. The planner searches the plan space by either adding new story node to the plan or using already existing nodes in the plan to satisfy a set of open requirements. With each new node, that nodes requirements are also added to the list of open requirements to be fulfilled. A standard partial ordered planner would search until a plan is found that connects starting point to the goals with no open requirements left.

When search algorithms attempt to connect start and end points of narrative space, they select exact existents from the starting point that are used in the plan. In the case of deferred planning, the planner is modified to produce

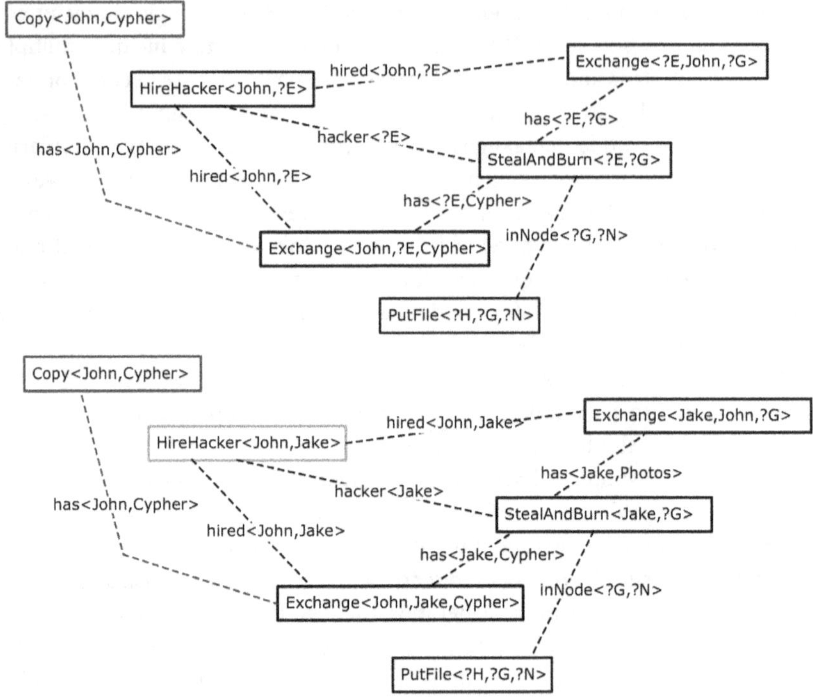

Fig. 2. Two states of a sample partial plan with anchor node denoted in blue and activated node denoted in green. Variable?E is bound after activation of the first story node.

partially bound plans where exact bindings for variables are not determined in advance. Instead, the planner creates plan-scope variables as needed. This means that the story is planned in an abstract form and specific instantiations of actors and actions are determined later on. These variables are bound just-in-time during activation phase. The only exception to this rule are the variables that are connected to the anchor event.

Exact existents are unknown during the planning phase. The deferred planner cannot make direct connections with the current world state like other planners. That is why deferred planner needs a mechanism to know when to stop planning and accept a partial plan. The concept of anchor events is put forward to provide this mechanism. An anchor event is a character action that is the triggering point of a narrative. A heuristic method selects the anchor event from existing events in the fabula. This heuristic method should suit the particular narrative intent of the designer. For instance, selecting a player's action as an anchoring event would result in a story connected to that action. Therefore, this heuristic puts emphasis on the consequences of the player's choice. Alternatively, the designer could favor actions of some key in-game characters, like the people in a king's court. In which case, the narrative would revolve around palace drama and intrigue.

Before accepting the plan to be valid, the plan has to connect to the current fabula. Like domain and goals, this anchor event is provided to the planner as an input prior to the start of planning. At current version of the system, the planner does not select the anchor event. The deferred planner would attempt to establish a connection with the current partial plan to an anchor event. This connection is established, when a causal link connects to rest of the partial plan with the anchor event during the planning phase. However, even with this connection established, there may not be a valid instantiation of the partial plan. A valid partial plan should have at least one set of appropriate variable mappings with the current fabula. The deferred planner continues the search until this condition is met. This will ensure that when the plan is activated there is a valid story available for the narrative to start with.(Fig. 2a). If the planner cannot establish a valid connection with the anchor event, planning step fails. Afterwards, the process can be repeated with another event selected as the anchor.

B. Activation Phase

Once a partial story plan is created, the planner will start activating story events and assigning related tasks to the story agents. These tasks are causally linked with next story event, therefore they are required for the story to progress. Whether these assigned tasks are started immediately or not depends on the behavioral programming of story agents in the simulation. In current implementation deferred planner was tested with a simulation with agents following simple templates of actions. However, it is also possible to have a secondary AI mechanism to resolve these tasks. Regardless of the method used, deferred planner keeps track of which story events are activated, completed or canceled.

When a story event is activated, the deferred planner would check for any unbound variables remains in that event. If such variables exist, the planner would make a unification check with the current fabula to determine which existent should fill the role. Chosen existent is assigned to the previously unbound variable. Since this variable is on plan-scope, story nodes and links with this variable are updated to reflect the change (Fig 2b).

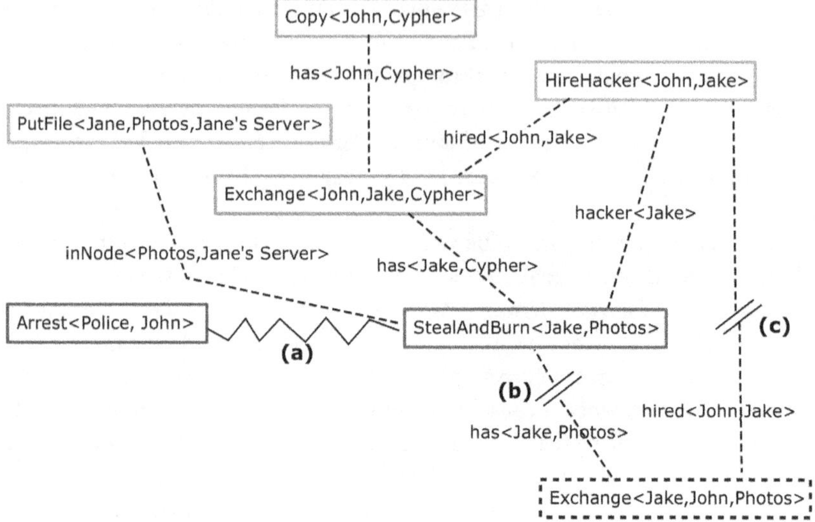

Fig. 3. Story plan is disrupted (a) via new event in simulation breaking causal links in current plan. (b) Causal links starting from the disrupted event are removed before modification process starts. (c) Other causal links that share same variable-bindings are reset to unbound variables.

Since a valid plan has at least one possible instantiation of a story, activation has a safety net to fall on. However, this activation is deferred on purpose. The fabula gets richer with every emergent event after the story started. Deferring activation, lets the planner pick existents from this richer fabula. If those existents created due to player actions, their inclusion would make the story more relevant to the player. It is also possible to run multiple story plans in parallel. In this case, an event from one plan can change the fabula that would, in turn, have the possibility to appear in the other plan. As such partial plans can be interwoven to tell multiple stories at the same time.

C. Modification Phase

In the case of a task cancellation or a requirement violation, deferred planner enters into a modification process. A task cancellation may arise when a character decides not to comply with the planners tasks. Alternatively, an emergent change in the game world may break a requirement of a causal link. Due to Deferred Plannings design principles, planner favors enabling emergent changes rather than fixing the story in a way that leads to the same outcome.(Fig 3).

In order accomplish this, deferred planner applies following operations. First, the causal links between completed story events and future events are reevaluated. In other words, the partial plan is divided into two graphs; one containing events that already occurred in the simulation, other containing projected future events that are not yet activated. The previously completed events in the story plans are kept in the planning process. As their conditions are already satisfied, they only serve for establishing causal links with new story events.

However the projected future events may no longer occur due to the disruption of the plan, hence they are marked as optional. These optional events can be added back into the new plan during search later on. Their conditions are not necessary to fulfill until this happens. Consequently, their conditions are not added to the open requirements initially. If after modification another node establishes a causal link with an optional node, that node's prerequisites are added back to open requirements. By turning old events to optional, deferred planner gives an opportunity for the emergent change to break previous plan create a new one. At the same time, by leaving optional nodes in the graph, deferred planner also offers an opportunity for these previously projected events to occur where appropriate. The resulting partial plan may include both new and old events.

Secondly, the causal links starting from the disrupted event are no longer valid. These causal links are removed from the modified plan. During removal of these causal links, variables on those links are set to unbound in all projected events. This operation can also break causal links between other completed and projected events that contain same variable binding. Similar to the planning phase, these unbound variables are assigned a plan-scope identifier. This assignment propagates through the causal link chain and can exist in multiple nodes. These unbound variables can be rebound to different existents. This gives an opportunity for the planner to reconcile with the conflict and search for new connections with new existents. Resulting graph would contain both completed and optional projected events, some of which are still causally linked, albeit with unbound variables. At this point, the partial plan looks similar to a search state in the middle of the planning phase, with no anchor event. So, a new anchor event is given as input. Then, deferred planner repeats planning phase to find a valid partial plan. In the current implementation, the event that broke the previous plan is selected as the new anchor event. With this heuristic, the planner puts emphasis on the impact that event caused in the narrative.(Fig 4).

IV. Discussion

The deferred planning is an attempt to approach narrative generation systems from a different angle. The systems goal is to incorporate principles of emergent

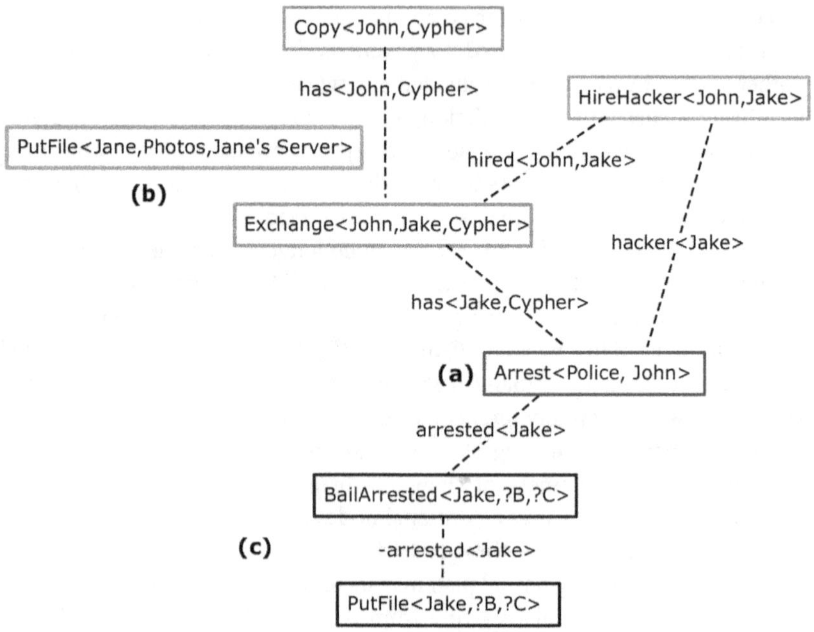

Fig. 4. The new Story-plan after modification. (a) The event that disrupted the previous plan becomes the new anchor point. (b) Completed events stay inside the new plan even if there is no current causal links connecting them to future events. Therefore they will be available for future modification and activation phases. (c) Newly added story events does not converge back to the original plan.

narratives into planbased narrative generation systems. As a top-down process, the deferred planning is not an emergent narrative system by definition. However, it does provide a flexible story plan model to enable changes in the plan through emergent events from a simulation. Therefore the deferred planning offers a plan based alternative solution for systems that have narrative aspirations similar to emergent narratives.

Narrative planners are favorable in cases where fulfilling requirements of a narrative is essential to the experience. This makes planners particularly useful in training scenarios where a specific condition (occurrence of an event, use of an action, a specific concept displayed) has to be fulfilled in order for that narrative to accomplish its purpose. Story plans are evaluated based on these conditions, both during and after generation process, to ensure cohesiveness of the narrative and its fulfillment of authorial goals. As mentioned before having authorial control over the experience is also an important aspect of game development.

On the other hand, while many games do have clear narrative direction, they also offer agency to the player for exerting power over the game world. Unless restricted significantly, the player in a video game has the ability to disrupt the previously planned story. Such disruptions can be remedied by repairing, replanning or preplanning possible interruptions in advance [9]. Even so, goal driven planners have a tendency to redirect narrative into the same outcome to fulfill goals. Therefore it presents a risk of the player feeling cheated and having his agency stripped of power and meaningful impact.

While narrative planners can replicate directed approach of storytelling, not capitalizing on the capacity of explosive permutations resulting from player interaction in a narrative generation system is a wasted potential for video games. As opposed to narrative planners, emergent narrative systems facilitates interaction between story agents in a simulated environment. The story agents and the player can act freely within the mechanisms set in the simulation without previously imposed story plan. However, having no story plan to validate also entails that resulting narrative may not satisfy authors narrative goals. The meta-narrative approaches like double-appraisal [18] and distributed drama management [19] make sure that the narrative at least will have a desirable dramatic trajectory. But even so, if an author wants to convey a specific narrative message, then the burden is on how the narrative domain and simulated mechanics are designed. These components have to be carefully adjusted to shape the possibility space of the emergent experience and make sure the narrative does not stray away from the agenda. As the narrative domain grows such adjustments become increasingly complex. Moreover, the changes in mechanisms should not impair the enjoyment value of the players experience. This concern imposes another layer to consider in using emergent narratives in video games. To put it another way, an emergent narrative designer has to juggle between player's overall enjoyment from the ludic gameplay and narrative intent when adjusting mechanics of the simulation. As demonstrated by the discussion on why distributed drama management should still be considered an emergent narrative, keeping a characters internal decision-making mechanisms protected from outside influences is one of the core tenants of emergent narrative ethos to protect bottom-up nature of the process [15]. The deferred planning does not follow this tenant, as it assigns tasks to the characters in order to lead the narrative into a desired trajectory. However, it's flexible story plan does leave room for improvised events. When working along with a behavior rich simulation it can bring those emergent happenings into the story plan. By leveraging interchange between top-down and bottom-up processes, deferred planning offers open-ended story plans that are authorable in the fashion of a plan-based system.

To test the systems applicability in video game use cases, the deferred planner was developed in tandem with a text based video game project. In the current state of the work, deferred planning was implemented inside a game engine with a small custom made corporate espionage narrative domain with hacking and matrix concepts from cyberpunk genre. This domain was chosen to compliment the video game. A set of qualitative assessments were made to ascertain its feasibility as a narrative generation system. These feasibility requirements were determined through analysis of related works, common issues present in narrative generation systems, and how the proposed system responds to these issues.

First of these requirements was providing strong narrative agency. The narrative agency is one-half of the narrative paradox where players' actions can disrupt requirements of a cohesive narrative. The deferred planning aspires to provide an unconstrained experience like emergent narratives. The test implementation did not have a designated player character of its own, rather let player control all characters in the game in a similar fashion to Prom Week [4]. Hence, it was possible for the tester to act as he desires without any regard for current story plan. The deferred planner successfully incorporated the player's actions in its process and modified the story plan without limiting the players ability to exert power over narrative progression. Moreover, the player was also able to completely disrupt previous narrative goals and lead narrative to different outcomes due to the design of the system.

However, providing strong agency also comes with its own issues. When a causal link is broken, projected story nodes become optional. However, as illustrated in the sample figure, when the player breaks a causal link it is possible none of the optional nodes may be used in the new narrative. In this case, the deferred planner creates story nodes at random and risks possibility of producing weak stories that have very little connection with previous narrative events. The deferred planning currently does not offer a feature to resolve this issue.

In light of this shortcoming, we can assert that deferred planning may not guarantee narrative cohesion throughout the process. Yet, both events before and after the disruption of the partial plan are cohesive when viewed individually. Therefore, the deferred planner only partially covers the narrative coherence requirement. This might be an acceptable compromise for some use cases. At any rate, this issue points to an area for improvement.

The aforementioned case may be mitigated with the use of domain specific knowledge in design. A planning request starts with an anchor event and a set of story nodes to form a structure. A narrative designer can set up pairs of anchor events and story node sets as cached story templates. In doing so, whenever a causal link is broken, the planner would not only change the anchored action but also add these new story nodes before restarting the search steps. As a result, the modified plan would include a coherent narrative structure even if all previous story nodes are obsolete. Nevertheless, such measures are heavily dependent on the individual narrative domain and craftsmanship of the narrative designer.

V. Conclusion

The initial findings showed that the deferred planning can be a valid alternative for narrative generation. The system is currently being evaluated by expert users and project will continue with player testing after its full integration with the video game project. The system itself is in its infancy and further research on its performance against other methods in a quantitative study is warranted. However, the current priority of the research is to field test the system in a game production pipeline. As a new method, deferred planning is open for improvement in many respects; like performance, authoring techniques, and overall integration with the simulation world. A real world application of the system and feedback from expert users is quintessential to plot the future of the research. It is my belief that with further development, deferred planning can present a production ready system that can be used in a variety of video game projects.

References

[1] R. Aylett, "Narrative in Virtual Environments - Towards Emergent Narrative," in *Narrative Intelligence: Papers from the 1999 Aaai Fall Symposium*. The AAAI Press, 1999, pp. 83–86.

[2] M. &. Mateas and P. Sengers, "Narrative Intelligence," in *Narrative Intelligence: Papers from the 1999 Aaai Fall Symposium*. The AAAI Press, 1999, pp. 1–10.

[3] M. Mateas and A. Stern, "Procedural Authorship: A Case-Study Of the Interactive Drama Fac¸ade," in *Digital Arts and Culture: Digital Experience: Design, Aesthetics, Practice (DAC 2005)*.

[4] J. McCoy, M. Treanor, B. Samuel, A. A. Reed, M. Mateas, and N. Wardrip-Fruin, "Prom week: Designing past the game/story dilemma." in *FDG*, 2013, pp. 94–101.

[5] B. Magerko, "Evaluating preemptive story direction in the interactive drama architecture," *Journal of Game Development*, vol. 2, no. 3, pp. 25–52, 2007.

[6] F. Charles, J. Porteous, and M. Cavazza, "Changing characters' point of view in interactive storytelling," in *Proceedings of the 18th ACM international conference on Multimedia*. ACM, 2010, pp. 1681–1684.

[7] S. G. Ware and R. M. Young, "Intentionality and conflict in the best laid plans interactive narrative virtual environment," *IEEE Transactions on Computational Intelligence and AI in Games*, vol. 8, no. 4, pp. 402–411, 2016.

[8] M. O. Riedl and V. Bulitko, "Interactive narrative: An intelligent systems approach," *Ai Magazine*, vol. 34, no. 1, p. 67, 2012.

[9] M. O. Riedl and A. Stern, "Believable Agents and Intelligent Story Adaptation for Interactive Storytelling," in *Proceedings of the 3rd International Conference on Technologies for Interactive Digital Storytelling and Entertainment*. Springer, 2006, pp. 1–12.

[10] R. S. Aylett, S. Louchart, J. Dias, A. Paiva, and M. Vala, "FearNot! An experiment in emergent narrative," in *Intelligent Virtual Agents: 5th International Working Conference*. Springer Berlin Heidelberg, 2005.

[11] J. H. Murray, *Hamlet on the Holodeck: The Future of Narrative in Cyberspace*, 1st ed. Newyork, NY: Free Press, 1997.

[12] M. O. Riedl and R. M. Young, "An Intent-Driven Planner for MultiAgent Story Generation," in *Proceedings of the Third International Joint Conference on Autonomous Agents and Multiagent Systems - Volume 1*. Washington, DC: IEEE Computer Society, 2004, pp. 186–193.

[13] S. G. Ware and R. M. Young, "CPOCL: A Narrative Planner Supporting Conflict." in *Proceedings of the Seventh Aaai Conference On Artificial Intelligence and Interactive Digital Entertainment*. Menlo Park, CA: AAAI Press, 2011, pp. 97–102.

[14] D. Thue, S. Schiffel, R. A. Arnason, I. S. Stefnisson, and B. Steinarsson, "Delayed Roles with Authorable Continuity in Plan-Based Interactive Storytelling." Springer, Cham, 2016, pp. 258–269.

[15] S. Louchart, J. Truesdale, N. Suttie, and R. Aylett, "Emergent Narrative Past, Present and Future of An Interactive Storytelling Approach," in *Interactive Digital Narrative: History, Theory and Practice*. Newyork, NY: Routledge, 2015, pp. 185–200.

[16] I. Swartjes, E. Kruizinga, and M. Theune, "Let's Pretend I Had a Sword: Late Commitment in Emergent Narrative," in *Interactive Storytelling: First Joint International Conference On Interactive Digital Storytelling*. Springer, 2008, pp. 264–267.

[17] I. Swartjes and J. Vromen, "Emergent Story Generation: Lessons from Improvisational Theater," in *Intelligent Narrative Technologies: Papers From The 2007 AAAI Fall Symposium*. The AAAI Press, 2007, pp. 146–149.

[18] R. Aylett and S. Louchart, "If I Were You: Double Appraisal in Affective Agents," in *Proceedings of the 7th International Joint Conference on Autonomous Agents and Multiagent Systems - Volume 3*. International Foundation for Autonomous Agents and Multiagent Systems, 2008, pp. 1233–1236.

[19] "Distributed Drama Management: Beyond Double Appraisal in Emergent Narrative," in *Interactive Storytelling*. Springer Berlin Heidelberg, 2012, vol. 7648, pp. 132–143.

[20] N. Suttie, S. Louchart, R. Aylett, and T. Lim, "Theoretical considerations towards authoring emergent narrative." in *ICIDS*. Springer, 2013, pp. 205–216.

[21] D. S. Weld, "An introduction to least commitment planning," *AI magazine*, vol. 15, no. 4, p. 27, 1994.

9. Cloth Tearing Simulation

Emre Onal
Middle East Technical University emrek1@gmail.com

Veysi İsler
Middle East Technical University isler@ceng.metu.edu.tr

Abstract— Among different physical simulation topics, cloth simulation is one of the most popular subjects in computer graphics. There are many different studies published on different aspects of cloth simulation, but there are not many studies focused on the tearing of cloth. Existing studies related to this topic have only dealt with some aspects of the problem [1] and have not provided general solutions. In this study, we provide a generic solution for different aspects of the problem of tearing cloth. Some of the points we focus on in this study include: providing realistic tearing effect, preserving polygonal area consistency and texture integrity after the process of tearing, and handling the tear properly for either outer physical impacts or inner manipulations like allowing a user to drag the cloth interactively. The technique proposed in this paper works with non-uniform cloth structures. It makes it easier to adopt the solution proposed here for many different simulation systems. The processing cost of the technique is quite small, so it is also appropriate for real-time simulation systems.

Keywords— ***Cloth Tearing; Cloth Simulation Spring Physics Fracturing materials***

I. Introduction

The motion of cloth has unique features when compared to the motion of other objects in the real environment. That is why cloth animation is a topic treated as a different area on computer graphics separately. There are many different studies on this topic [2, 3, 4], but most of them are related to movement of cloth, interaction with other objects and performance optimizations. There are not enough studies related to the tearing of cloth. There are existing studies on fracturing materials, but since the nature of cloth physics differs from other kinds of objects, such methods are not directly applicable to handling the tearing of cloth.

There are different methods used for modelling cloth in simulations [5, 6]. They have different advantages and drawbacks. We review different aspects of these methods and extend them to constitute the required changes to handle tearing in the cloth structure. This may result from an outer physical effect or directly by a user interaction.

Detecting where and when a rupture will occur, finding the path that the tear will propagate and making the related structural changes constitute the main parts of the problem. At the same time, preserving texture integrity after the tearing is also important.

Another problem is to find a solution which could work on non-uniform cloth structures. It is a more generic solution for cloth modelling and also provides performance gain by allowing different resolutions on different parts of a cloth. We developed a model which uses triangular meshes and spring constraints in traditional continuum sheet model together.

Damaged and undamaged parts of the cloth respond to an impact in different ways. While triggering a tear on an undamaged region, we face strong resistance, whereas damaged parts of a cloth show less resistance against tearing. For some materials with unique characteristic features, this could be different. However, this is the case for a majority of cloth-like materials. We take into consideration these issues and apply different constraints according to the condition of the cloth to reflect a more realistic result.

II. Background and Related Work

A. Background

Continuum sheet model with springs on a regular grid [3] is a well-known and widely used cloth model. It has some constraints to simulate realistic cloth behavior.

This model gives visually satisfactory results, but it is not designed to handle structural changes in the cloth like tearing. The connections between particles are not designed to be separated. Also, in this model all parts of cloth need to

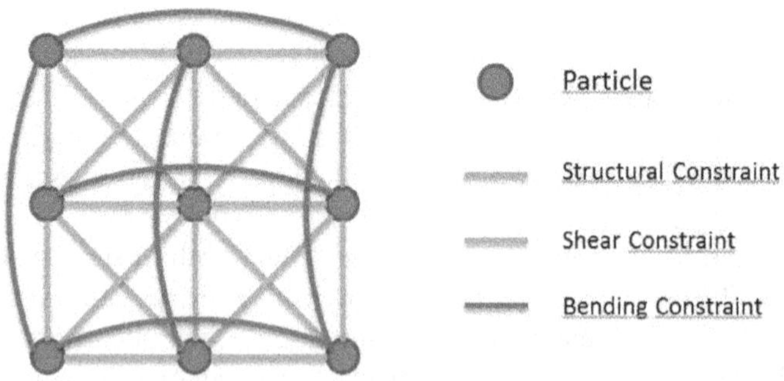

Figure 1: Structural, shear and bending constraints on the cloth model.

be modelled with the same resolution. There are some other studies which uses irregular triangular meshes [7, 8]. Using irregular triangles makes it possible to model different parts of the cloth with different resolutions. The main advantage of this approach is the performance gain.

B. Related Work

There are not many studies directly related with the tearing of cloth. In an earlier study, Terzopoulos and Fleischer [9] proposed some methods on modelling inelastic de-formation. In one of the examples they showed fracture propagation on surfaces with a net falling over an obstacle. The fibers on the cloth were subject to fracture limits. The fibers were broken when the ball fell onto the obstacle and the net was torn, but this model was not a cloth with a polygonal surface.

Metaaphanon et al. [1] worked on the cloth tearing problem but in a specific condition for woven clothes. They used both the standard continuum sheet model and a yarn-level model. First, the cloth was completely modeled according to the standard continuum sheet model, then the area around the torn line was modeled according to the yarn-level model, and tearing occurred in this yarn-level model. Their main focus was to simulate the behavior of threads on the torn lines with their yarn-level model. In many studies related to cloth, the tearing of cloth is mentioned as a future study [10, 11], but there is yet to appear a complete study that proposes a successful generic solution for tearing in cloth.

Hellrung et al. [12] designed a system for cracking and shattering objects. Zhaosheng Bao and Jeong-Mo Hong [13] proposed an algorithm to handle the fracture of stiff and brittle materials in which the objects are treated as rigid bodies, but such studies are not directly applicable to cloth tearing because of the difference between the nature of the cloth and the rigid materials. O'Brien and Hodgins had some studies on crack initiation and propagation [14], and they evolved the existing techniques used for simulating flexible objects. That study was about fracturing brittle materials only and was not applicable to cloth-like materials like the other studies mentioned above, but this method was also extended to support ductile fractures by adding a plasticity model to the former finite-element method used in their former study [15]. Fracturing a brittle material cannot be used to represent a cloth tear, but it would be possible to use a ductile object fracture to look like a cloth tear by configuring the material properties, however this will not be appropriate for a whole cloth simulation system.

III. Methodology

A. Cloth Model

In our cloth model we use springs and irregular triangular meshes. Our method relies heavily on data relation and benefit from this while reconstructing the cloth structure during rupture. The data structures for particles, springs and triangles hold data related to each other. A spring has the data of its neighbour triangles and a particle has the data of the springs that is connected to itself. A triangle holds texture positions in addition to springs and particles constructing itself. These data are used in different steps like; searching the tear path, reconstructing the cloth structure, maintaining texture integrity after tearing, calculating an interpolated normal vector for a particle which is used for lighting by searching the surrounding triangles, and detecting the proper particles that will be used for bending and shear constraints.

Since we change the structure to a non-uniform model, we cannot use the shear and bending constraints as they are in the standard continuum sheet model. The two ends of bending and shear springs are calculated using indices on the grid, but in our structure we do not have a regular grid and indices anymore. Also, springs crossing each other irregularly would make it harder to process the segmentation of regions on the cloth. Not to sacrifice the visual quality maintained by those constraints, we developed a different approach for them.

Kelager et al. published a study about a triangle bending constraint model [16] in which the difference between the normals of two adjacent triangles were used to determine the bending constraint between those triangles. The nature of that study is appropriate for our triangle model since we know the two neighbour triangles for any spring, and we also know the surrounding spring vectors for any triangle. However, we found a better solution that would meet our needs. Since we know the neighbour triangles for a spring, we can reach the opposite corners of those triangles. We create a spring connected to these two opposite corners like the spring shown with the red dotted line in Figure 2. The repulsion force of this spring meets the bending constraint between the two triangles. In Figure 2, this bend-shear spring applies a repulsion force until it reaches the length it has when the two triangles are on the same plane. An additional benefit of this spring is that the attraction force of this spring also meets the shear constraint. This way we are able to meet the two constraints at the same time with a low cost.

The non-uniform placement of particles makes it difficult to find the right particles to apply bending and shear constraints. Also, it is not possible to detect the springs to eliminate when a rupture occurs and configure them again according to the newly changed structure. In our solution, we relate this

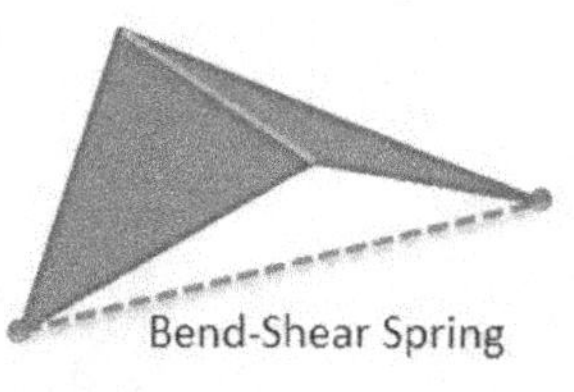

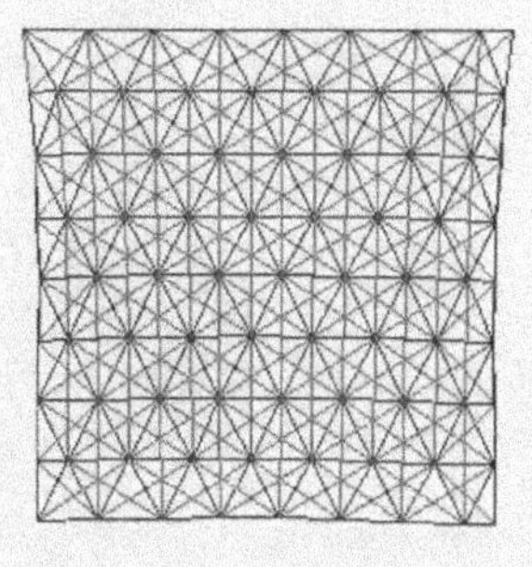

Figure 2: Bend-shear springs.

bend-shear spring with the spring on the intersection edge of the two triangles, as on the orange spring in Figure 2 on the left. This way we can keep track of these bend-shear springs and eliminate and recreate them when needed after a tearing occurred. On the right we see all the bend-shear springs in a cloth drawn with red, and the structural springs drawn with black.

Bend shear springs are created dynamically during the tearing. We use the length of the surrounding springs of the two adjacent triangles and the cosine rule to be able to calculate the length of a bend-shear spring we create between these two triangles. The length is the distance between the tips of the two adjacent triangles when they reside on a plane with no stress.

B. Tearing

Basic Tearing

During the simulation, as a result of the movement of the particles, the length of the springs change. We assume that when the length of the spring exceeds a certain threshold, the cloth would need to be ruptured around that area. In Figure 3, the cloth is dragged through the blue point and there is tension on the red line. After we stretched a little bit more, the deformation threshold is exceeded and the cloth is torn on that line.

Here the neighbour triangles of the stretched spring are eliminated and new triangles are created in place of them by dividing the former triangles into two.

We prepare new springs according to the new triangles and use them for the creation of those new triangles. In Figure 4 the red springs are the remainder of the spring, which is subjected to high tension before the tearing. The green springs divide the former triangles into two. The two facing couples of green

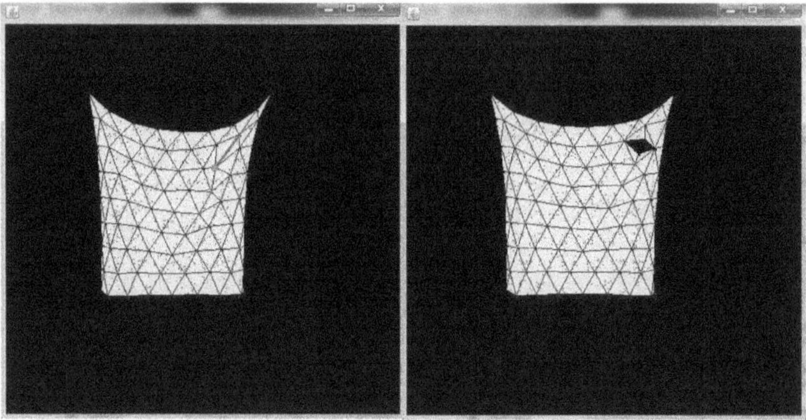

Figure 3: Tension on the cloth and the basic tear.

springs are on the same line of the texture of the cloth, but we need two separate springs for each texture line at those torn edges because they should not be connected anymore. That means both of the triangles on the opposite side of the tear should have their own springs so as not to be connected to the triangle on the other side of the tear line. In Figure 4, since m3 and m4 do not share an adjacent spring, they can move independently. On the other hand, since m3 and m5 share the same red spring, they are connected and they do not move independently.

There is a small calibration that needs to be mentioned. In this example, the spring on the intersection edge is elongated and ruptured because the length of it exceeded the tearing threshold. We create two new springs with the half length of the original spring indicated as red springs in the middle of Figure 4. If we do not change the positions of the particles at the middle of the tear at the orange point, the newly created springs will also be created and elongated more than the tearing threshold, so they will be ruptured again and again. To prevent this effect, we need to shorten the length of those springs so we move the tips at the middle of the tear closer to the other edge of the spring and create it that way so the new length will not be long enough to be ruptured again.

Weak Points

When you start to tear a cloth, for most kind of materials, the ruptured parts become weaker. When tension increases around these weak points, the tear

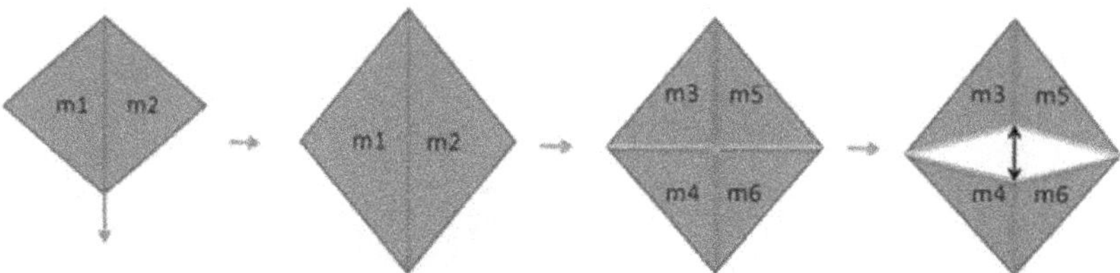

Figure 4: Basic tear.

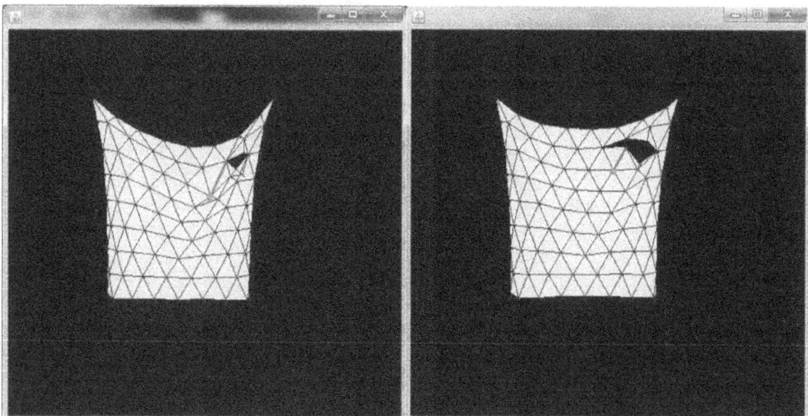

Figure 5: Tension and tear around a weak point.

tends to continue through these ripped parts and make the hole larger, rather than creating another hole next to the former one.

In Figure 5, when there is high tension on a string which is connected to a weak point, our former method would create another hole near the first, whereas in real life examples the tear would continue along the same tear line. Cloth can show less resistance at these weak points, and a tension which is not strong enough to tear an undamaged cloth would be enough to rupture a weak point, so we apply a smaller tearing threshold for springs that are connected to weak points. Here, when the tension on the red line exceeds a certain threshold, the tearing will occur, but since one of the ends of this spring is on a weak point this time, the neighbour triangles of the stretched spring will not be divided into two as in the first basic method in Figure 4. Instead, we select one of the springs connected to this weak point and detach the two neighbour triangles.

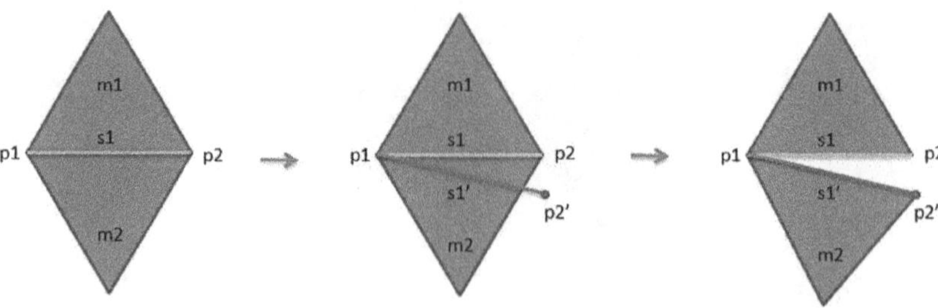

Figure 6: Detaching the triangles connected to the weak point p2.

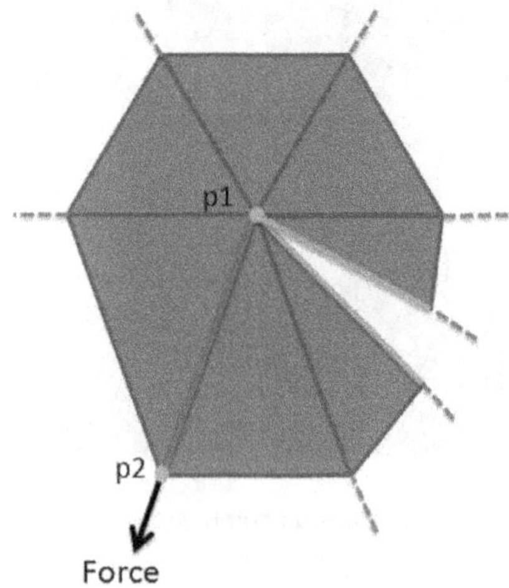

Figure 7: Possible paths for a tear on a weak point.

Figure 6 demonstrates the recreation of structures after the spring selected for the tear path.

An important problem at this step is to determine the tear path. For example, in the case illustrated in Figure 7, there is tension on the red spring and there are eight triangles connected to the weak point, which is under pressure. Two of them reside on a torn edge, indicated as green on the figure. One of the springs is the spring with the tension, indicated as red. There are five springs left that the tear may continue to grow on.

Here the selection of the spring is the problem. There is not a unique solution at this point, and the result may differ according to the characteristics of

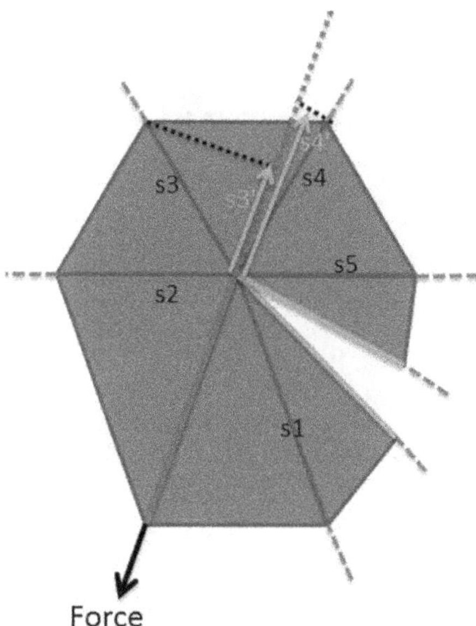

Figure 8: The projection of spring forces on the axis of tension.

the material of the cloth. We propose a heuristic, which is similar to most of the cases we see in the real world and produces visually pleasing results.

The main idea in our solution is to find the closest spring to the perpendicular axis of the force direction on the right side. The idea is to find the path which maximizes the torque. We can observe this in Figure 8. The tension on the red spring is increased and so the weak point in the middle is pulled through the red spring. In the first step, we find the spring that shows the highest resistance against the force. We find it by calculating the projection vectors of spring forces on the force axis and detecting the highest one on the opposite direction of the force.

Here s3' and s4' are the projections of force vectors of the springs s3 and s4. We did not show all of them for sake of clarity. For this example, we find that s4 is the spring that has the highest resistance against the tension on the red spring.

After this step, the remaining springs are divided into two groups. s1 and s5 are on the right side of the tension, s2 and s3 are on the left side of the tension. Since the tear connected to the weak point is on the right side of the tension, s1 and s5 are not under stress, tearing one of them is not logical, so we need to select either s2 and s3 for the tear path. We find the correct side by checking the connectivity between triangles. The red spring and s4 are not connected through the triangles on the right side, but they are connected through s2 and s3 on the left side.

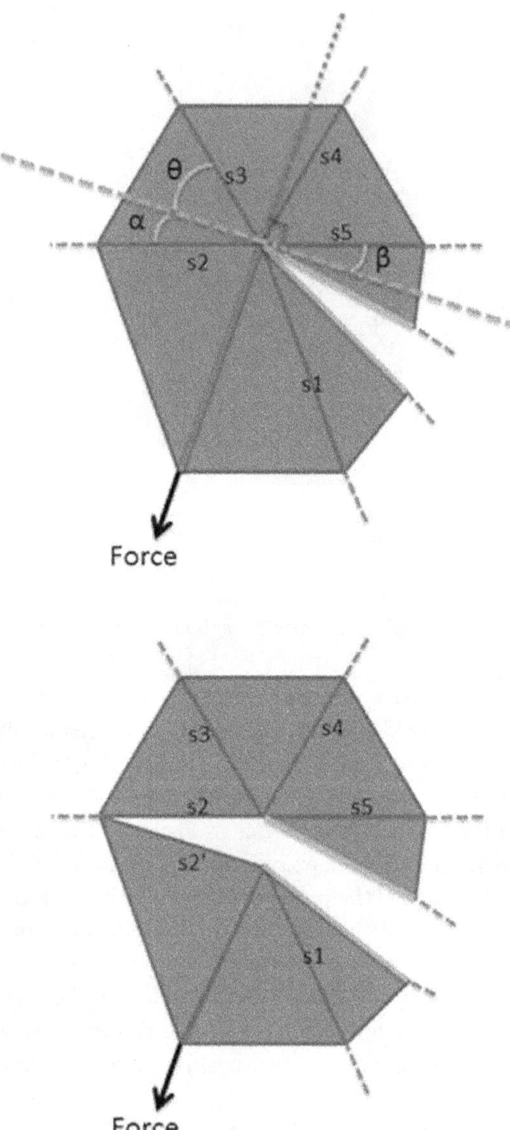

Figure 9: The angles between the spring forces and the axis perpendicular to tension axis.

After detecting the right springs to control for tearing, we found that they are s2 and s3. Our aim is to find the one which has the smaller angle between its force vector and the axis perpendicular to the tension axis, indicated as the blue dotted line in Figure 9. For this example, between s2 and s3, s2 has the smaller angle, so we choose it for the tear path. The angle between the force vector of s5 and the axis perpendicular to the tension axis is shown as here. Even though it could be smaller than this, we do not take it into consideration since it is on other side of the tension axis.

After a weak point is ruptured and as the tear continues on a weak path, the cloth structure and weak points change. For Figure 6, after the triangles connected to the weak point, p2, are detached, p2 or p2' are not weak anymore. The change of weak points can be seen in Figure 10, and are shown as blue points on the wire-frame view.

C. One Point Connections

There is an exceptional case which needs to be addressed. In some cases, the connections between two different parts of the cloth may consist of only one point. The tear would have come to the edge of the cloth or two different tears may have come across at a point. Here that point is a weak point, but the algorithm we use for weak points searches for a spring to continue the tear. However, in this case the opposite parts of the cloth on the tension axis should have been detached directly without considering the springs.

We detect the situation by comparing the number of the springs which have a connection to the spring with the tension and the total number of springs connected to the point of interest. We create a new particle, p' instead of p, and apply a small displacement along the axis of tension to be able to prevent the tear after the detachment by shortening the length of the spring with the tension. We use the same threshold as we use for weak points in one point connection cases, but it is possible to use a smaller threshold here. This is also a subject which could change according to the characteristics of the cloth material.

Texture Integrity

Another important part of the problem is to preserve texture integrity after tearing. In a continuum sheet model, particles are placed on the corners of a regular grid and cloth texture is applied to this plane. As the simulation runs, the positions of these particles change. These positions are used while drawing the polygons. In order to draw the correct texture portion on a polygon, we have to know the texture coordinates at the corners of that polygon. In a sheet model

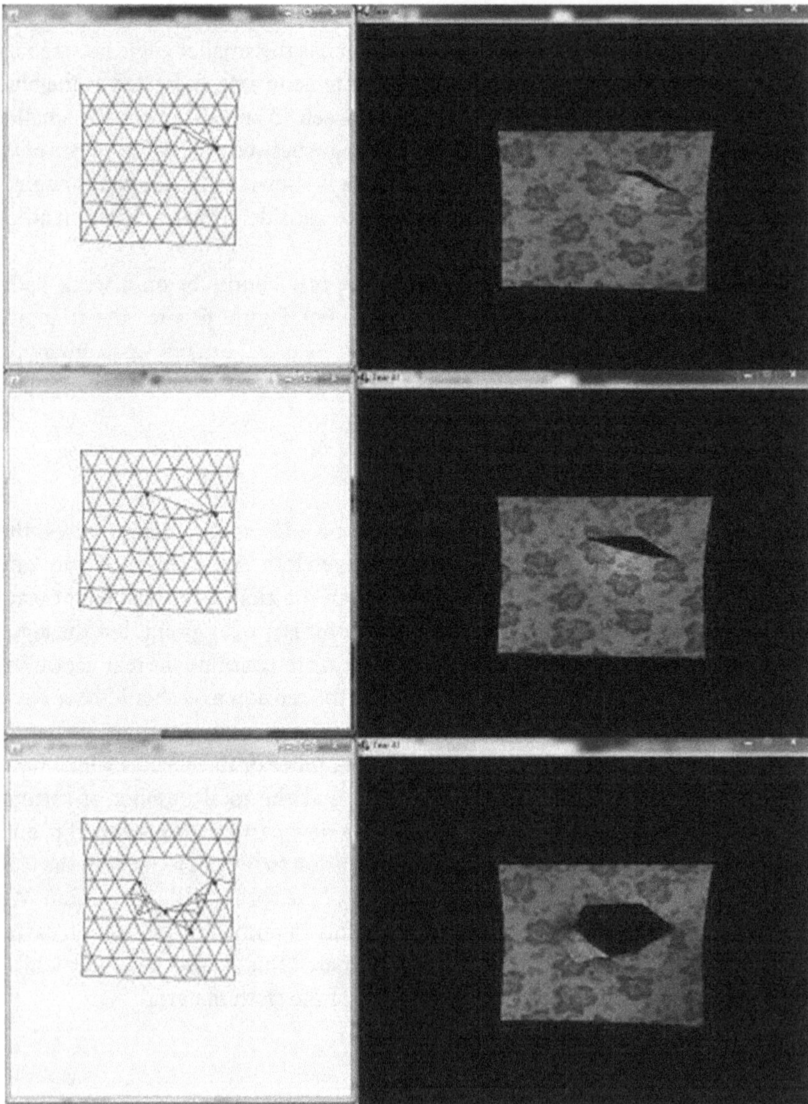

Figure 10: Weak point propagation.

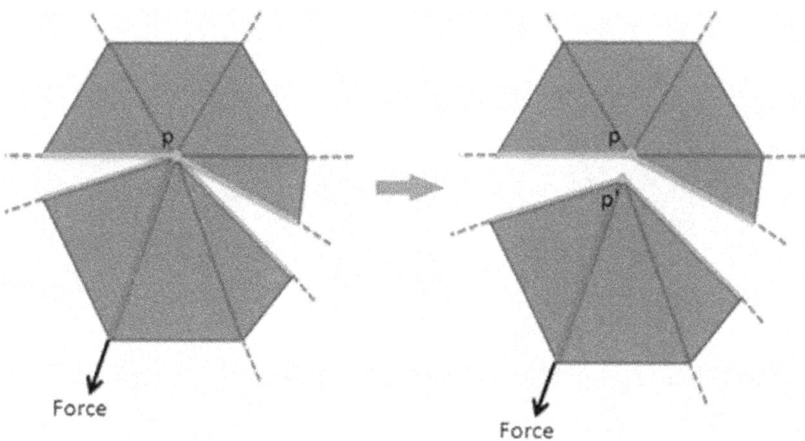

Figure 11: Tear on a one point connection.

which uses a regular grid, it is possible to reach the texture coordinate on a particle using the indices. Since we use non-uniform structure we cannot use such an approach. Also, it is not suitable for a system in which the model is subject to structural change.

At this point, we take advantage of the data stored in triangle structure. At the initialization step, we store the texture positions on the vertices of the triangle in itself. This way we do not lose this data for a triangle during the simulation. When two adjacent triangles are separated, as in Figure 12, they do not share the same particle object on the separated side after the tearing. Since the texture position is stored in triangle objects, we still know the texture position on p3 and p3'.

In the basic tear model, the triangle is divided into two. The old triangle is eliminated and two new triangles are created instead of the old one, as in Figure 13. We need to calculate the texture positions for these newly created triangles. The position value from the former triangle is used for the undamaged parts of the new triangles. For the ruptured edge, we find the texture position in the middle using the texture position on the tips of the ruptured spring.

In Figure 13, the middle of the texture positions of p2 and p3 is calculated and used at the edges p4 and p5 on the newly created triangles.

Here is an example from our implementation in Figure 14. After we pulled and tore the cloth, we dragged the red point back to the place it was originally connected to. We use a texture which enables us to identify the portions on the

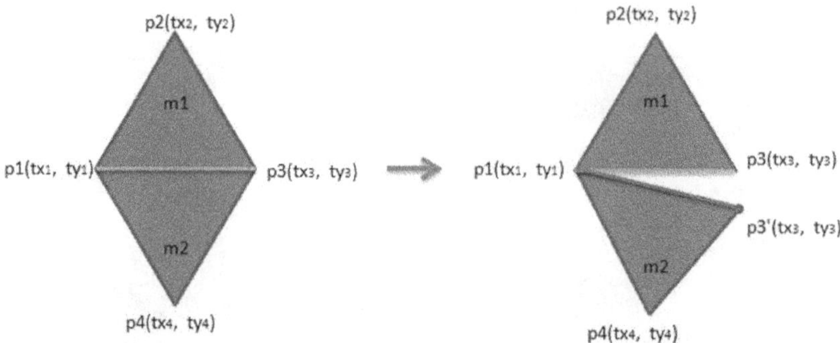

Figure 12: Texture position assignment after detaching one side of two triangles.

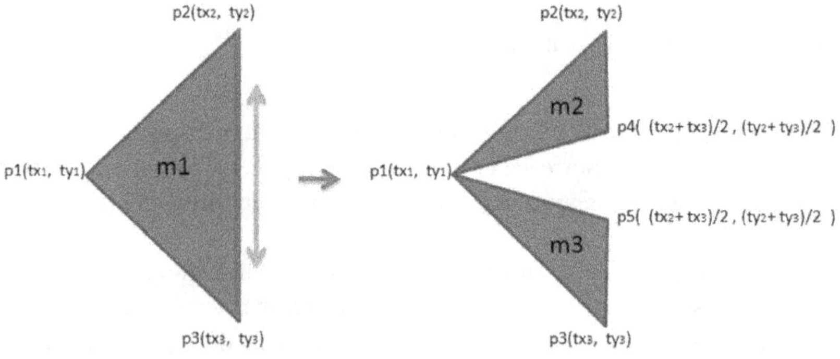

Figure 13: Texture position assignment after dividing a triangle.

cloth, and we see that after the tearing, texture integrity is maintained correctly; no area is lost and texture positions are preserved correctly at the torn parts.

IV. Experiments and Results

We developed a cloth simulation implementation and applied our method in this model. Here we used a simple rectangular cloth model (see Figure 14), but our proposed method is applicable to complex irregular cloth models. The cloth is attached from the up-per edge. You can drag the cloth from any point with the mouse interactively, and the lengths of the springs on the cloth change. This way the tension on the cloth changes, and if it exceeds a certain threshold, those parts of the cloth are torn.

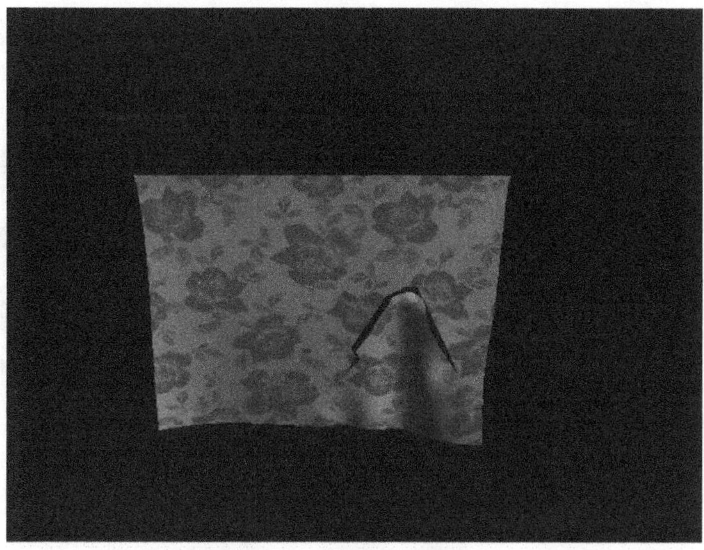

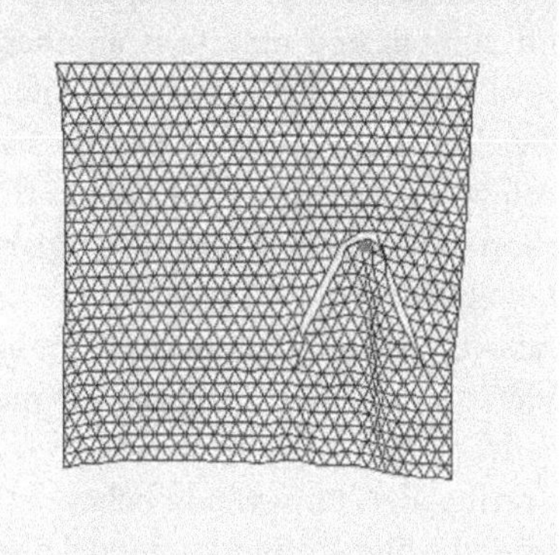

Figure 14: Textured, torn cloth.

A. Physical Interaction with Outer Objects

We also tested the success of our method to see how it works when the cloth interacts with other objects. There are two key points,

(i) size of the impact area, and (ii) speed of the impact. The pressure is spread among a greater number of particles as the impact area increases and the pressure applied for each particle decreases. Thereby, an impact on a smaller area is more likely to cause a tear on the cloth, whereas an impact on a larger area tends

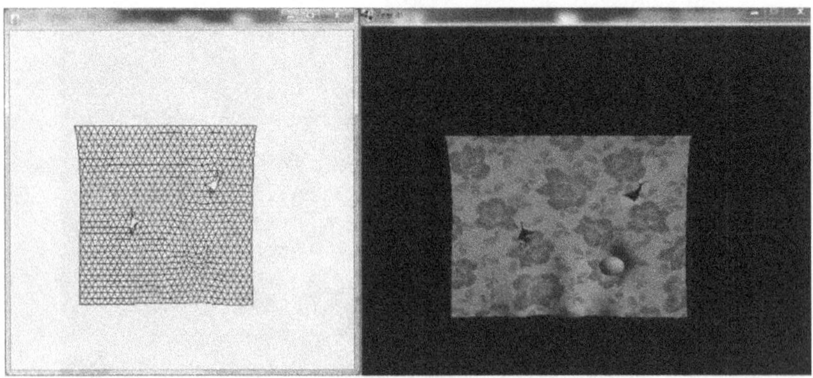

Figure 15: ball with a radius of 0.5 is thrown to the cloth with a speed of 3.

to push the cloth without penetrating. In a similar way, faster impacts are more likely to tear the cloth while slower movements of the same object would not be able to tear it because of the difference in pressure applied.

Handling collision with complicated objects is another extensive research area and outside the scope of this study. In our implementation, we used spherical balls to collide with the cloth, but our model proposes a generic solution to the tearing problem regardless of the shape of the objects that collide with it, so long as the collision detection is handled properly, it is capable of working for an interaction with any kind of object.

In most of the images above, we used the images of an implementation with a low resolution cloth model to be able to explain the technique clearly. We used a high resolution cloth model for the implementation presented in this chapter where we tested how the tearing algorithm would behave in an interaction with a physical object. In addition, the high-resolution model provided a better collision response and showed that our method has no performance problems with detailed cloth models.

In this implementation, we throw balls at the cloth of different sizes and at different speeds. In the first example we observed that impacts applied on a small area can tear our cloth model. The thrown ball has a radius of 0.5 units, which can be considered as a small size for this example, and the speed of the ball is 3 units. In Figure 15 it is shown that there are three balls thrown at the cloth. Two of them tore the cloth and the last one is about to tear.

In the second example we increased the ball radius to 1. When we threw the ball at the lower part of the cloth, it passed beneath the cloth without tearing. Cloth slid on the ball as in Figure 16

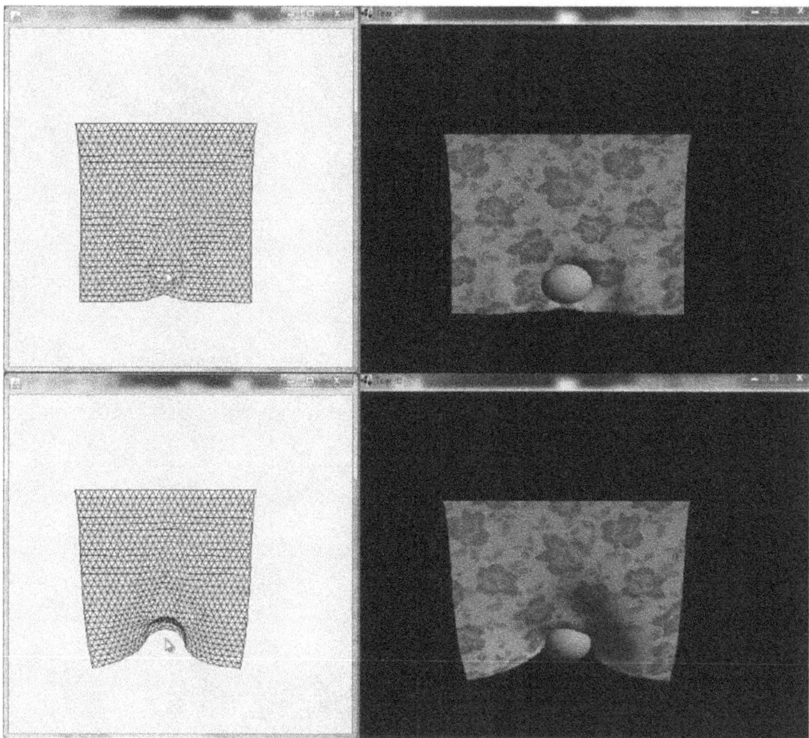

Figure 16: A ball with a radius of 1 is thrown to the lower part of the cloth with a speed of 3.

If we throw the ball at the upper part of the cloth instead, the ball faces some resistance, resulting from strain that is caused by the weight of the lower part of the cloth. Here the size of the pressure area and the speed of the ball is at a level which can apply an impact that can tear the cloth before the cloth would be able to slide on the ball. This is shown in Figure 17.

As a third example, we threw a large ball with a radius of 2 and with a speed of 3. The cloth slid over the ball without being torn. The interaction surface is large, so the pressure is spread around. A greater number of springs responded to the impact so the tension at each spring decreases and they can resist the impact without being torn, as in Figure 18.

As a final example, we increased the speed and threw the same ball with the radius of 2 at a speed of 10 this time. Because of the speed of the ball, the tension in the springs increased so fast that the cloth cannot resist the impact and is torn, as in Figure 19.

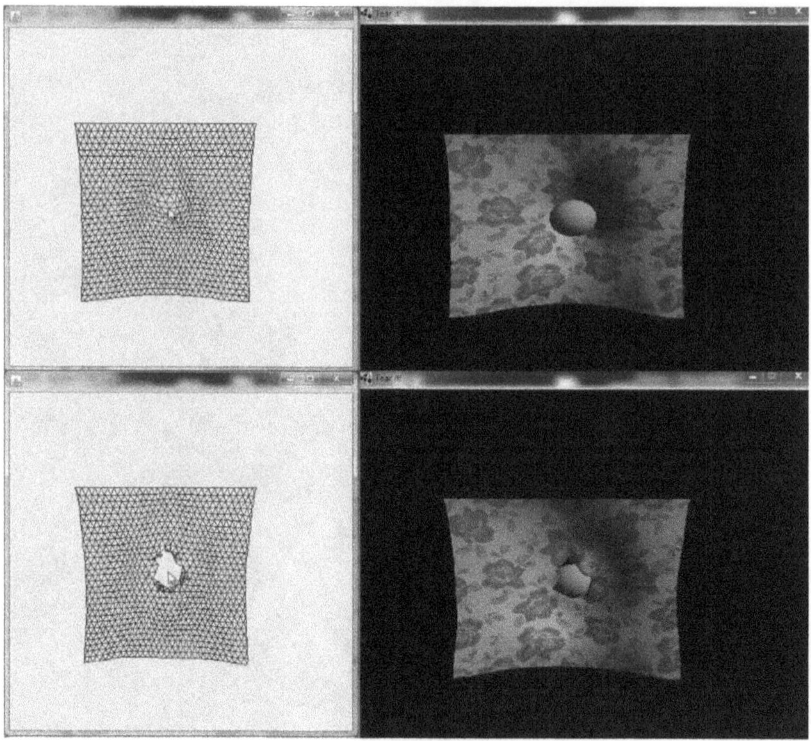

Figure 17: A ball with a radius of 1 is thrown to the upper part of the cloth with a speed of 3.

B. Tear Response to Force Direction

Another experiment we carried out focused on the change of the tear propagation according to the direction of the force. According to the heuristic we developed for weak points, the tear path should be perpendicular to the direction of the force. It means that the tear should continue somewhat in the direction of the pull. We can see that the method could perform visually convincing results as in Figure 20.

C. Profiling Results

One of the most important criteria for evaluating the success of a method used for modelling an interactive physical simulation is real time performance.

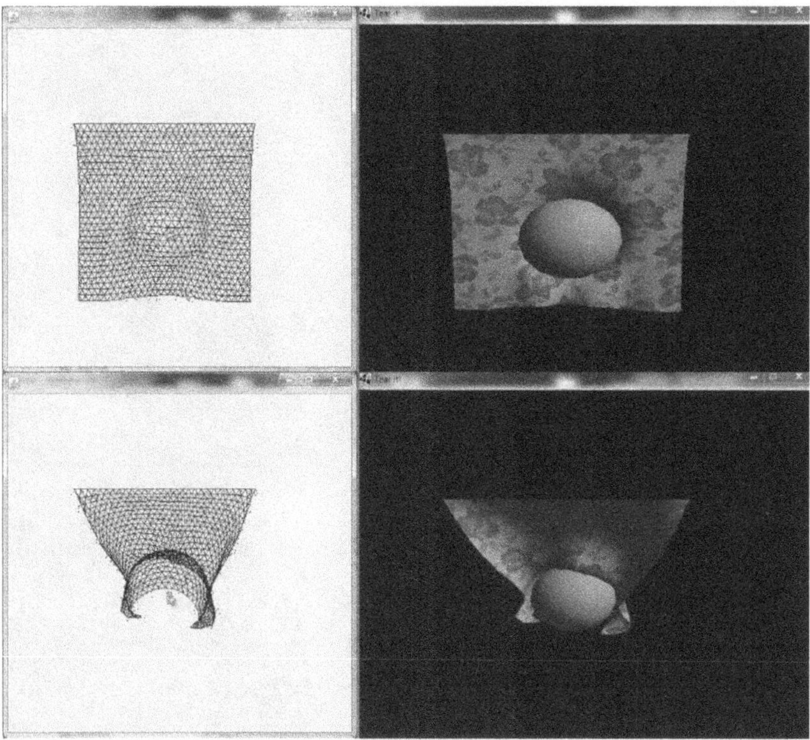

Figure 18: A ball with a radius of 2 is thrown to the cloth with a speed of 3.

Running cloth simulations in real time is not that much of a problem for the processing power of today's computers. We extend the current cloth simulation model with new capabilities without compromising the performance. Necessary calculations about tearing are processed only if a high tension is detected on a part of the cloth. Since this is a one-time process, it does not affect the overall performance of the simulation, so our method works without a problem on real time systems. In order to assess performance, we ran a simulation for 35 seconds. During the simu-lation, we tore different parts of the cloth constantly and observed the method durations. Initial parameters of the simulation:

Resolution of the cloth model used in simulation is 30x30, which makes 900 particles. There are 2581 structural springs There are 2465 bend-shear springs. There are 1682 triangles.

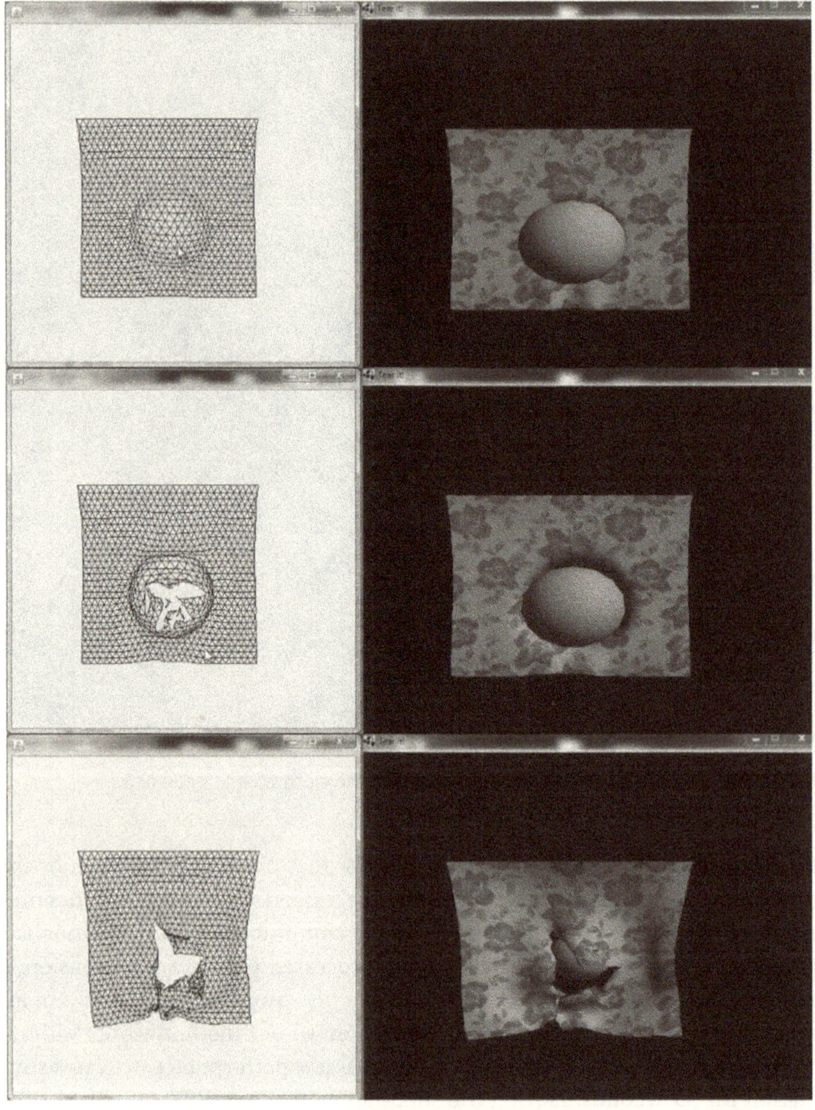

Figure 19: A ball with a radius of 2 is thrown to the cloth with a speed of 10.

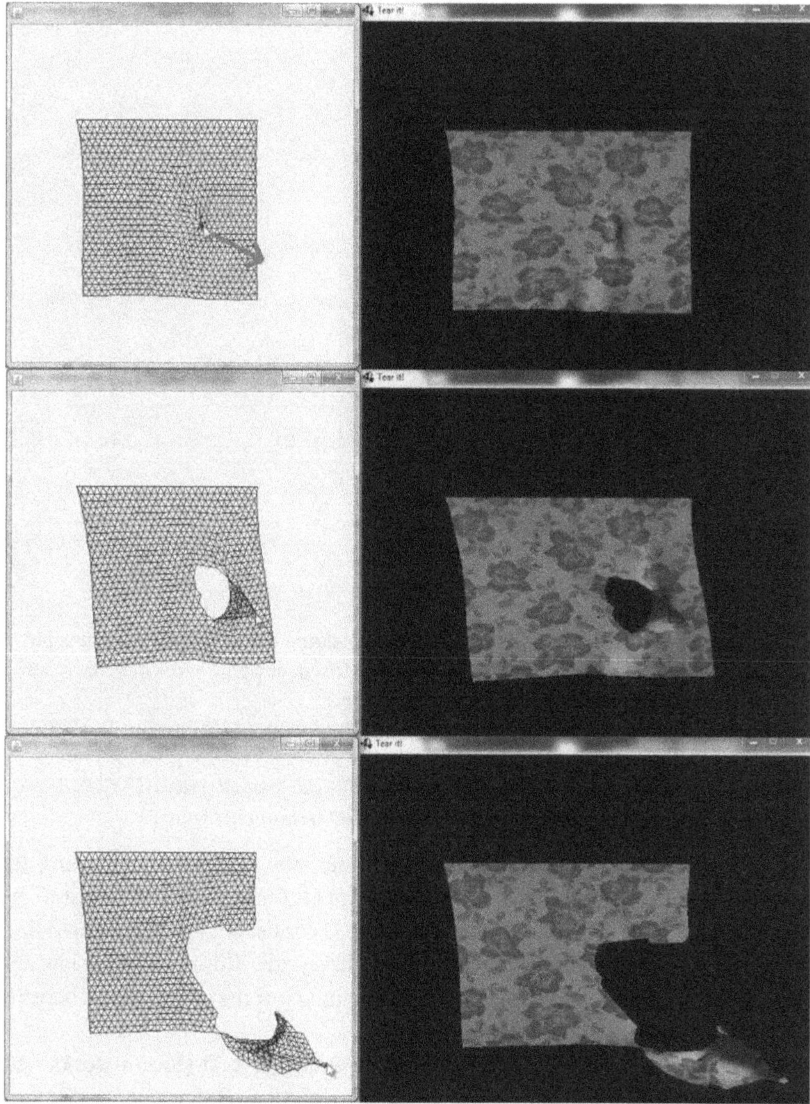

Figure 20: An example of the response of the tear to the direction of the force.

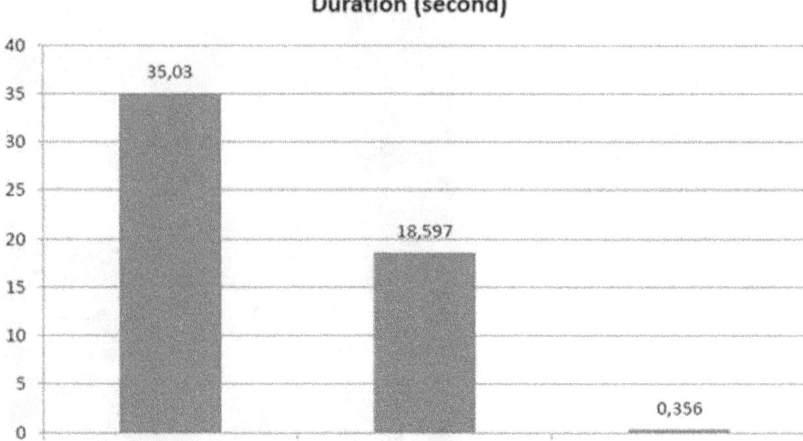

Figure 21: Textured, torn cloth.

These values change during the simulation as the tearing occurs. The forces that apply on each particle are the spring force, gravity, air friction, damping and wind. The specs of the computer that we run this simulation on are:
Operating System: 64-bit Windows 7 Ultimate Processor: AMD Phenom II X2 555 3.20GHz Ram: 4GB DDR3
GPU: NVIDIA GeForce GTX560 (Core Clock: 820MHz, Memory Size: 1GB GDDR5)
The profiling result obtained using Netbeans profiler is shown in Figure 21.

The whole simulation takes 35 seconds, but our concern is regarding the Tearit.update() method. This is the place where all the required calculations for the simulation are done, and it takes about 18 seconds. On the highlighted line, we can see our tear() method. It takes 356 milliseconds. This is relatively low and does not have a significant effect on the performance of the simulation. There are 510 tearing process during this simulation.

Moreover, our model can reduce the computational cost since it works with non-uniform models. It is possible to model a cloth with fewer particles with a non-uniform model. Unnecessary triangles can be eliminated where a high level of detail is not necessary. This way the number of particles in the system is reduced, resulting in performance gain.

V. Conclusion

A. Contributions

There have not been many studies around the tearing of cloth. The existing ones have only dealt with some aspects of it, and they have not provided a general solution to this problem. Our aim in this paper was to fill this gap and provide a general solution to this problem, dealing with all aspects of it.

We tried to recreate a realistic tearing effect while partitioning the triangles on the cloth and generating the tear path by taking advantage of physical rules to achieve satisfactory results.

The presented method preserves the polygonal area consistency after the tearing by avoiding the elimination of any polygonal area at the process of tearing. At the same time, texture integrity is maintained successfully, regardless of any structural changes by the help of the triangle structure we used in our model.

Another important aspect of this study is that it is able to react successfully to external physical impacts. The reaction changes realistically according to the pressure area and the speed of impact. It is also capable of responding to direct user interactions successfully. The tear propagation changes according to the direction of force and this increases the sense of physical reality. We showed examples of these in the experiments and results section.

As much as the proposed method provides results with a good level of realism, it is also able to provide these results at an insignificant cost.

There is an important factor that makes the contributions mentioned above more valuable. The cloth model we used in our method is a generic model which could be adapted easily to many applications. It is based on spring physics, which is used widely used for cloth simulations, and also supports non-uniform cloth structure with the help of irregular triangular meshes.

B. Future Work

There are many studies about collision handling with cloth. A study which observes the interaction of different kinds of objects with this cloth model would be an interesting study, like using a knife for cutting a cloth.

Spring model is used for cloth simulations regularly, but it is also used for modelling some volumetric rigid objects [17] Strict bending constraints are applied to the springs in some of those models to constitute volumetric shape. It may be possible to extend this model to support cracks on a volumetric object.

Another area of expertise, creating tearing sounds during the process of tearing the cloth, would be a nice complementary study to this study. However, the resolution of polygons determines the moment of tear and this would be a difficult problem for developing a successful algorithm for generating sounds in a proper way.

Another original addition to this study would be implementing a tessellation methodology to this algorithm. It is mentioned that the method presented in this study is applicable to multi-resolution cloth models. The parts modelled with coarser resolution could be torn in detail by subdivision and produce more realistic results even for a low resolution cloth model.

References

[1] Metaaphanon, N., Bando, Y., Chen, B.-Y., AND Nishita, T. 2009. Simulation of tearing cloth with frayed edges. *Comput. Graph. Forum* 28, 7, 1837–1844.

[2] De Aguiar, E., Sigal, L., Treuille, A., AND Hodgıns, J. K. 2010. Stable spaces for real-time clothing. In *ACM SIGGRAPH2010 papers*, ACM, New York, NY, USA, SIGGRAPH '10, 106:1–106:9.

[3] Provot, X. 1995. Deformation constraints in a mass-spring model to describe rigid cloth behaviour. In *Graphics interface*, Canadian Information Processing Society, 147–147.

[4] Bridson, R., Fedkiw, R., AND Anderson, J. 2002. Robust treatment of collisions, contact and friction for cloth animation. In *Proceedings of the 29th annual conference on Computer graphics and interactive techniques*, ACM, New York, NY, USA, SIGGRAPH '02, 594–603.

[5] NG, H. N., AND Grimsdale, R. L. 1996. Computer graphics techniques for modeling cloth. *Computer Graphics and Applications, IEEE* 16, 5, 28–41.

[6] Nealen, A., Mller, M., Keiser, R., Boxerman, E., AND Carlson, M. 2006. Physically based deformable models in computer graphics. *Computer Graphics Forum* 25, 4, 809–836.

[7] Baraff, D., AND Witkin, A. 1998. Large steps in cloth simulation. In *Proceedings of the 25th annual conference on Computer graphics and interactive techniques*, ACM, New York, NY, USA, SIGGRAPH '98, 43–54.

[8] Narain, R., Samii, A., AND O'Brien, J. F. 2012. Adaptive anisotropic remeshing for cloth simulation. *ACM Transactions on Graphics* 31, 6 (Nov.), 147:1–10. Proceedings of ACM SIGGRAPH Asia 2012, Singapore.

[9] Terzopoulos, D., AND Fleischer, K. 1988. Modeling inelastic deformation: viscolelasticity, plasticity, fracture. *SIGGRAPH Comput. Graph.* 22, 4 (June), 269–278

[10] Jain, N., Kabul, I., Govindaraju, N. K., Manocha, D., AND LIN, M. 2005. Multi-resolution collision handling for cloth-like simulations. *Computer Animation and Virtual Worlds* 16, 3–4, 141–151.

[11] Haggstrom, O. 2009. Interactive real time cloth simulation with adaptive level of detail.

[12] Hellrung, J., Selle, A., Shek, A., Sifakis, E., AND Teran, J. 2009. Geometric fracture modeling in bolt. In *SIGGRAPH 2009: Talks*, ACM, New York, NY, USA, SIGGRAPH '09, 7:1–7:1.

[13] Bao, Z., Hong, J.-M., Teran, J., AND Fedkiw, R. 2007. Fracturing rigid materials. *IEEE Transactions on Visualization and Computer Graphics 13, 2 (Mar.)*, 370–378.

[14] O'Brien, J. F., AND Hodgins, J. K. 1999. Graphical modelling and animation of brittle fracture. In *Proceedings of the 26th annual conference on Computer graphics and interactive techniques*, ACM Press/Addison-Wesley Publishing Co., New York, NY, USA, SIGGRAPH '99, 137–146.

[15] O'Brien, J. F., Bargteil, A. W., AND Hodgins, J. K. 2002. Graphical modeling and animation of ductile fracture. *ACM Trans. Graph.* 21, 3 (July), 291–294.

[16] Kelager, M., Niebe, S., AND Erleben, K. 2010. A triangle bending constraint model for position-based dynamics. In *Workshop on Virtual Reality Interaction and Physical Simulation VRIPHYS.*

[17] Criswell, B., Lentine, M., AND Sauers, S. Avatar: Bending rigid bodies.

10. A Dynamic Exit Choice Method for Real-Time Indoor Evacuation Scenarios

Oner Barut
Hacettepe University onerbarut@cs.hacettepe.edu.tr

Murat Haciomeroglu
Gazi University murath@gazi.edu.tr

Cumhur Y. Ozcan
Hacettepe University cumhuryigitozcan@cs.hacettepe.edu.tr

Hayri Sever
Hacettepe University sever@hacettepe.edu.tr

Abstract— Most of the evacuation simulations focus their attention to the accuracy of the simulation. Therefore, their models fail to provide real-time performance which is indispensable for almost all video games. In this paper, we propose a real-time crowd simulation model specifically designed for indoor evacuation scenarios to be used by entertainment systems. While agents in our model perform local collision avoidance using an agent-based steering technique, the optimal exit gates and the global paths to these gates are determined efficiently by the proposed system. We represent our simulation environment as a uniform grid and we perform Dijkstra path finding for all exit gates, which computes shortest paths from all exit gates to all cells. Then, for each grid cell, we choose the exit gate having shortest path within all gates and assign this gate to the cell, which means all agent in the grid cell will evacuate the area using the selected exit gate. Later on, we reverse the previously found paths between every grid cell and their associated exit gates in order to calculate guidance velocity of the agents in the cells. Our model takes advantage of the limited goal point availability by only performing one Dijkstra path finding per exit gate to further increase the efficiency of the simulation. Test results indicate that our method not only manages to balance the agent load on the exit gates despite non-uniform position distribution of the agents in the test scenarios, but also achieve lower average and maximum evacuation times than mostly used nearest gate selection approach.

Keywords— simulation; real-time scenarios; crowd simulation

I. Introduction

Crowd simulation can be introduced as imitating motion and behavior of a large number of real entities such as humans or animals as close as possible to reality.

This replication includes navigating vast number of virtual characters from their initial positions towards their goal positions in a virtual environment. In addition, a real-time simulation should be as efficient as possible to run on a consumer level hardware, especially in games.

Although both movie and video game industries intensively use the crowd simulation models, evacuation planning is another important application area of the crowd dynamics. Evacuation planning can be defined basically as measuring and improving the time required to move all the agents inside a building, facility, vehicle, etc. to a safe place in an emergency situation. In addition to being too expensive and time consuming, using humans for evacuation experiments may be very dangerous and be concluded with casualties. Therefore, evacuation simulations, which perform these experiments on a virtual environment with virtual agents, are commonly used.

Evacuation simulation of crowds has been studied over the past decades. Numerous evacuation models have been proposed to perform accurate computer simulations. Kuligowski et al. [1] published a comprehensive review of existing models. As a summary, according to the current literature, cellular automaton models [2], lattice gas models [3], social force models [4], fluid dynamic models [5] and agent-based [6] models are employed to navigate the agents in evacuation simulations. These models may be divided into two groups; microscopic and macroscopic in terms of scale. Microscopic models deal with the individual agents while macroscopic models consider the agents in the crowd as a whole. Another possible categorization of these models, which labels them as discrete and continuous, is based on time and/or space.

In evacuation simulations, one the most fundamental parts is determining optimal exit gates for each agent to perform the immediate evacuation of all the agents in simulation environment. Kuligowski et al. [1] performed a detailed comparison of the exit selection strategies of the existing evacuation simulation systems. According to their study, many evacuation simulations employ shortest distance metric between agent positions and exit gate positions in order to decide the minimum-cost exit gate for every agent. One of the major drawbacks of this approach is that, when agents in the evacuation scenario are not distributed uniformly in the simulation environment, a large number of agents will decide to navigate towards the same gate. In that case, some of the exit gates will be congested while the others are not used intensively. As a result of the unoptimal exit gate decisions of the agents, average evacuation time and maximum evacuation time of the simulation tend to be higher. Also in a video game, agents would seem unintelligent.

On the other hand, Kuligowski et al.'s [1] work also indicates that there are evacuation models that consider not only the distance but also width, congestion level, awareness, signage of the exit gates in order to plan the evacuation route of the agents. The subject evacuation models either determine the minimum-cost exit gates conditionally or seek for the minimum-cost path through the exit gates bearing in mind the above parameters. Recently, Ehtamo et al. [7] performed the exit route choice based on a game theoretic approach, which tries to maximize the utility of all agents, and Guo et al. [8] employed a logit-based method for exit choice.

Although there are successful evacuation models in the literature, these techniques are computationally too intensive for a real-time simulation. Our technique is designed for real-time performance therefore; the proposed model is not intended for accuracy but creating plausible evacuation scenarios to be used by entertainment industry.

II. System Overview

In the proposed model, we developed an evacuation system for indoor simulation environments with a limited number of exit gates. We employed a uniform grid structure, which consists of square cells that covers the entire simulation area. Each grid cell contains the information about an optimal predetermined exit gate that has minimum cost to evacuate all agents currently in that cell. Every cell also contains the whole path information to the determined minimum-cost exit gate.

The proposed method utilizes Reciprocal Velocity Obstacles (RVO) [9] model for local navigation of virtual agents. In every simulation step, agents query their current cell and head towards the minimum-cost exit gate, which their current cell has already detected (see Section III). For this purpose, agents use the path information stored in cells and determine a preferred velocity vector considering first three cells along their path. They calculate average center positions of the first three cells (except from the one the agent is currently at) and construct a vector directing from the center of the current cell to the average center of the following cells. Then agents normalize their vector and scale it by their intended speed to obtain preferred velocity vector. By using the preferred velocity vectors of each agent as input, RVO calculates a final velocity vector for each agent as output. The agents use these collision-free velocity vectors while navigating towards the exit gates. A general overview of the proposed system is illustrated in Figure 1.

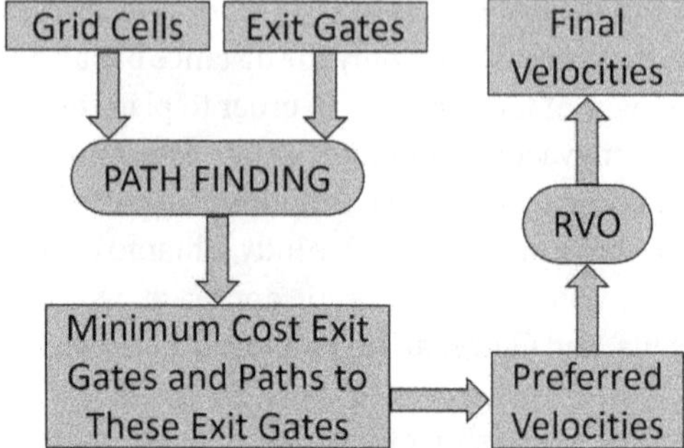

Fig. 1: System overview of the proposed technique that uses path finding for determining exit gates. A path finding process is performed per exit gate. Consequently, minimum-cost exit gate of each grid cell and path to this gate from the cell are determined. Benefiting from the path information of each cell, a preferred velocity (heading towards the determined minimum-cost exit gate) is calculated for all agents in the cell. RVO is used in order to calculate final velocities of all agents via utilizing their preferred velocities.

III. Determining Exit Gates

The proposed technique determines the minimum-cost exit gate for all cells by utilizing Dijkstra path finding algorithm. Basically, instead of searching for paths from cells to exit gates, each gate determines a path from itself to every grid cell. Considering we have a very small number of exit gates relative to the number of grid cells in almost all evacuation scenarios, it is obvious that the proposed technique will require very few pathfinding operations independent of the simulation environment size. However, it should be also noted that the number of path finding operations is linearly proportional to the number of exit gates the evacuation environment has.

The presented evacuation system performs pathfinding operations for all of the exit gates once per each second. Each gate determines a cost (C) to each grid cell based on the cost function (Equation 1). Each gate compares its cost (for each cell) with the cost of other gates. If the cost from that gate to the current cell is smaller than the other gates, the current cell updates its cost and path information. After finishing the path finding operation for all exit gates, each cell is aware of its minimum-cost exit gate and the path going towards that exit gate.

$$C = C_T + C_W \tag{1}$$

Benefiting the cumulative nature of the Dijkstra cost function, travel cost (C_T) that corresponds the cost of passing from the current cell, which Dijkstra has arrived, to the neighboring cell of the current cell is given in Equation 2. This cost function calculates travel time in seconds, which will be required while passing from the current cell to the neighboring cell. In Equation 2, $\left|\overrightarrow{P_C P_N}\right|$ indicates the distance between the current cell (P_C) and one of the neighboring cells (P_N), S_A indicates the average travel speed of the agents on the simulation.

$$C_T = \frac{\left|\overrightarrow{P_C P_N}\right|}{S_A} \tag{2}$$

Travel cost computed using Equation 2 is used in Dijkstra path finding operations to decide minimum-cost paths from exit gates to grid cells. However, to determine minimum-cost exit gates for all grid cells, another cost, named waiting cost (C_W), is also included. Waiting cost takes into account the evacuation time (in seconds) of all agents who are on the grid cells that has already determined the same gate as minimum-cost exit gate. For this purpose, each exit gate stores total number of agents that will be evacuated via itself and when a grid cell determines an exit gate as minimum-cost exit gate, the exit gate updates its total agent number by aggregating the number of agents inside the grid cell. The cost function, which computes the waiting cost, is given in Equation 3. In Equation 3, N_A indicates the total number of agents that will be evacuated from the same gate and S_E indicates the evacuation speed (agents per second) of the exit gate. Instead of performing a dynamic computation of the evacuation speed variable of the exit gates, we heuristically choose a constant value (one agent per seconds) for performance considerations.

$$C_W = \frac{N_A}{S_E} \tag{3}$$

One can think that grid cells might determine two different minimum-cost exit gates among two consecutive path finding progresses in a one-second period and such a case might cause oscillations of agents although it is not observed in our test scenarios. One possible solution to that would be introducing a penalty cost when a grid cell tries to change its previously determined minimum-cost exit gate. Thus, potential oscillations of moving agents will be avoided successfully and with insignificant additional computational cost.

IV. Test Scenarios & Results

In order to evaluate the proposed evacuation system, we have generated two test environments, which represents common indoor evacuation cases. Both test environments consist of a square shaped closed environment (32 × 32 square grid cells with 4 meters edge length). Agents' intended speeds are assigned randomly between 1 – 3 m/s.

The first test environment is a large hall that has four equally sized exit gates, which are positioned at each corner of the environment. The agents are placed being close to the upper left exit gate in a way that they form a square shape as large as a quarter of the environment. An illustration of the simulation environment is given in Figure 2a.

The second test environment is a stage hall that has a rectangular stage leaning to the middle of one of the edges of the environment. The simulation environment also has three exit gates, which is identical in terms of size. Two of these exit gates are positioned at the edges of the environment next to the stage. The last exit gate is located on the edge opposite of the stage. The agents are placed in front of the stage such a way that the agents are located equally distant from either of the two edges next to the stage. An illustration of the simulation environment is given in Figure 3a.

We employed two different test scenarios on the simulation environment described above. For the first scenario, we assumed that agents will choose the nearest gate (see Figures 2b and 3b) which is a commonly used method in many evacuation systems. For the second one, the proposed method is used (see Figures 2c and 3c). We also employed two different metrics while evaluating the scenarios: The average evacuation time of the agents and the maximum evacuation time of the agents.

In Tables I and II, average and maximum evacuation times (in seconds) of the agents are presented for large hall and stage hall test environments respectively. Considering these results, it is obvious that the second scenario, which corresponds to the proposed evacuation system, finalizes the evacuation of virtual agents in the simulation environments quicker than the opponent scenario in terms of both the average and the maximum evacuation times.

As emphasized before, we constructed our test environments having non-uniform agent distributions among the exit gates. Because of this fact, during the simulation of the first test scenario in both test environments, the exit gates close to the agents' initial positions were very congested while there were almost no agents heading towards distant exit gates (see Figures 2b and 3b for each test environment respectively). On the other hand, in the second test scenario in both test environments (see Figures 2c and 3c respectively), the agents used the gates more intelligently and balanced the agent load among the exit gates roughly equal which leads to a decrease in both average evacuation time and maximum evacuation time. In short, the agents that use the proposed system look smarter (which is often desired in games) by using the exit gates more organized.

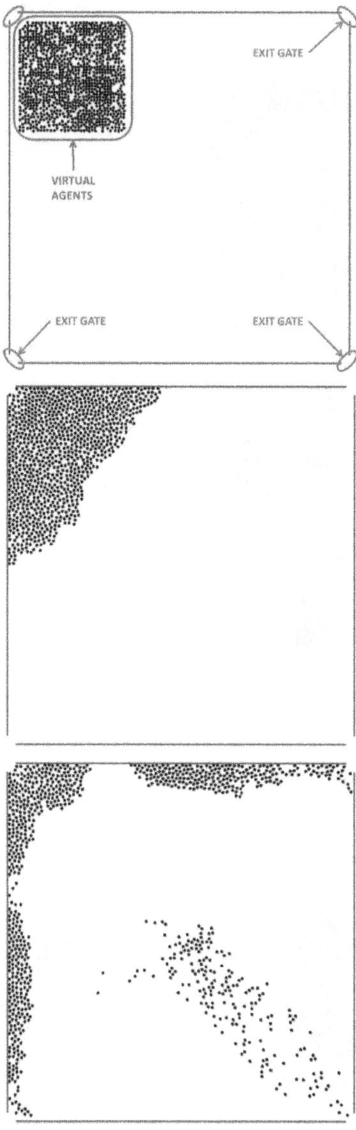

Fig. 2: (a) Top view of the initial simulation environment, which has four exit gates and virtual agents standing close to the upper left exit gate. (b) Snapshot taken shortly after the beginning of the first scenario in which agents choose nearest gate (the one at the upper left corner). (c) Snapshot taken shortly after the beginning of the second scenario that determines minimum-cost exit gates via path finding.

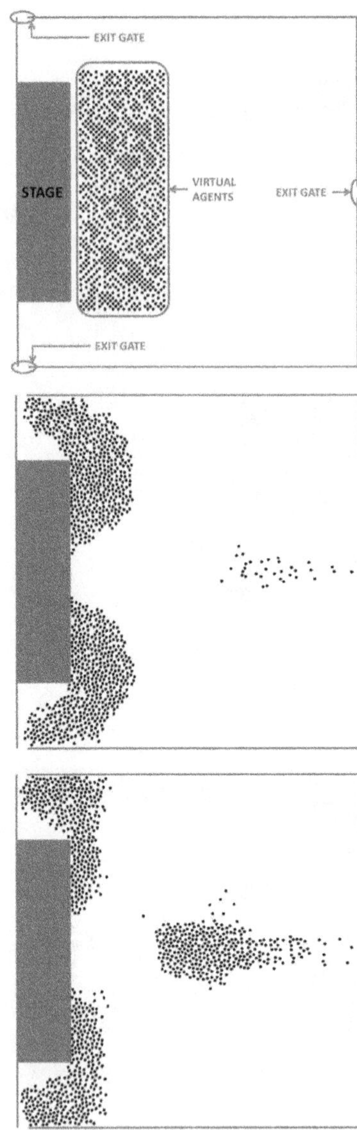

Fig. 3: (a) Top view of the initial simulation environment, which has three exit gates and virtual agents standing right front of the stage. (b) Snapshot taken shortly after the beginning of the first scenario in which agents choose nearest gates. (c) Snapshot taken shortly after the beginning of the second scenario that determines minimum-cost exit gates via path finding.

TABLE I: Average and maximum evacuation times of both scenarios in large hall (in seconds)

Agent #	Nearest Gate		Path Finding	
	Average	Maximum	Average	Maximum
500	122.75	254.57	67.40	121.92
1000	230.08	465.34	95.54	178.01

TABLE II: Average and maximum evacuation times of both scenarios in stage hall (in seconds)

Agent #	Nearest Gate		Path Finding	
	Average	Maximum	Average	Maximum
500	73.23	133.74	68.40	115.68
1000	126.42	250.97	101.03	198.14

To assess the computational overhead, which the path finding progress introduces, we measured average completion time of the simulation steps (excluding the rendering time) during both of the test scenarios containing 1000 virtual agents in stage hall simulation environment. According to the results obtained, the presented evacuation system requires 4.15 % extra time per simulation step to perform 3 reverse path finding operations (one per exit gate) on a grid consist of 32 × 32 cells.

V. Conclusion

In this paper, we proposed a real-time evacuation simulation technique. The main goal of the proposed technique is intelligently planning the paths of the agents during an evacuation scenario with a minimal cost by taking advantage of the limited number of exit gates. The presented technique performs path finding from the exit gates to the grid cells and by reversing the found paths, it not only obtains the paths from cells through the minimum-cost exit gates but also assigns the minimum-cost exit gates to these cells. While determining the minimum-cost paths, we consider not only the distance but also the possible congestion of the exit gates. Agents are navigated towards the minimum-cost exit gates thanks to the RVO, which is a well-known and commonly used microscopic steering and collision avoidance algorithm.

The test results indicate that although we employed test environments which have non-uniform agent distributions, our model successfully optimizes the evacuation time. Using each gate effectively also increases the plausibility of the simulation (more intelligent-looking agents) as can be observed from the Figures 2 and 3.

Another important advantage of our method is that computational overhead of minimum-cost exit gate determination progress is minimized because of performing a path finding operation per limited number of exit gates once in every second. Hence, our technique will achieve similar evacuation time results as nearest exit gate selection approach with a negligible computational overhead when the test environments have uniform agent distributions.

Finally, testing the recommended evacuation system with different number of exit gates and various grid sizes with the aim of estimating the scalability is not covered in this study and is left for the future work.

References

[1] E. Kuligowski and R. Peacock, "A review of building evacuation models," NIST Technical Note 1471, Tech. Rep., July 2005.

[2] Z. Daoliang, Y. Lizhong, and L. Jian, "Exit dynamics of occupant evacuation in an emergency," *Physica A: Statistical Mechanics and its Applications*, vol. 363, no. 2, pp. 501–511, 2006. [Online]. Available: http://www.sciencedirect.com/science/article/pii/S0378437105008447

[3] M. Fukamachi and T. Nagatani, "Sidle effect on pedestrian counter flow," *Physica A: Statistical Mechanics and its Applications*, vol. 377, no. 1, pp. 269–278, 2007. [Online]. Available: http://www.sciencedirect.com/science/article/pii/S0378437106012428

[4] A. Seyfried, B. Steffen, and T. Lippert, "Basics of modelling the pedestrian flow," *Physica A: Statistical Mechanics and its Applications*, vol. 368, no. 1, pp. 232–238, 2006. [Online]. Available: http://www.sciencedirect.com/science/article/pii/S037843710600118X

[5] R. M. Colombo and M. D. Rosini, "Pedestrian flows and nonclassical shocks," *Mathematical Methods in the Applied Sciences*, vol. 28, no. 13, pp. 1553–1567, 2005. [Online]. Available: http://dx.doi.org/10.1002/mma.624

[6] X. Pan, C. Han, K. Dauber, and K. Law, "A multi-agent based framework for the simulation of human and social behaviors during emergency evacuations," *AI & SOCIETY*, vol. 22, no. 2, pp. 113–132, 2007. [Online]. Available: http://dx.doi.org/10.1007/s00146-007-0126-1

[7] H. Ehtamo, S. Helivaara, T. Korhonen, and S. Hostikka, "Game theoretic best-response dynamics for evacuees' exit selection," *Advances in Complex Systems*, vol. 13, no. 01, pp. 113–134, 2010. [Online]. Available: http://

www.worldscientific.com/doi/abs/10.1142/S021952591000244X

[8] R. Guo and H. Huang, "Logit-based exit choice model of evacuation in rooms with internal obstacles and multiple exits," *CHINESE PHYSICS B*, vol. 19, no. 3, MAR 2010.

[9] J. van den Berg, S. Guy, M. Lin, and D. Manocha, "Reciprocal nbody collision avoidance," in *14th International Symposium on Robotics Research*, Sep. 2009.

11. Keep Brushing! Developing Healthy Oral Hygiene Habits in Young Children with an Interactive Toothbrush

Lal Gamze Bozgeyikli, Evren Can Bozgeyikli, Andrew Raij

Abstract - *Cravy Brush* is an interactive toothbrush attachment that helps young children (3–8 years old) develop a daily habit of brushing teeth. The attachment displays an interactive game that provides real-time visual and auditory feedback. In addition to feedback, *Cravy Brush* also provides virtual rewards when children brush properly. Alarm feature calls for children to brush their teeth twice a day at desired times. We collect background data regarding tooth brushing sessions, which can be presented later on to parents. In this paper, design and development of *Cravy Brush* and lessons learned are presented. As a starting point, we conducted a pilot test with one child for a week's duration. Feedback from both parents and the child revealed that *Cravy Brush* made the tooth brushing experience more fun and was effective at encouraging the child to brush her teeth regularly for the dentist-recommended two minutes.

Keywords - Interactive toothbrush; tooth brushing game; children; persuasive technologies; behavior change

I. Introduction

Daily preventative dental hygiene habits, such as brushing teeth, are of great importance for people of all ages. When teeth are not taken care of daily, they can develop cavities and decay [1]. Such decay is often painful, and may require dental procedures, which are also painful and stressful for adults and children. Besides the self-confidence of a healthy smile, bad oral health is supposed to be correlated with some serious diseases such as heart attack, though no scientific evidence of direct effect is found yet.

Good oral health is proportional to brushing teeth at the correct frequency, duration and technique [2]. For effective oral hygiene, people of all ages should brush teeth twice a day for two minutes [3]. Unfortunately, in a study conducted in 2009 [4], the average tooth brushing time of the general population was given as approximately 45 seconds, far less than the recommend two minutes. According to a recent study however, more than half of the children at the age of five do not brush their teeth twice a day and deficits in brushing, such as wrong technique and insufficient duration are observed [5].

To overcome insufficient tooth brushing at childhood, parental supervision while brushing is strongly recommended [6]. In addition, making the process fun by singing a favorite song of children or pretending to play a game is suggested by some front runner oral healthcare brands [7, 8]. Though it can be emotionally satisfying, tracking children's tooth brushing and motivating them by improvised theatricals may incur a burden on parents in the rush of their daily routines. These tooth brushing deficits of children are likely to persist into adulthood [9]. For good dental health that will last for the rest of their lives, children should learn healthy oral hygiene habits as early as possible.

In this paper, we describe the design of *Cravy Brush*, a toothbrush attachment that provides children with an entertaining interactive brushing experience. By making brushing more fun, with the game embedded toothbrush add on, *Cravy Brush* helps young children develop a positive lifelong habit of brushing teeth for the right frequency and duration. Our paper makes the following contributions:

- Design and development of an interactive toothbrush add-on that aims at helping children gain tooth brushing habit.
- Design and implementation of a children's tooth brushing game to make the process more fun for them.
- Parental and children's feedback from the one week pilot test.
- Lessons learned throughout this study, regarding designing interactive game experiences for young children.

II. Related Work

Systems that focus on reinforcing tooth-brushing habits exist both in the literature and industry. We first mention related work from academia and then discuss commercial products.

A. Dental Hygiene Systems from Academia

Chang et al. propose a playful system that encourages children learn correct tooth brushing techniques with the help of a video game [10]. In the game, mirror image of the child's dirty teeth is displayed on a screen and the child tries to clean the virtual teeth by brushing their own teeth. The prototype consists of a cable connected toothbrush equipped with led markers that are tracked by surrounding video cameras to identify tooth brushing motions. Brushing game which incorporates child's brushing strokes is displayed on a monitor in front of the child. The authors' system provided improvement in tooth brushing technique and duration of children according to their user study results. The work provides a complete system with scientific results but its requiring special hardware equipment such as cameras, monitor and LED markers makes the design impractical to use in daily life.

Another intelligent tooth brushing system is proposed by Flagg et al. which caters to elder individuals who may forget brushing their teeth [11]. Aim of the work is to develop a tooth brushing system that will remind and guide the user. Authors developed a vision based tooth brushing system that consists of a bi-color striped toothbrush and web cameras. However, the work suffered from limitations in depth recognition with image processing and needed further improvement to effectively serve as a prompting system.

B. Dental Hygiene Systems from Commercial Entities

There are many commercial products focusing on making the tooth brushing experience more fun and habitual for children. An example is "Tooth Tunes" by Arm & Hammer Inc. which plays a preloaded song for two minutes when the button on the toothbrush is pressed [12]. The product focuses on encouraging the child to brush their teeth for a complete two minutes. Although it is an innovative product, since the music plays for two minutes and then stops, it makes the reward at the process. In the end, the child does not get any reward, which may turn out to be a discouraging factor for the completion of two minutes. Tooth brushing movements of the children are not incorporated into the brushing experience, the song keeps playing even when the brush is kept still after the button has been pressed.

A popular application for children that is recently released is "Disney Magic Timer" by Oral-B [13]. The application can be downloaded to smartphones or tablets. When started, the application counts down for two minutes and reveals a picture gradually as the time passes. The colorful characters seem spot on to encourage the children but the downside is that there is no mechanism to ensure that the child has been brushing their teeth for the reward to appear, just starting the application is enough. Brushing movements of children are not incorporated into the application.

Another commercial product for children that focuses on completing the two minutes brushing duration is "Brush Buddies Talking" by Ashtel Dental [14]. The product is designed in the form of different animals talking for two minutes while the child brushes their teeth. Information about effective oral care is given during these talks. At the end, verbal positive feedback is also given to encourage the children. The downside of the product might be considered as focusing on the auditory feedback only, while omitting the visual. Again, brushing movements are not incorporated into the design, pressing the button activates the two minutes talking session.

Though not interactive, another commercial product example we find worth mentioning is Jordan's toothbrushes that are specially designed for children [15]. These toothbrushes are designed in the form of toddler toys in order to encourage children to use them. Although the design of the brushes seem really attractive, only physical form is used for encouragement.

An interactive commercial product for adults is "Oral-B Professional Care Smart Series 5000 with Smart Guide Electric Toothbrush" [16]. Toothbrush has a separate smart guide that can be put onto sink countertop or attached on the wall. Smart guide provides visual information on how much time has spent brushing each four quadrant of the mouth. After the user completes two minutes, a smiley face appears as an appreciation. Janusz et al. performed a study on effectiveness of the mentioned brush [17]. Adults using the smart brush brushed their teeth more thoroughly according to their user study results. As another improvement, users brushed four quadrants of their mouths more evenly in duration. The product uses high technology to sense the brushing movements and give interactive feedback as an upside. But since the display is monochromatic as the product is designed for adults, it may not be encouraging for children who are used to more colorful displays.

All of these mentioned examples have their own merits and contributions. Our work differs in providing an interactive digital game for children that gives real time visual and auditory feedback with a practical invisible product design that does not require any environmental installments. Children can use *Cravy Brush* with their favorite toothbrushes since it is designed as an attachable add-on for any toothbrush. In the video game, there are bacteria that need to be destroyed which provides interactivity and engagement, while making sure that the child completes two minutes brushing time. Child's brushing strokes are recognized with their 3D orientation. Brushing of upper and lower teeth are incorporated into the game in the form of bacteria destroyed. If the child stops brushing, the system recognizes this and responds accordingly to encourage children to keep brushing. We employ a rewarding mechanism to encourage children come back and brush their teeth regularly. Surprise rewards are also employed to increase motivation. *Cravy Brush* can be programmed to call for the child to brush their teeth at desired times of the day, serving as a reminder. Finally, the system captures tooth brushing data of the children which gives parents the opportunity of reflection and appropriate behavior adjustment afterwards.

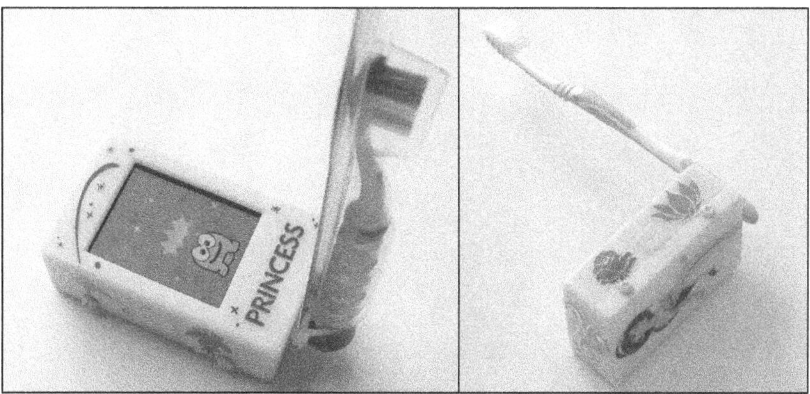

Figure 1. *Cravy Brush* prototype attached to two different children's toothbrushes. Left: Front view displaying the interactive game. Right: Back view of the case.

III. Cravy Brush

Cravy Brush is a small attachment for toothbrushes that features an interactive children's game on a colored LCD screen and gives auditory feedback via loudspeakers. Our target audience are children of age between three and eight. Currently, high fidelity proof of concept prototype of *Cravy Brush* is built up that consists of a mini smartphone covered with waterproof material and put inside a specially designed 3D printed case. This prototype can be attached to any toothbrush with a flex cuff. A general view of *Cravy Brush* prototype that is attached to two different children's toothbrushes can be seen in Figure 1.

A. Prototype Design and Implementation

While designing *Cravy Brush's* high fidelity proof of concept prototype, emphasis is given on invisibility, compatibility and compactness. Making the prototype invisible means that it does not require any environmental installment and can be used practically without too much effort. The reason behind this is not to repel the user. Compatibility means that the user can use the product with any existing toothbrush. This is important because we do not want to take away any comfort from the children. If they have a favorite toothbrush, they can still keep on using it. Compactness means that the add-on is as small as possible while containing all necessary elements of visual and auditory feedback. The reason behind using a mini smartphone for the prototype is its compactness while

Figure 2. 3D printed case of *Cravy Brush* prototype to hold the mini mobile phone inside. Left: Charging hole cover. Middle: Bottom part of the case. Right: Top part of the case.

providing many technical capabilities such as LCD screen, three axis accelerometer and loudspeakers. Current dimensions of the attachment prototype are 5.7x9.5x3.1 cm. For the final product, a specially designed set of mini display, loudspeakers and embedded batteries will be used. The end product is planned to make a smaller attachment of approximate dimensions of 5.5x3.5x2.0 cm.

For the prototype, a case to hold the mini smartphone was designed and produced with a 3D printer. While designing the case, it's being as small and stable as possible was of great importance. We also tried to keep the design aesthetic and simple, following the guidelines in the work of Consolvo et al. [18]. The case consists of three parts as can be seen in Figure 2. Two parts hold the phone and the third part is to enable charging of the phone inside. At the bottom of the case, there are bridge holes for flex cuff to pass. This way, *Cravy Brush* can be attached to a toothbrush tightly. In order to prevent turning of the toothbrush in its own axis inside the flex cuff, a saw toothed offset is added to the middle of these bridges.

After 3D printing, the model is decorated with stickers to give it an attractive look for children. Before the phone is put into the case, it is wrapped with waterproof plastic as a precaution layer. After several trials, the angle of attachment that enabled the child to best see *Cravy* while brushing turned out to be 90 degrees upward at the back of the brush facing inside, as can be seen in Figure 1.

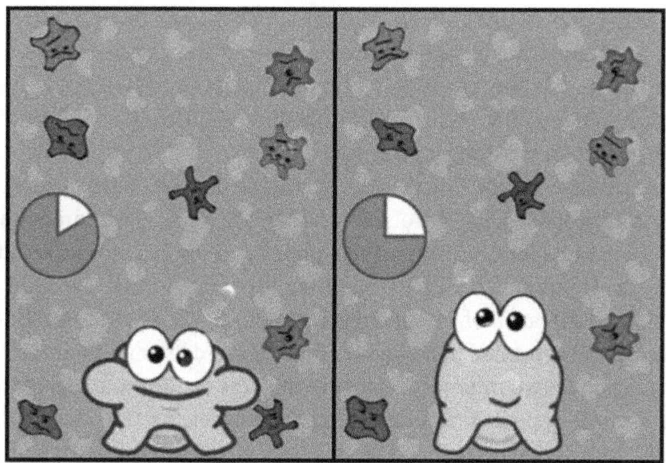

Figure 3. Screenshots of the game showing *Cravy* chewing, bacteria around, clock and foam particles. Left: 20 seconds have passed since beginning of the brushing session. Right: 30 seconds have passed since beginning of the brushing session.

B. Game Design and Implementation

In order to help children develop regular tooth brushing behavior for the right duration, a children's game is designed that features a cute monster called *Cravy*. *Cravy* eats bacteria. In order to make *Cravy* eat bacteria, the child has to brush their teeth. Twice a day, *Cravy* craves for bacteria and calls for the child to help him eat them by brushing their teeth. Two general screenshots of the game which shows *Cravy*, bacteria, timer and foam particles are presented in Figure 3. The child has to feed *Cravy* two times a day. The idea is to create a bond between the child and the character besides providing them a game, so that they would keep coming back to feed it.

There are 10 bacteria at each game session, five at the bottom and five at the top. For every 10 seconds of upper and lower teeth brushing, a bacteria at the corresponding side is eaten by *Cravy*. The reason behind that is to make the child brush their upper and lower teeth evenly in duration, one minute each. For good dental health, it is suggested to brush upper and lower teeth for one minute [19]. When the brush is turned upside down, *Cravy* also falls down to the other side to give the child a sense of rotation and engagement. It is known that as the user's actions affect the gameplay, they are more engaged into games [20]. Hence, we tried to make the children's movements affect the gameplay while avoiding overwhelming them with too many variables. User's tooth brushing strokes with their 3D orientation are recognized using the data coming from the three axes accelerometer inside the smart phone. If the maximum difference between the acceleration values of any axis exceeds a threshold value over a predefined period of one second, then brushing is recognized.

The enemy bacteria characters in the game can also be seen as a form of data abstraction as proposed by Consolvo et al. [18]. They tell the children how much time they need to spend brushing their upper and lower teeth in an attractive way. 10 seconds buffer is added for brushing each part so that the child will get enough time to defeat all of the enemies without getting stressed out. But on the other hand, we also wanted to make sure that the child brushes their teeth for the proper duration of two minutes. To ensure that, the resulting screen of the game showing rewards appears only after the child brushes for 120 seconds. So even if the child destroys all of the bacteria in a perfect 100 seconds, they will have to brush until it sums up to 120 seconds to be able to see the resulting screen. There is a child friendly visual timer on the screen showing how much time remains as a form of feedback, shown in Figure 3.

C. Feedback and Rewards

In the resulting screen, stars are presented to provide positive feedback based on the child's tooth brushing performance. Three stars are given for defeating all of the bacteria and brushing for a complete two minutes. Two stars are given for brushing for two minutes but having any number of bacteria remaining on the screen. Finally, one star is given for not completing two minutes brushing time. Rewarding system is designed to encourage children at first place. So even if they fail to complete the proper brushing time, no punishment is used, instead a star is given to encourage them. The idea behind this approach is to motivate them to keep trying by using a positive reinforcement as suggested in [18]. Other than the stars, there is a badge visual that appears only if the children have brushed their teeth in the previous 12 hours period. The reason for this is to make the child keep brushing regularly in order to see the badge on the screen and feel a sense of accomplishment. Our final type of reward is rare items that are in the form of attractive visuals for children such as princess crown, balloon and cake slice. Rare items appear with a 20 % probability if the child got three stars and the badge at the current brushing session, meaning that they had brushed their teeth in the previous 12 hours period and they had brushed their teeth for two minutes in the current session, giving upper and lower teeth one minute each approximately. It is shown that unexpected rewards provide more motivation

Figure 4. Screenshot of the game showing the resulting screen. Left: Rare item of cake slice. Middle: Three stars meaning that the child destroyed all of the bacteria and brushed for two minutes. Right: Badge meaning that the child brushed their teeth in the previous 12 hours period.

in individuals as compared to expected ones [21, 22]. Hence, these rare items are designed to serve as a variable to surprise the children. All of the mentioned reward types can be seen in Figure 4. After the brushing is completed, *Cravy* enters into score state until the next brushing session. While in the score state, *Cravy* celebrates by jumping continuously with accompanying sparkle particles. If there are remaining bacteria, they also remain on the screen as a form of feedback. Stars, badge and rare items are kept at the center of the screen.

The reason behind this long score state is to give the child opportunity of reflection on their performance. *Cravy* still responds to children movements in the score state. If the child turns the brush upside down, *Cravy* also jumps onto the other side of the screen, continuing jumping there.

To give a clear understanding of different game states, a flowchart which summarizes each possible state of the game and their relations to each other is presented in Figure 5. In the idle state, the game is ready for playing. Two

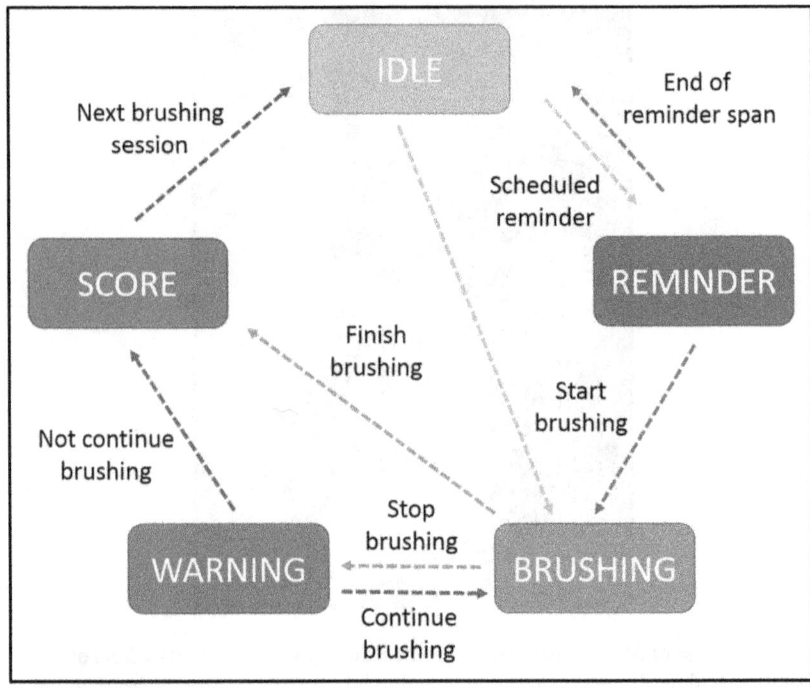

Figure 5. Flowchart presenting each possible game state and their relations to each other.

times a day at the predetermined times, *Cravy Brush* enters into the reminder state in which *Cravy* calls for the child to brush their teeth. When the child starts brushing, it enters into the brushing state and the interactive game of feeding *Cravy* with destroyed bacteria begins. If the child stops brushing during the game, warning state gets activated. In the warning state, background music is stopped, *Cravy* stops chewing and gives verbal prompt to the child to keep brushing. Then, if the child continues brushing, brushing state gets reactivated until two minutes time is completed. If the child does not continue brushing after stopped once or completes two minutes brushing time successfully, score state gets activated. Score state features rewards and *Cravy* celebrating the previous session. *Cravy Brush* remains in the score state for a programmable duration that is currently set to 12 hours. After that, the circle is completed by the activation of idle state, meaning that *Cravy* is ready for another brushing session.

D. Audio Effects and Voices

In the game, sound effects play when bacteria gets destroyed and during the celebration at the end. If the child stops brushing, *Cravy* gives verbal feedback to make them continue brushing ("Keep Brushing!"). Cheerful and upbeat background music is played as long as the child keeps brushing, to make the brushing process more fun. If the child stops brushing, the music stops. The stopped music communicates that the child should start brushing again.

Cravy Brush has an alarm feature to remind the child to brush their teeth at the programmed times of the day. As Fogg states in his behavior model for persuasive design [23], individuals may need to be triggered to perform a desired action. Timing of the trigger is also important for activities that need to take place at specified times of the day such as tooth brushing. When in the alarm mode, *Cravy* calls out for the child to brush their teeth. Instead of using an alarm sound solely, *Cravy* verbally asks the child for help. The idea behind that is to create a more attractive form of trigger for children.

E. Data Collection

Another important feature of *Cravy Brush* prototype is automated background data collection. Currently, we are collecting data in raw format which can be presented to the parents after being edited into an easily readable form. This can enable the parents to see when and for how long their children brushed their teeth. Unintentionally, people have a tendency to significantly overestimate the time they brush their teeth as a recent study by Terezhalmy et al. reveals [24]. Data collection is a helpful way of being aware of actual times spent on activities as proven by various studies [25, 26]. With automated data collection, parents can help their children make necessary adjustments on their tooth brushing behaviors. This reduces unintentional overestimation errors. For these reasons, in the end product, an automated data collection user interface is planned to be implemented.

IV. User Evaluation

A female, four year old child used the *Cravy Brush* prototype for a week's duration. Before conducting the user study, we decorated the casing with stickers of the child's specific interests to increase the prototype's attractiveness. Her parents used to brush her teeth every day before the user study, hence she had no prior self tooth brushing experience. For the testing period, she brushed her teeth with *Cravy Brush* every day. Her parents video recorded the brushing sessions and provided us with them. At the end of the study, interviews were conducted with parents and the child which involved answering some qualitative questions and providing feedback on the brushing experience.

A. Parental Feedback

Parents of the testing child were pleased with *Cravy Brush* overall. They said that their child was excited about using *Cravy Brush* during the whole testing duration. They also stated that they would want their child to use such a product every day. Her parents said that after trying *Cravy Brush* she started to brush her teeth on her own, which helped her to become a bit more independent. They also said that this achievement made the child excited and happy.

Although it was not designed as a training product, *Cravy Brush* helped the child learn how to brush her teeth on her own, for the first time, without getting any parental help. The parents stated that after stopping using the *Cravy Brush* prototype, the child still brushed her teeth by herself without any help.

Another feedback of the parents was that the prototype's being a bit heavy for little children. Although a mini cell phone of 120 grams weight is used for the prototype, it still challenged the child to hold the brush comfortably with one hand sometimes. Another negative feedback was regarding the placement of the display. Parents informed us that the child held the brush in a weird way while trying to see what was happening on the screen. This may degrade the brushing quality which is antipodal to the whole idea behind the study.

A positive feedback was on how responsive the child was to the *Cravy* character. Every time *Cravy* talked, she consistently responded, talking to him as if he was a real character, as suggested by Isbister and Nass [27]. Regarding the rewarding system, parents said that the child was excited and motivated about getting three stars. At her first few trials, she did not manage to get three stars, which disappointed her a bit. But then, she consistently started to get three stars, which made her happy. Other than the three stars, she also looked for the badge icon on the screen after brushing, getting happy as she saw it. She was also disturbed by the remaining germs, if any, asking why they were still there after the brushing was complete. We consider these as positive indicators of the successful aspects of our design.

Other than using it as a toothbrush, the child continued playing with *Cravy Brush* after brushing her teeth. She played with *Cravy* as if it were another doll in her collection. *Cravy* "talked" with other dolls, "walked" up and down the dollhouse stairs and done many more imaginative things.

At the first few days of user study when the child was not so comfortable with brushing her own teeth, she picked up *Cravy* without parents' prompting and shook it to hear the in game music and sounds. After getting comfortable with brushing her teeth on her own, one day while sitting in the living room the child grabbed *Cravy Brush* and started brushing her teeth without any toothpaste. These use cases show us that the child perceives tooth brushing experience with *Cravy* as a game rather than a necessary daily routine.

Parents stated that she liked being taped while brushing, asking them to take videos of her while brushing her teeth or doing other activities during the day. She associated being video recorded with the act of tooth brushing using *Cravy Brush*. This feedback gave us the idea of turning the product into an augmented reality experience which shows the child and interactive characters together on the display or as a projection, as a future possible improvement.

Parents' final feedback was about the child's general bed time routine. They said that before the user study, the child had a tendency to drag out the bedtime process consisting of her hygienic care and story reading, to stay up longer and play. But they happily said that this did not happen after using the *Cravy Brush* prototype since the child was too excited to use it at nights. They stated that it helped them to make the bed time a little bit easier. This feedback gave us the idea of expanding *Cravy Brush* as an entire bed time routine companion to help parents in their child's bed time process, as a future development.

B. Child's Feedback

We requested the parents to ask four simple questions to their child for us, after completing the user study. They asked these questions to the child and video recorded her answers. In addition to giving each question explanatory answers, the child provided additional feedback also. First question was *"What did you like about Cravy?"* The child's answer was *"How he lets me know about all of the stuff when I am done and then I know when I am done."* The second question was *"Would you like to use Cravy every day?"* The child answered *"Yeah! Cravy is cool."* The third question was *"Do you like brushing your teeth with Cravy?"* and the child quickly answered *"Yeah."* The fourth and final question was *"Is there something you don't like about Cravy?"* The child's answer was *"Um, no. There isn't anything I don't like."* As additional feedback the child also said that she liked the colors of *Cravy* and the stickers on it. As the parent asked her if she would like *Cravy* as much without the stickers on it, she said that she likes it with the stickers on it.

V. Lessons Learned

In this section, we would like to share lessons we have learned throughout this study from its beginning up until now, considering that they might help other researchers in their studies regarding child friendly interactive designs involving games.

First of all, we learned that when designing games for children, their testing sessions can give developers ideas they have never thought of before. So, starting the testing early in the process might help more. In our case the most obvious example we never thought of before was the child's playing with the prototype as a normal toy outside of the tooth brushing sessions.

Turning the prototype into something attractive for the child using different colors and decorative aids like stickers helped a lot in our case. The child embraced the prototype quickly although it was too plain in its outer shape form as compared to other commercial children's products. Colorful particles displayed in the game also attracted the child's attention.

During the development phase, we have adjusted the parameters by self-trials, pretending to brush like a child. But when the prototype is given to the child, it remained ineffective since it did not recognize her tiny brush strokes. The child was brushing her teeth with really little strokes and at a small pace. So, the prototype was giving too many prompts, since it was not recognizing any change in movement above the specified threshold. It was a discouraging factor for the child getting prompts to keep brushing while she was already brushing. We quickly made an additional calibration according to our observations from the videos and in this calibration, parameter values were given considerably too small as compared to our previous calibrations. The recalibration solved false prompt giving problem and we learnt that when designing games for little children, pretending to play like them as adults remains ineffective. Making them try the prototype as early as possible and adjusting the parameters accordingly can be a more clever way which would save time.

We saw that an interactive companion encourages the child to fulfill her daily tooth brushing duty in a fun way. The game helped spending two minutes time without getting bored. The child was motivated to destroy all the bacteria and see the resulting screen, which ensured that she brushed for two minutes. We observed that providing the child interactive small missions of destroying bacteria one at a time kept her busy while brushing and resulted in more proper duration of time spent. We believe that such interactive companions featuring games can be effectively used to reinforce daily regular health routines that are not normally so fun to fulfill. The child-avatar bond can extend outside the health applications into other playful interactions that are not related to the health behavior of interest. Children treat the avatar as another toy in their collection. This also implies that a child's existing bonds with toys, dolls, and action figures could be leveraged in health apps and games.

We designed *Cravy Brush* software to be always on. It allows the child to brush their teeth two times a day. After the brushing sessions, it remains in the score state which features *Cravy* celebrating. This came with a more frequently charging burden for the parents but also resulted in the child's playing with the prototype throughout the day. So, we think that while designing products for children, keeping them accessible any time they want to play with, even if at a more restricted state, is important to make the bond between the child and the product stronger.

While designing it, we wanted the *Cravy Brush* to be attached to the tooth brush instead of standing still somewhere else. The reason behind that was to encourage the child to embrace the tooth brush, associating it with playing a video game. However, this resulted in some difficulties during brushing since the child was curious about what was going on in the screen and holding the brush in an orientation that gave her a proper viewing angle. This is a potential degrading factor for tooth brushing quality, which is not desirable. But as an advantage on the other hand, it resulted in the child's playing with the toothbrush on which *Cravy Brush* was attached as if it was one of her other toys.

As another lesson learned, making the character talk was more important than we gave credit for during the design. The child tirelessly gave answers to the character when *Cravy* prompted her to keep brushing, even though it was not a question at all. We think that while designing products for little children, making characters talk can considerably increase the child's immersion and make the experience more interactive and fun for them. Young children can form powerful bonds with interactive avatars that communicate verbally and respond to a child's real-world behavior (in our case, the motion of a toothbrush.) The formation of this child-avatar bond is likely critical to the development of new health habits in young children. Health app and game developers targeting this population should consider integrating avatars with these characteristics.

As a final insight, young children have little experience with health behaviors, and thus doing these behaviors correctly (in some cases, for the first time) requires the child to focus entirely on performing the behavior. An interactive tool like *Cravy Brush* may in fact distract the child and detract from performing the behavior correctly. Careful thought is needed to ensure the design of health behavior apps and games for young children manage their attention properly.

VI. Conclusion and Future Work

In this paper, design and development of *Cravy Brush*, an interactive tooth brush attachment for children, is presented. The aim of the study is to help children gain tooth brushing habit for the right frequency and duration by making the tooth brushing experience more fun for them. Results of the user study conducted with a four years old child revealed that *Cravy Brush* was effective in these targeted aims. Feedback provided at the end of the user study indicated that there are many features of the prototype that need to be improved, which gave us clear directions for future work.

As future improvements, we are planning to work on a lighter prototype which may be attached to somewhere other than the toothbrush. Instead of solely focusing on tooth brushing, we want the improved version to cater to many more bedtime routines such as taking a bath, getting dressed, flossing, brushing teeth, brushing hair, reading a story, using the potty, getting and staying in bed. This way, we believe that it will serve as a helpful small assistant for parents, which leads the children through the entire bedtime routine. In the future versions, it's being designed as a mobile playable toy is important to enable the child play with it throughout the day as one of their other normal toys. Making the in game animations more dynamic and adding more talking to characters would help making the experience more fun and interactive for children. A parental user interface for data reviewing is also planned to be implemented, considering several advantages of reflection on the collected self-data. We are also considering to build up an augmented reality version, featuring the child and the virtual characters overlaid.

Until now, we have tested the prototype with one child for a long duration. For an effective and generalizable user study, number of participants should be increased. So in the future, we are planning to test the prototype with more children to discover even more ways to improve and suit *Cravy Brush* to the little children's needs.

References

[1] Bellini, H.T., Arneberg, P. and von der Fehr, F.R. Oral hygiene and caries. A review. *Acta Odontol Scand. 1981*, 39(5):257–65.

[2] Choo, A., Delac, D.M. and Messer, L.B. Oral hygiene measures and promotion: Review and considerations. *Australian Dental Journal 2001*, 46(3):166–173.

[3] Attin, T. and Hornecker, E. Tooth brushing and oral health: how frequently and when should tooth brushing be performed? *Oral Health Prev Dent. 2005*, 3(3):135–40.

[4] Creeth, J.E., Gallagher, A., Sowinski, J., Bowman, J., Barrett, K., Lowe, S., Patel, K. and Bosma, M.L. The effect of brushing time and dentifrice on dental plaque removal in vivo. *J Dent Hyg. 2009*, 83(3):111–6.

[5] Franzman, M.R., Levy, S.M., Warren, J.J. and Broffitt, B. Tooth-brushing and dentifrice use among children ages 6 to 60 months. *Pediatric Dentistry 2004*, 26(1):87–92.

[6] Dean, J.A. and Hughes, C.V. Mechanical and chemotherapeutic home oral hygiene. *Dentistry for the child and adolescent (8th ed.)*. Mosby, St. Louis, MI, USA, 2004.

[7] Maintaining Your Kids Oral Health at Every Stage: Tips and Tools from Dr. Salzer by Oral-B. http://www.oralb. com/topics/maintaining-your-childs-oral-health-at-every-stage.aspx.

[8] Colgate Oral and Dental Health Resource Center. Oral Health for Children. http://www.colgate.com/app/CP/US/EN/OC/Information/Articles/Oral-and-Dental-Health-at-Any-Age/Infants-and-Children/Toddler-Child-Transitional-Care/article/Oral-Health-for-Children.cvsp.

[9] Becker, G.S. Habits, Addictions, and Traditions. *Kyklos 1992*, 45(3):327–45.

[10] Chang, Y.C., Lo, J.L., Huang, C.J., Hsu, N.Y., Chu, H.H., Wang, H.Y., Chi, P.Y. and Hsieh, Y.L. Playful toothbrush: ubicomp technology for teaching tooth brushing to kindergarten children. In *Proceedings of the SIGCHI Conference on Human Factors in Computing Systems (CHI '08)*, ACM Press (2008), 363–372.

[11] Flagg, A., Boger, J. and Mihailidis, A. AN INTELLIGENT TOOTHBRUSH: Machines for Smart Brushing. *RESNA/ICTA 2011: Advancing Rehabilitation Technologies for an Aging Society 2011*, Toronto, Canada.

[12] ARM & HAMMER Tooth Tunes Musical Toothbrushes. http://www.spinbrush.com/toothtunes.

[13] Oral-B. Disney Timer App. http://www.oralb.com/stages/disney-timer-app.

[14] Brush Buddies Talking. http://ashteldental.com/oralcare/ distributor/brush-buddies-talking-category.html.

[15] Award for Design Excellence - Jordan. http://www. jordan.no/en/dental-care/News/Award-for-Design-Excellence.

[16] Oral-B ProfessionalCare SmartSeries 5000 with SmartGuide Electric Toothbrush. http://www.oralb.com/products/professional-care-smart-series-5000/.

[17] Janusz, K., Nelson, B., Bartizek, R.D., Walters, P.A. and Biesbrock, A. Impact of a Novel Power Toothbrush with SmartGuide Technology on Brushing Pressure and Thoroughness. *The Journal of Contemporary Dental Practice 2008*, 9(7):1–8.

[18] Consolvo, S., McDonald, D.W. and Landay, J.A. Theory-driven design strategies for technologies that support behavior change in everyday life. In *Proceedings of the SIGCHI Conference on Human Factors in Computing Systems (CHI '09)*, ACM Press (2009), 405–414.

[19] Crest - How to Brush Your Teeth. http://www.crest.com/dental-hygiene-topics/how-to-brush-your-teeth.aspx.

[20] Lok, B., Naik, S., Whitton, M. and Brooks, F.Jr. Effects of handling real objects and self-avatar fidelity on cognitive task performance and sense of presence in virtual environments. *Presence: Teleoper. Virtual Environ.* 12, 6 (2003), 615–628.

[21] Steinberg, E.E., Keiflin, R., Boivin, J.R., Witten, I.B., Deisseroth, K. and Janak P.H. A causal link between prediction errors, dopamine neurons and learning. *Nature Neuroscience 2013*, 16, 966–973.

[22] Tucker-Ladd, C. Psychological Self-Help. *Methods for Changing Behaviors.* http://www.psychologicalselfhelp.org/.

[23] Fogg, B.J. A behavior model for persuasive design. In *Proceedings of the 4th International Conference on Persuasive Technology (Persuasive '09).* ACM Press (2009), 1–7.

[24] Terezhalmy, G.T., Bsoul, S.A., Bartizek, R.D. and Biesbrock, A.R. Plaque removal efficacy of a prototype manual toothbrush versus an ADA Reference Manual toothbrush with and without dental floss. *J Contemp Dent Pract. 2005*, 6(3):1–13.

[25] Wolf, G. The Data-Driven Life. http://www.nytimes.com/2010/05/02/magazine/02self-measurement-t.html?pagewanted=all&_r=0.

[26] Li, I., Dey, A. and Forlizzi, J. A stage-based model of personal informatics systems. In *Proceedings of the SIGCHI Conference on Human Factors in Computing Systems (CHI '10)*, ACM Press (2010), 557–566.

[27] Isbister, K. and Nass, C. Consistency of personality in interactive characters: verbal cues, non-verbal cues, and user characteristics. *Int. J. Hum.-Comput. Stud. 2000*, 53, 2, 251–267.

12. Enhancing Gamepad FPS Controls with Tilt-Driven Sensitivity Adjustment

Gazihan Alankuş
Department of Computer Engineering Izmir University of Economics Izmir, Turkey gazihan.alankus@ieu.edu.tr

Alp Arslan Eren
Department of Computer Engineering Izmir University of Economics Izmir, Turkey alp.eren@ieu.edu.tr

Abstract— Playing a first person shooter (FPS) game using a gamepad is inherently more difficult than playing it with a mouse. In this study, we propose a novel input method for gamepads to improve the FPS aim task. We take a user-centric approach to identify unused secondary motions to be indicative of players' intents. We consider the gamepad tilt as a secondary motion that users naturally perform while playing with the gamepad, and interpret it as the intent of the user to move the FPS aim faster. We use a tilt sensor to detect the amount of tilt, and use it to increase the sensitivity of the gamepad analog joystick when the gamepad is tilted. We compare this to two other input methods: a pure gamepad and a tilt-aim scheme. Our experiments show that our approach is statistically better than the tilt-aim approach, and is comparable with the pure gamepad approach.

Keywords—video games gamepad; first person shooter; tilt input; sensitivity adjustment; human-computer interaction

I. INTRODUCTION

First person shooter (FPS) games are one of the most popular genre of games with a big market and an increasing number of titles released every year [1]. FPS games typically provide a fast-paced gameplay experience, in which the user interacts with the game world from the point of view of the avatar. FPS games are popular both in PCs and in game consoles; however, console gamers are typically disadvantaged in FPS games because the gamepad is not an ideal input device for FPS games.

Conceptually, the inputs that an FPS game requires typically consist of a 2D vector for walk and strafe, a 2D position in polar coordinates for aiming at the objects in the environment, and various triggers for interacting with the aimed object (use, take, shoot, etc.) Acquiring targets by aiming quickly at the correct object is very important in fast-paced FPS games. Therefore, the input device used for the aim should be appropriate for providing the desired 2D position in polar coordinates. In PCs, the mouse is used for aim and its 2D position on the mouse pad provides a good approximation for the 2D position in polar coordinates—the continuous on-screen visuals provide sufficient feedback for the player to determine and update the desired target position for the mouse. This enables players to aim at targets quickly and correctly in PCs.

Gamepads are the standard controllers for popular game consoles, most notably Xbox One and PlayStation 4. The two analog joysticks of gamepads enable users to input continuous 2D vectors within a 2D unit disc. To aim in FPS games, players control the 2D position in polar coordinates using the right analog joystick. However, unlike the mouse, the analog joystick does not let the user directly input a 2D position. Rather, the user inputs a 2D velocity in time, and uses it to arrive at the desired 2D position. Compared to the mouse, this is an indirect way of controlling the aim. Rather than directly providing the desired position, the user has to provide a series of velocities that will cause the aim to arrive at the desired position.

This indirect input for aim puts console FPS players at a disadvantage. As a result, game companies typically do not mix PC and console FPS gamers in online games [2] and add features such as aim-assist (auto-aim) to help console players. We believe that this inequality between PC and console gamers provides Human-Computer Interaction researchers with an opportunity to improve the FPS gaming experience in consoles and help many console players enjoy FPS games more.

Some modern gamepads have integrated inertial controllers that can track the orientation of the gamepad in space, e.g., Sixaxis and DualShock 4 controllers from Sony contain three-axis accelerometers and gyroscopes. However, using the gamepad orientation to control FPS aim directly has not created successful interactions. Therefore, FPS games use these sensors for simpler tasks such as shake detection and leaning the character around walls.

In this study, we propose a novel input method to make use of inertial sensors of gamepads to enhance the FPS aiming experience. We take a user-centric approach and make use of secondary motions that users automatically perform while playing FPS games with gamepads. Our input method increases the sensitivity of the analog joystick in response to the tilt angle of the gamepad. We compare this approach with the state of the art in a within-subjects study and show that it is better than using the gamepad orientation for controlling aim directly. Our results also suggest that this method may enhance aiming with the gamepad.

II. Related Work

A. Gamepads for Aiming

Researchers previously reported issues related to position input with the analog joysticks of gamepads. Natapov et. al. [3] studied the use of the gamepad for point-select tasks. They compared the gamepad to mouse and Wii remote IR pointing, and showed that "analogue stick targeting is not a good method for performing point-select tasks". They later replaced the right analog joystick with a trackball to enable direct position control [4], and showed that this resulted in better user performance in games [5].

Several researchers have studied the use of the gamepad in FPS games, and have concluded that the direct position input of the mouse is better than the gamepad in the essential FPS task of target acquisition [6]–[10]. Nevertheless, players are still using the gamepad in consoles to play FPS games. This may be because it provides a noise-free and lag-free way to play games, unlike some other novel sensor technologies. Ardito et al. showed that for positioning tasks, gamepads are preferable to tilt-based accelerometer inputs [10]. Similarly, Zaranek et al. showed that for FPS target acquisition, gamepads are preferable to the Kinect and the Move input devices [11] that are both susceptible to noise and lag since they rely on computer vision. Therefore, gamepad is still preferred in consoles for FPS games, and improving the gamepad experience in FPS games can enhance the experiences of many console gamers.

B. Tilt Inputs in Games

Low-cost three axis MEMS accelerometers and gyroscopes are used for detecting tilt as input in devices such as mobile phones, tablets and the Wii remote controller. A number of modern gamepads also contain tilt sensors (e.g., Sixaxis and DualShock 4). Therefore, techniques that enhance the gamepad experience with tilt sensors can be practically used with gamepad controllers that already contain these sensors.

Tilt input has been used extensively in mobile games. Multiple researchers have shown that in simple mobile games, tilt controls competes with touch controls in terms of game success and user preferences [12]–[15]. However, a fundamental investigation of Fitts' law for tilt controls show that tilt is not a precise method of input [16] and therefore may not be suitable for games that require precision. Further research shows that using tilt as velocity input (similar to gamepad analog joystick) is less precise than using tilt for position input in which the orientation is mapped directly to position [17].

C. Tilt Inputs in FPS Games

The research on tilt inputs in FPS games and similar VR navigation tasks is limited. For the fundamental aiming task of FPS, tilt inputs were shown to be better than discrete button inputs [18], but worse than the gamepad analog joystick as well as the mouse [10]. These findings indicate that tilt is not an appropriate input for aiming. This is also reflected in the games industry. Tilt was found to be "not sensitive enough" by developers of Resident Evil 5 [19] and accelerometers are not integrated in Xbox 360 controllers because of similar reasons [20].

Some mobile FPS games have implemented aiming by moving the device (Cube for iOS, MosKillTo for iOS, Call of Duty Strike Team for iOS), which are likely to suffer from the shortcomings of tilt-based aiming. A notable example is Uncharted: Golden Abyss for PlayStation Vita [21], in which the user fine-tunes the aim by moving and tilting the device after drawing the weapon. The larger motions of the aim and the navigation of the avatar are performed using the analog joystick and without the device motion. Isolating the tilt inputs to be used only for precise aiming and not for general aiming appeared to help improve tilt-based interaction for this game.

Many mainstream FPS games for the PlayStation Sixaxis controller have opted not to use tilt input for aiming. They preferred to use the motion inputs for secondary actions such as leaning near walls (Battlefield 4), having to keep the gamepad still while aiming with the sniper rifle (Killzone 2), shaking the gamepad to change attire, looking around under a dumpster lid, and shaking the enemy with a magical doll (Metal Gear Solid 4).

There is a notable exception that is also related to our approach of using tilt for sensitivity adjustment. Killer Freaks from Outer Space (which was later released as ZombiU) makes use of the combination of high-precision accelerometer, gyroscope and magnetometer in WiiU to use the controller's orientation for FPS aiming [22]. The game detects the speed of the user's motions to adjust the sensitivity of aim so that slower motions would be used for precise aiming and faster motions would be used for larger motions of the aim. This is similar to the well-known approach of mouse pointer acceleration. While it relates to our approach in that it also adjusts FPS aim sensitivity according to user input, it is a simple adjustment using the main mode of motion input (i.e., adjusting tilt sensitivity using tilt speed). Ours is different in that we use the user's otherwise unused secondary motions to improve the non-motion based input method of aiming with the gamepad analog joystick.

III. Approach

We take a user-centric approach to find an effective way of using tilt sensors to enhance FPS aiming with the gamepad. We use the secondary motions that users already perform while playing with gamepads, and adjust the aim sensitivity accordingly.

Most input devices require a specific set of motions to play a game (e.g., moving the mouse with the hand, moving the analog joystick of a gamepad with the thumbs, etc.). When users are engaged deeply with a game, they sometimes tend to perform other motions that are not required and are ignored by the input device (e.g., moving the head while dodging attacks in an FPS, tilting the gamepad when failing to steer a car during a sharp turn, etc.). While a behavioral study is necessary to better understand these commonly-observed secondary motions, we believe that they contain useful information that reflect the intentions of the user. Specifically, we assume that users tilt the gamepad when they want the avatar to move faster than the maximum velocity possible with the analog joystick. We believe that these unused motions can be used to enhance the game experience.

In this study, we make use of these otherwise ignored secondary motions to better reflect the user's intentions in the game. For example, the intentions of the user may be different in the following two behaviors: (1) moving the analog joystick all the way to the right without moving the gamepad itself, (2) moving the analog joystick all the way to the right and also tilting the gamepad to the right. We assume that in case (2), the user intends to move the avatar faster compared to case (1). Therefore, we use the amount of gamepad tilt to make the avatar really move faster in case (2). By using the amount of tilt to adjust the magnitude of the velocity input, we aim to better reflect the user's intentions in the game. As the user tilts the gamepad more, we effectively increase the sensitivity of the analog joystick input and move the crosshair with higher velocities, allowing the user to aim quickly to targets that are away from the crosshair. Holding the gamepad in a level pose reduces the sensitivity of the analog joystick and allows the user to aim precisely on targets that are close to the crosshair. This way, we aim to reflect the user's intentions more closely in the game.

We implement this sensitivity adjustment for the analog joystick in a simple way. We track the sensitivity separately for X and Y axes. We use the roll angle of the gamepad to adjust the sensitivity for the X axis of the analog joystick and used the pitch angle similarly for the Y axis. We used the equations below to compute the velocity input by the analog joystick.

$$v_x = \left(s_x^0 + s_x^1 \frac{|\psi|}{\pi} \right) v_x^0$$

$$v_y = \left(s_y^0 + s_y^1 \frac{|\phi|}{\pi} \right) v_y^0$$

In the equations above, v_x^0 and v_y^0 are the components of the raw velocity input from the analog joystick, ψ is the roll angle, ϕ is the pitch angle, s_x^0 and s_y^0 are the sensitivity when the gamepad is level, s_x^1 and s_y^1 are the sensitivity increases when the gamepad is turned 180 degrees, and v_x and v_y are the velocity components that is provided as an input to the game. We use the absolute values of the angles so that sensitivity increases both for right and left when the gamepad is tilted to the left.

To determine the values of sensitivity coefficients for our implementation, we conducted a simple pilot test in which we tried different values and subjectively assessed the playability of the game with different sensitivity values. As a result, we chose $s_x^1 = 6s_x^0$ and $s_y^1 = 6s_y^0$ as the sensitivity coefficients to be used in our usability tests.

With this scheme, the crosshair's motion velocity was controlled both by the analog joystick and the gamepad tilt. The analog joystick provided the direction and the base magnitude of the velocity. The degree of tilt scaled the magnitude of this velocity so that the crosshair moves faster when the gamepad is tilted.

IV. Method

In a within-subjects design, we compared our novel input method of using tilt to adjust gamepad sensitivity to two other input methods: conventional gamepad input and using gamepad tilt to aim.

A. The game

We used the FPS constructor tool for the Unity 3D game engine to create a simple FPS game that can be controlled with the different input methods that we aimed to compare. In the game, the user controls an avatar with a rifle and views the game world from the avatar's point of view (see Figure *1*). The goal is to kill all the enemies that are spawned simultaneously from four different locations in four waves in a square map surrounded by walls. The next wave started after the player kills all the enemies in a wave. We did not let the avatar die in the

Figure 1. The game that we used in the experiments.

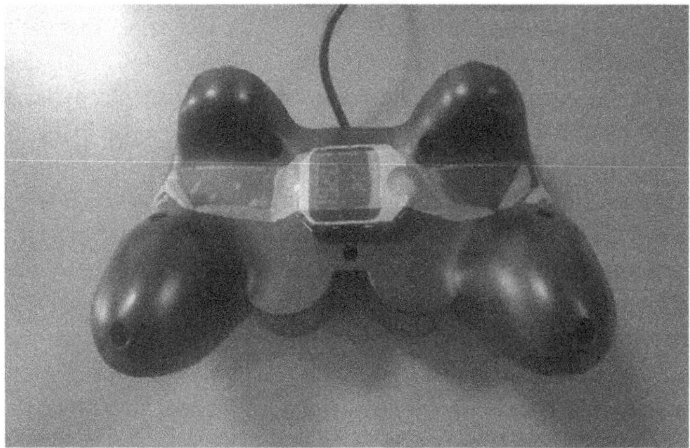

Figure 2. The gamepad and the accelerometer.

game to avoid shorter sessions and to enable fair comparisons between the input methods. Therefore, the game session continued until the player killed all the enemies in the last wave.

B. Hardware

We used a conventional PC gamepad with an analog joystick for the input device. This gamepad did not contain integrated sensors to detect tilt. To track the tilt, we used a Texas Instruments ez430 Chronos Sports Watch that has a three-axis accelerometer. We removed the body of the watch from its wristband, and taped it on the bottom of the gamepad (see Figure 2). The watch communicated wirelessly with the PC via its USB dongle and provided approximately 20 three-axis accelerometer readings per second. Using a lowpass filter, we tracked the direction of gravity and computed the pitch and roll angles that represented the tilt of the gamepad compared to the default orientation that we captured at the beginning of the game.

C. Conditions

Our experiment had three conditions: Gamepad, Tilt and Sensitivity.

Gamepad: This condition reflected the conventional way of playing FPS games with a gamepad. The player used his or her thumbs to operate the two analog joysticks. The left analog joystick moved the avatar without changing the aim direction (walk/strafe) and the right analog joystick changed the aim direction of the avatar.

Tilt: This condition reflected the state of the art for FPS games that use the tilt to control the aim direction. The amount of tilt in the pitch and roll angles determined the velocity for the aim direction in the vertical and horizontal planes (i.e., the character constantly turned right while the user is tilting the gamepad to the right).

Sensitivity: This condition represented our novel input method for FPS games, in which we track the gamepad tilt to make use of the user's secondary motions while playing the game with the analog joysticks. As explained in Section 0, we use the amount of tilt to scale the magnitude of the velocity input of the right analog joystick that is used for aiming.

The three conditions differed only with respect to how the aim in the game is controlled with user input. In all conditions, the player used the R1 button with the right index finger to shoot, and the left analog joystick to walk and strafe.

D. Participants

We recruited 13 participants through announcements in university lectures. We used two of these participants in our iterations on the input method and experiment design. We used the rest of the 11 participants (one female, ages 20–25) in our within subjects experiment. Participants' experiences in gaming, FPS games and gamepads varied.

E. Sessions

Sessions took place in a quiet office environment. Each user participated in only one session. After signing a consent form, the participant sat in front of a laptop computer and used the gamepad to play the three game conditions in succession. We counterbalanced possible learning effects with a Latin Square design. We conducted semi-structured interviews and questionnaires at the beginning of the session, after the user played each condition, and at the very end of the session. Before playing each condition, we briefly explained how the inputs worked and the game allowed the user to practice for one minute. Afterwards, the user played the game in which enemies attacked in four waves. New waves started after all enemies in the previous wave were dead. The game ended when the user killed all the enemies in the last wave. After the game, the user filled in a questionnaire about the input method that he or she just used. At the end of the session, we conducted a semi-structured interview in which the user commented on his or her experiences with the different input methods.

F. Data

In the beginning of the session, we collected data about the user and his or her experiences with video games and gamepads. We collected the different platforms that the user plays games in, whether the user plays FPS games, whether they play FPS games online, whether they played an FPS game with the gamepad and if they did, how much they liked it.

While playing the game, we tracked how long it took the user to complete the game (dependent variable, DV 1). We assumed that a short game duration would reflect the success of the input method as the efficiency of target acquisition would result in the user killing the targets faster. We also collected how many times the user was shot, i.e., how much damage the user received (DV 2). We assumed that this would reflect the difficulty of using the input method as the user may get shot more because of not being able to evade attacks due to the mental load that the input method requires, and also because of not eliminating the enemies quickly. In addition, we kept track of users' missed shots (DV 3) to reflect their failure to acquire targets, which may be caused by the input method.

After each game, the user filled the questionnaire in Table 1. (DV 4–9) After the third game, at the end of the session, we asked users about their choices in input methods (DV 10) and users ranked the three input methods that they have experienced.

V. Results

A. Statistical Analysis

We analyzed each of the 10 dependent variables (see **Error! Reference source not found.**) in an analysis of variance with input method (Gamepad vs. Tilt

TABLE I. AFTER-GAME QUESTIONNAIRE

DV	Question	Answer
DV 4	Was it easy to play the game?	Likert (1-5)
DV 5	Was it easy to get used to this input method?	Likert (1-5)
DV 6	Was it easy to kill nearby enemies?	Likert (1-5)
DV 7	Was it easy to kill far enemies?	Likert (1-5)
DV 8	Was the game fun to play?	Likert (1-5)
DV 9	Do you think you would be successful in the game once you got used to this input method?	Likert (1-5)

vs. Sensitivity) as a within-subjects factor. We verified the sphericity assumption with Mauchly's test of sphericity for all variables, except for damage received and fun factor ($p>0.05$). We used the Huynh-Feldt correction for these variables in the following analyses. Using repeated measures analysis of variance (RM-ANOVA) for each variable, we found that the main effect of the input method was significant for the following variables: game duration (DV 1), damage received (DV 2), and perceived ease of use (DV 4). For each of these variables, we performed post-hoc comparisons using the Fisher's LSD test to further uncover the effects of the input method. Below we analyze each of these variables in detail.

Game duration (DV 1). RM-ANOVA showed that the main effect of the input method was significant with $F(2, 20)=10.186$, $p<0.005$, $\eta 2=0.505$. Post-hoc comparisons using the Fisher's LSD test indicated that players spent significantly more time to finish the game in the Tilt case (M=199.18, SD=44.546) compared to both the Gamepad case (M=158.64, SD=29.659, $p<0.005$) and the Sensitivity case (M=171, SD=33.791, $p<0.05$). This may indicate that Tilt was more difficult to use than both Gamepad and Sensitivity.

Damage received (DV 2). RM-ANOVA with Huynh-Feldt correction showed that the main effect of the input method was significant with $F(1.207, 12.067)=4.532$, $p<0.05$, $\eta 2=0.312$. Post-hoc comparisons using the Fisher's LSD test indicated that players received significantly more damage ($p<0.05$) in the Tilt case (M=69.45, SD=61.876) compared to the Gamepad case (M=38.27, SD=26.661). This may mean that Tilt required more mental load compared to Gamepad. The players received even lower damage on average in the Sensitivity case (M=34.18, SD=23.302), however the differences are not statistically significant.

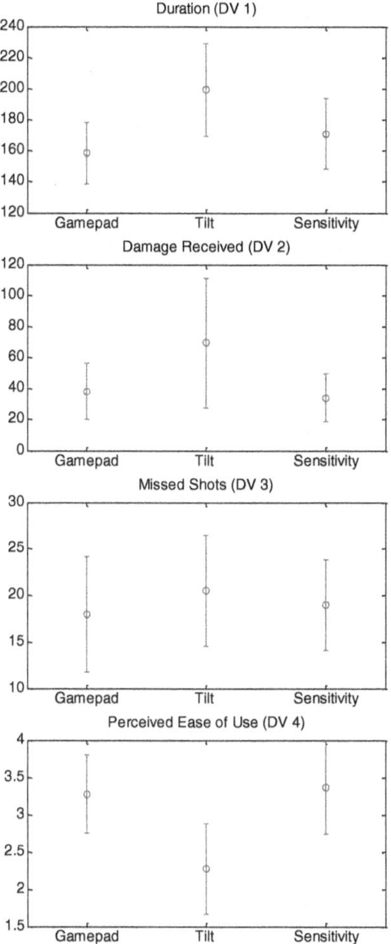
Duration (DV 1)
240
220
200
180
160
140
120
Gamepad
Tilt
Sensitivity
Damage Received (DV 2)
120
100
80
60
40
20
0
Gamepad
Tilt
Sensitivity
Missed Shots (DV 3)
30
25
20
15
10
Gamepad
Tilt
Sensitivity
Perceived Ease of Use (DV 4)
4
3.5
3
2.5
2
1.5
Gamepad
Tilt
Sensitivity

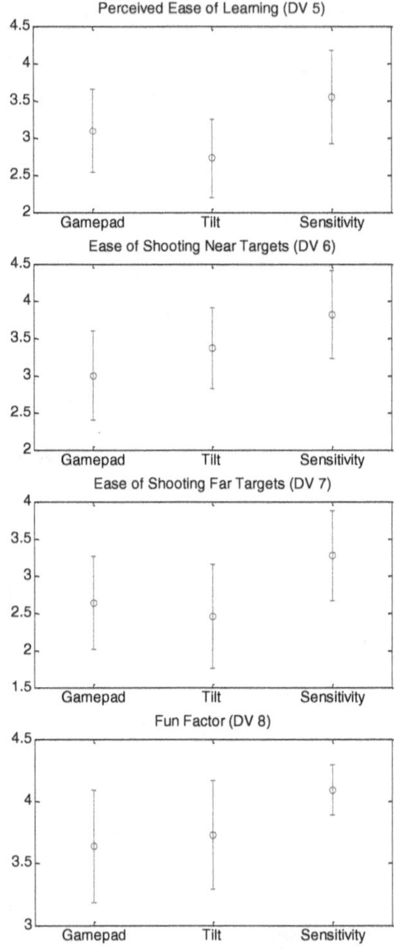

Perceived ease of use (DV 4). RM-ANOVA was significant with $F(2, 20)=7.151$, $p<0.005$, $\eta2=0.417$. Post-hoc comparisons using the Fisher's LSD test indicated that players found both the Gamepad case ($M=3.27$, $SD=0.786$, $p<0.001$) and the Sensitivity case ($M=3.36$, $SD=0.924$, $p<0.05$) significantly easier than the Tilt case ($M=2.27$, $SD=0.905$). This may indicate that controlling aim with the tilt may be less intuitive for players compared to the other two input methods that use the analog joystick. In addition, users found Sensitivity to be slightly easier than Gamepad on average, but the difference was not statistically significant.

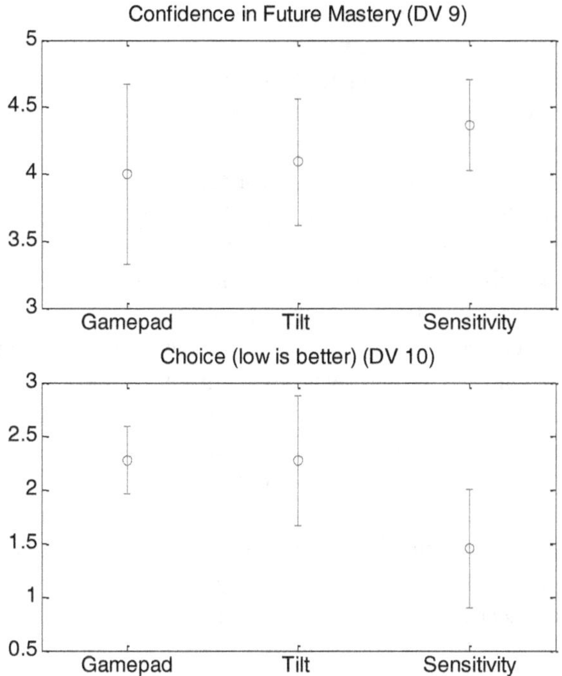

Figure 4. Estimated distributions of dependent variables. Error bars indicate 95 % confidence intervals.

B. Statistically Insignificant Observations

The statistically significant results above favor Gamepad and Sensitivity over Tilt; however, they do not make a distinction between Gamepad and Sensitivity. We believe that the limitations of our sample size may have caused this. In order to make the best use of our data, we investigate the means of the rest of the variables to gain further insight between the two cases.

We observed that means of the questionnaire results all favor Sensitivity over the two other input methods. Average perceived fun factor is the largest for Sensitivity, followed by Tilt and then Gamepad. On average, players ranked Sensitivity better in their choice of input device compared to Gamepad and Tilt. Average perceived ease of learning is the highest in Sensitivity, followed by Gamepad and Tilt. Average perceived ease of shooting near targets is the highest for Sensitivity, followed by Tilt and then Gamepad. Average perceived ease of shooting far targets is the highest for Sensitivity, followed by Gamepad and then Tilt. Average confidence in future mastery was the highest for Sensitivity, followed by Tilt and then Gamepad. While the differences are not statistically significant, these results all favor Sensitivity, including perceived ease of use that we analyzed in the previous subsection.

The data related to users' gameplay favor Gamepad and Sensitivity. Average game duration is the lowest for Gameplay, followed by Sensitivity and Tilt. This may mean that Gamepad let the users eliminate the enemies faster. Average number of missed shots parallels this observation, as it is the lowest for Gameplay, followed by Sensitivity and Tilt. Together, the means of these two variables may indicate that users acquired targets more accurately and eliminated them faster with Gamepad, followed by Sensitivity, and then Tilt. However, damage received favors Sensitivity over Gamepad and Tilt, as the average damage received is the least for Sensitivity and is followed by Gamepad. This may mean that the mental load of the input method was the least with Sensitivity.

While the results in this section are not statistically significant, they support our qualitative observations during the user tests. Users generally enjoyed the Sensitivity input method, however; since it was a new way of interacting with a game, the learning effect may have hindered their performances.

VI. Discussion

A. Discussion of Results

Our results indicate that it is possible to make use of tilt input to positively enhance the aim inputs in FPS games played with gamepads. The Tilt and Sensitivity input methods that we implemented to enhance the gamepad experience were simply additions to the default controls of the gamepad, i.e., the user could still play the game with the analog joystick and have a similar experience if he or she did not tilt the gamepad. However, the additional input of gamepad tilt to control the aim in the Tilt case not only did not enhance the interaction, but also significantly hindered the user experience according to the results of our statistical analysis. Whereas using the tilt to adjust the speed of aim in the Sensitivity case did not have such a detrimental effect.

We believe that having both tilt and the analog joystick control the aim simultaneously in the Tilt case created a competition between the two inputs. As the user tried to control the aim with the analog joystick, the inevitable motions of the gamepad introduced unwanted noise in the aim. This led some users to stop using the analog joystick and rely on the tilt to control the aim, which has been shown previously to be inferior to gamepad inputs [10].

In contrast, having tilt and analog joystick inputs affect aim in different but complementary ways in the Sensitivity case did not hinder the user experience. Tilting the gamepad to enable the analog joystick to move faster in the Sensitivity case resulted in a similarly successful game experience as the plain gamepad in the Gamepad case according to the results of our statistical analysis. Our qualitative observations and differences in means of questionnaires responses indicated that users received this novel input method well. When asked to rank their preferred input methods, eight out of 11 participants ranked the Sensitivity case as their first choice in the questionnaire.

B. Relation to Previous Work

Our results are in agreement with previous work that showed tilt inputs to be less precise compared to gamepad inputs [10]. Similarly, in our experiments, the Tilt case that used the gamepad tilt to control the aim resulted in slower completion of game tasks (longer game sessions), increased received damage, and was perceived to be more difficult compared to other two input methods.

Our work is the first to propose an alternative use for the gamepad tilt to enhance the analog joystick input in gamepads. Previous studies aimed to replace the role of the analog joystick with tilt, whereas we introduced a method to use tilt to enhance it. While we focused on FPS games, we believe that this idea can also be adapted to other types of games played with the analog joystick.

C. Better Interpreting Users' Intentions

A possible shortcoming of our approach is that it depends heavily on our assumption of how to interpret users' secondary motions while playing with a gamepad. As presented in Section 0, we assumed that the attitude of the gamepad (i.e., how much it is tilted) reflects users' intentions of aim speed. However, there may be parameters other than the value of the tilt angle that also contain information about the user's intent. For example, tilting the gamepad faster may mean that the user wants to move the aim faster. The speed in which the user changes the attitude, among other parameters of motion, is ignored in our assumption of the meanings of secondary motions. More accurate interpretations of secondary motions that may be uncovered by behavioral studies may help create better control schemes.

D. Enhancing Existing Input Methods with Novel Sensors in a User-centric Way

As new sensors are appearing in input devices, developers and researchers are finding ways of using them to interact with games. However, the straightforward way of integrating such sensors into games may not always be the best choice. As in the case of tilt for aim, using the tilt to replace the analog joystick for entering velocity was not a good choice, possibly because the analog joystick retracts to the neutral position while tilt must be neutralized manually by the user. Using tilt to directly input a position rather than a velocity can be a better choice [17]; however, this is not suitable for FPS games as they require continuous and long motions that can wrap around in the polar coordinates. This would require the user to spin the gamepad around and keep in poses that are very different than the neutral pose. In addition, precision and input delay of accelerometers can also play a role in hindering the success of tilt for aiming.

Rather than fixating on and forcing the straightforward ways of using these devices, developers and researchers should take a step back and reconsider the abilities, requirements and limitations of these novel sensors as well as the expectations, behaviors and abilities of users that will use them. In this study, we adopted a user-centric approach and considered the naturally occurring behaviors of users as they played FPS games with the gamepad. Then, we found a way to use the tilt sensor to address the unused secondary motions of the users to better reflect their intentions in the game. Our experiments showed that this was a better approach compared to the straightforward use of tilt for aim. Through a similar user-centric approach, we can find ways to use other novel input devices to enhance the gaming experience for users. For example, even though the Kinect may not be preferred as the main input device to play fast-paced games that require quick and precise reactions, it can be used in conjunction with a gamepad to enhance the game experience.

E. Limitations

The present study has a number of limitations that should be considered while interpreting the outcomes. Firstly, the sample size in our experiments was small (11) which limits the statistical power of our study. Therefore, we opted to use the Huynh-Feldt correction instead of Greenhouse-Geisser when sphericity assumption was violated, and we used the Fisher's LSD test without the Bonferroni adjustment for multiple comparisons. Nevertheless, we found meaningful and statistically significant results that supported our qualitative observations.

Another limitation of our study is that the FPS game that we created was a simple one and represented general gameplay, rather than focusing on cases in which the advantages and disadvantages of the considered input methods would be stressed. The noise associated with regular gameplay may have prevented some of the results from being statistically significant. A better game design would be to have isolated tasks within a game that can help contrast the input methods. Comparing the data about such sections of the game rather than whole game sessions may uncover more information about the effects of the different input methods.

Another limitation is the learning effect related to novel input methods. Before each game, we included a one-minute practice session in which the user practiced and learned the input method. Instead, teaching users how to use the input methods more effectively through tutorials could help contrast the differences in the input methods.

VII. Future Work

The novel input method that we introduced in this study may lead to new possibilities for future studies, related to the improvement of the present study, using motion sensors to enhance existing methods of playing games, and better addressing the velocity-position mismatch in gamepads and FPS games.

The experiments that we presented in this study provide a convincing argument for using tilt not for aiming directly in FPS games but for adjusting the sensitivity of the analog joystick to aim in FPS games. However, the experiments do not provide statistically significant evidence about whether this novel input method or the basic use of gamepads is preferable. As a part of our future work, we plan to improve our game and experiment design in a way that will stress the differences between these input methods, and run a study with a larger sample size and more statistical power.

Another line of future work that stems from this work is to find practical uses for new motion sensors to enhance existing methods of playing games. For example, Zaranek et al.'s study [11] indicates that gamepads are preferable to the Playstation Move and Microsoft Kinect in FPS games. We believe that using a similar methodology from this study, we can find ways of making use of users' secondary motions while playing with the gamepad to enhance the game experience with these devices. For example, the Kinect can track the user's head and may help him or her dodge bullets by moving the head.

While we did not address it in this study, there is an inherent mismatch between the velocity input of the analog joystick and the position input that FPS aiming needs. Natapov et al. [4] address this by using a trackball instead of an analog joystick. We would like to search for less invasive ways of enhancing the gamepad experience with other sensors to facilitate easier position input and improve the game experience.

VIII. Conclusion

In this paper we introduced a novel way of using a gamepad and a tilt-sensing three-axis accelerometer to enhance the way that users play FPS games with the gamepad. Our input method relies on the user-centric observation that users perform secondary motions of tilting the gamepad when they are not satisfied with the current velocity of motion they can achieve with the analog joystick. By using the amount of tilt to scale the velocity provided by the analog joystick and effectively increasing its sensitivity when the user tilts the gamepad, we attempt to realize the user's intent for tilting the gamepad. We compared this method to two other input methods: a plain gamepad input and using tilt directly to aim in FPS games. We showed with statistical significance that our method does not have the detrimental effects of using tilt to directly control the aim and is comparable with the plain gamepad input. Our experiments have also suggested that our method may be preferred by users and may put less mental load on the user; however, these results were not statistically significant.

References

[1] M. Hitchens, "A Survey of First-person Shooters and their Avatars," *Game Stud.*, vol. 11, no. 3, 2011.

[2] "PC gamers 'destroyed' console gamers in tests, says Voodoo PC founder." [Online]. Available: http://www.gamesradar.com/pc-gamers-destroyed-console-gamers-in-tests-says-voodoo-pc-founder/. [Accessed: 15-Aug-2014].

[3] D. Natapov, S. J. Castellucci, and I. S. MacKenzie, "ISO 9241-9 evaluation of video game controllers," in *Proceedings of Graphics Interface 2009*, 2009, pp. 223–230.

[4] D. Natapov and I. S. MacKenzie, "The trackball controller: improving the analog stick," in *Proceedings of the International Academic Conference on the Future of Game Design and Technology*, 2010, pp. 175–182.

[5] D. Natapov and I. S. MacKenzie, "Gameplay evaluation of the trackball controller," in *Proceedings of the International Academic Conference on the Future of Game Design and Technology*, 2010, pp. 167–174.

[6] K. M. Lenz, A. Chaparro, and B. S. Chaparro, "The Effect of Input Device on First-Person Shooter Target Acquisition," *Proc. Hum. Factors Ergon. Soc. Annu. Meet.*, vol. 52, no. 19, pp. 1565–1569, Sep. 2008.

[7] P. Isokoski and B. Martin, "Performance of input devices in FPS target acquisition," in *Proceedings of the international conference on Advances in computer entertainment technology*, 2007, pp. 240–241.

[8] C. Klochek and I. S. MacKenzie, "Performance measures of game controllers in a three-dimensional environment," in *Proceedings of Graphics Interface 2006*, 2006, pp. 73–79.

[9] J. Tsai and M. Stoklosa, "Evaluation of Input Devices for an FPS Program Used in Law Enforcement Tactics Training," *Proc. Hum. Factors Ergon. Soc. Annu. Meet.*, vol. 52, no. 27, pp. 2087–2091, Sep. 2008.

[10] C. Ardito, P. Buono, M. F. Costabile, R. Lanzilotti, and A. Simeone, "Comparing low cost input devices for interacting with 3D Virtual Environments," in *2nd Conference on Human System Interactions, 2009. HSI '09*, 2009, pp. 292–297.

[11] A. Zaranek, B. Ramoul, H. F. Yu, Y. Yao, and R. J. Teather, "Performance of modern gaming input devices in first-person shooter target acquisition," in *CHI'14 Extended Abstracts on Human Factors in Computing Systems*, 2014, pp. 1495–1500.

[12] P. Cairns, J. Li, W. Wang, and A. I. Nordin, "The influence of controllers on immersion in mobile games," in *Proceedings of the 32nd annual ACM conference on Human factors in computing systems*, 2014, pp. 371–380.

[13] K. Browne and C. Anand, "An empirical evaluation of user interfaces for a mobile video game," *Entertain. Comput.*, vol. 3, no. 1, pp. 1–10, 2012.

[14] S. Medryk and I. S. MacKenzie, "A comparison of accelerometer and touch-based input for mobile gaming," in *International Conference on Multimedia and Human-Computer Interaction-MHCI 2013*, pp. 117–1.

[15] W. Hürst and H. C. Nunez, "Touch Me, Tilt Me–Comparing Interaction Modalities for Navigation in 2D and 3D Worlds on Mobiles," in *Advances in Computer Entertainment*, Springer, 2013, pp. 93–108.

[16] I. S. MacKenzie and R. J. Teather, "FittsTilt: The application of Fitts' law to tilt-based interaction," in *Proceedings of the 7th Nordic Conference on Human-Computer Interaction: Making Sense Through Design*, 2012, pp. 568–577.

[17] R. J. Teather and I. S. MacKenzie, "Position vs. velocity control for tilt-based interaction," in *Proceedings of the 2014 Graphics Interface Conference*, 2014, pp. 51–58.

[18] P. Gilbertson, P. Coulton, F. Chehimi, and T. Vajk, "Using 'tilt' as an interface to control 'no-button' 3-D mobile games," *Comput. Entertain. CIE*, vol. 6, no. 3, p. 38, 2008.

[19] T. B. P. Tuesday and 3 June 2008, "Sixaxis not sensitive enough," *Eurogamer.net*. [Online]. Available: http://www.eurogamer.net/articles/sixaxis-not-sensitive-enough-for-fps-aiming-resi-5-producer. [Accessed: 25-Jul-2014].

[20] "Microsoft Talks About Why There Are No Tilt Or Motion Sensors in the Xbox One Controller." [Online]. Available: http://gamingbolt.com/microsoft-talks-about-why-there-are-no-tilt-or-motion-sensors-in-the-xbox-one-controller. [Accessed: 13-Aug-2014].

[21] "Uncharted: Golden Abyss video shows off gyro-aim controls," *ComputerAndVideoGames*, 15-Dec-2011. [Online]. Available: http://www.computerandvideogames.com/329930/uncharted-golden-abyss-video-shows-off-gyro-aim-controls/. [Accessed: 20-Sep-2014].

[22] Ruthie, "Ubisoft: Wii U makes FPSs easier to control," *Nintendo Wii U Go*, 02-Aug-2011. [Online]. Available: http://wiiugo.com/ubisoft-wii-u-makes-fpss-easier-to-control/. [Accessed: 15-Aug-2014].

Computer Graphics

13. Experimental Analysis of QEM Based Mesh Simplification Techniques

Ecem İren
Department of Computer Engineering Ege University
İzmir, Turkey ecem.iren@gmail.com

Murat Kurt
International Computer Institute Ege University
İzmir, Turkey murat.kurt@ege.edu.tr

Abstract— In this research study, effects of mesh simplification on visual quality are examined by using quadric edge collapse decimation method. In this context, we analyze simplifications of various objects by investigating the Peak Signal-to-Noise Ratio (PSNR) values, difference images, and compression ratios. Experiments are performed in MeshLab environment and it is shown that when model is chosen as complex, simplification error between reference and simplified models increases much more in comparison with simpler models. At the same time, if we use high compression ratio, higher simplification error is reached. It could be concluded that compression ratio affects the error linearly.

Keywords—mesh simplification; quadric error metrics; QEM; visual quality analysis; quadric edge collapse decimation

I. INTRODUCTION

Various applications in computer graphics need complex and detailed models for providing reality. For this reason, models are captured with high resolution but complexity of the model causes an increase in the computational cost. To solve this issue, producing simpler forms of such models has gained great importance. In this target, studies [1, 2, 3, 4] such as surface simplification and multiresolution modeling, which are creating the models with sufficient levels of details for rendering applications, have achieved popularity. Those studies are interested in simplifying surfaces by taking the polygonal model as input and obtaining a simplified model in the end. Due to preventing loss of information, the most of these studies [1, 2, 3] suppose that input model is composed of only triangles. Output model ensures some intended features such as a specific face count or a maximum tolerable error [2]. In this study, our aim is to observe impacts of mesh simplification on the visual quality and storage sizes. For this purpose, we will select quadric edge collapse decimation method for simplifying models. This method is also known as Quadric Error Metrics (QEM) based mesh simplification and we will use MeshLab [5] for evaluating QEM based mesh simplification. We will evaluate simplification with ten different objects and analyze results in terms of categories like data size, number of faces and PSNR differences between simplified mesh model and original model.

The paper is organized as follows: Section 2 mentions about some studies which analyzes different simplifications by categorizing them. Section 3 explains the QEM based simplification method in detail and Section 4 evaluates the simplification effects of that method on simplified graphical objects.

II. Related Work

Mesh simplification techniques propose different approximations to accomplish simplification process and they can be classified as:

A. Vertex Decimation Based Techniques

It is a technique described by Schroeder et al. [1]. They defined an algorithm that involves geometrical and topological operations in order to decrease the number of triangle faces. Their implementation chooses a vertex for removing adjacent faces and translates resulting hole to a triangle. Vertex decimation preserves topology of the model but this capability is not so important in multiresolution rendering systems. Also, this method works slowly [1, 2].

B. Vertex Clustering Based Techniques

This method decides the closeness of vertices and when some vertices are found close to any vertex, a new representative vertex is created and used to remove detected vertices as close previously. Clustering is divided into six stages and it starts with grading step in which each vertex is given a weight respect to its visual importance. After that, triangulation process is applied to transform polygons to triangles. Then, vertices are grouped into sets depend on their geometric similarity with clustering. In synthesis part, a representative vertex is calculated. Subsequent to this, duplicated triangles, edges and points are deleted in elimination part. Finally, normals of newly created triangles and edges are adjusted. Though vertex clustering runs fast and alters topology of the model, it does not give quality responses [1, 6]. In another study [7], a new mesh simplification algorithm was developed to handle faults of error accumulation. It has preprocessing stage in which each surface of the model is triangulated by connecting vertices and recorded those triangles in a table by giving numbers. Afterwards, each vertices are classified and selected for deletion operation with classical QEM algorithm. Finally, principal curvature technique is used due to the fact that it exhibits geometrical characteristics and curvature is insensitive to noise interference so it provides robustness.

C. QEM Based Techniques

Garland and Heckbert [2] proposed a surface simplification algorithm including iterative contraction of valid vertex pairs. Beside this, it benefits from quadric error metrics to keep track of approximate error while model is being simplified. At the end of the operation, result vertices of final model are hold in quadrics. Improved algorithm also could join unconnected regions in the model and supports non-manifold models by having a capability of not maintaining the topology. Furthermore, it protects main features of the model after simplification and performs its task very rapidly. In addition, they enhanced the algorithm [3] by adding a capability which could simplify surfaces with vertex properties such as texture and colors. Tarini et al. [4] presented an approach to quad mesh simplification responsible for the task of generating a low complexity quad mesh from a high complexity one. The algorithm depends on local operations which preserve quad structure. Furthermore, they presented a Triangle-to-Quad conversion algorithm which is used for obtaining the initial quad mesh from a given triangle mesh. Tang et al. [8] introduced a new mesh simplification algorithm related with QEM. The algorithm produces a new vertex from the midpoint of contraction edge. Since algorithm does not take feature of mesh model into account and also computation of new vertex is very complex with it, this algorithm is explored to improve original one. Thereby, it is considered simple but after some experiments, it's seen that results do not meet expected real-time processing. Yao et al. [9] developed a QEM algorithm based on discrete curvature. They made experiments on different models to prove that both geometry and topology structure and the features of the original models are absolutely retained by utilizing discrete curvature. Andersson et al. [10] proposed a restricted mesh simplification algorithm by utilizing edge contractions. They evaluated the method of iteratively contracting edges and boosted it by putting a constraint in which crossing edges will not generated from the contraction. By the way, the proposed approach works under a condition that the set of generated output points is needed to be a subset of the input set. For instance; the process of an edge contraction should be carried out on the one of its neighboring vertices. They also pointed out that some problems come in view during the edge contraction of triangulations such as final triangulation may not be produced in a planar form. Since the edge contraction is qualified as valid when the resulting triangulation is planar, they tried to analyze the troubles of specifying viable contractions and computing them. Moreover, Hoppe [11] proposed a new paradigm in order to simplify objects with appearance features. Firstly, a new quadric error metric which is capable of simplifying meshes with appearance attributes is identified. This metric captures both geometric error and attribute error. Following that, relating the quadrics with edge-driven data structures provides simplification of models with attribute discontinuities. With the aid of previously introduced two techniques called as memoryless simplification and volume preservation, results are further improved and get better. After some experiments on a variety of meshes with colors and normals it is seen that the new metric brings some advantages. It measures error that depends on geometric correspondence in R in a more intuitive way. In addition, it requires less storage area because of linearity between its space complexity and the number of attributes. Furthermore, the quadric matrix has a sparse structure which accounts for the algorithm to make an evaluation more quickly. Finally, it is stated that created simplified meshes show the same accuracy with the ones produced by the previously enhanced more expensive optimization by the same author.

III. Method

QEM based algorithm [1] depends on iterative contraction of vertex pairs. It is a generalization of iterative edge contraction. Vertex pair contraction is described by $(\mathbf{v}_1, \mathbf{v}_2)\Diamond\mathbf{v}$. An initial model is selected and some pair contractions are applied in order to simplify it. Until desired simplification rate is obtained, contraction operations are repeated. At the end of each contraction, a new simplified model is produced. Pair selection is important issue and valid pairs should be defined according to two rules:

- $(\mathbf{v}_1, \mathbf{v}_2)$ pair should create an edge.
- Value of $||\mathbf{v}_1, \mathbf{v}_2||$ should be less than a threshold parameter.

If threshold value is chosen as too high, it causes non-connected vertices to become paired. Otherwise, if it is selected as 0, algorithm acts like a simple edge contraction algorithm. So, it must be chosen carefully. After deciding all valid pairs, cost of each contraction should be computed. For calculating this, error at each vertex should be found with a symmetric 4×4 quadric matrix **Q**. Finally, error formula is written as $\Delta(\mathbf{v}) = \mathbf{v}^T\mathbf{Q}\mathbf{v}$. In order to achieve a contraction, position of result vertex must be determined and generally it is adjusted by selecting position of $\mathbf{v}_1$, $\mathbf{v}_2$ or $(\mathbf{v}_1 + \mathbf{v}_2)$ /2. While selecting the position, it is preferred the value which minimizes $\Delta(\mathbf{v})$. Also, a new quadric matrix is calculated as $\mathbf{Q} = \mathbf{Q}_1 + \mathbf{Q}_2$ for result vertex. After computing optimal position and quadric matrix for each valid pairs, the error cost of new vertex is identified as $\mathbf{v}^T(\mathbf{Q}_1 + \mathbf{Q}_2)\mathbf{v}$. Then all valid pairs are put into a minimum heap with their contraction costs.

Lastly, the pair which has least cost is removed from the heap and costs of all valid pairs are updated iteratively. At this point, constructing **Q** matrix of each vertex is a problem. In this method, error quadrics are derived from a way similar to the one suggested by Ronfard and Rossignac [12]. It is observed that each vertex is created from an intersection of a set of planes with this manner. Error of each vertex is associated with this set by finding sum of squared distance to its planes as follows where **p** is an element of its planes and $\mathbf{K}_p$ is equal to $\mathbf{pp}^T$:

$$\Delta(\mathbf{v}) = \mathbf{v}^T(\Sigma_p \mathbf{K}_p)\mathbf{v}. \tag{1}$$

Here each **p** (plane) is represented with $[a\ b\ c\ d]^T$ where $ax + by + cz + d = 0$ and $a^2 + b^2 + c^2 = 1$. Finally, $\mathbf{K}_p = \mathbf{pp}^T$ is illustrated as the following:

$$\mathbf{K}_p = \mathbf{pp}^T = \begin{bmatrix} a^2 & ab & ac & ad \\ ab & b^2 & bc & bd \\ ac & bc & c^2 & cd \\ ad & bd & cd & d^2 \end{bmatrix} \tag{2}$$

TABLE I. Statistics of the simplified three-dimensional models

Model	Metrics		
	#Faces	*#Vertices*	*Data Size*
Armadillo	345,944	172,974	3.9 MB
Bunny	69,451	35,974	2.89 MB
Dragon	871,414	437,645	32.2 MB
Golfball	245,760	122,882	2.66 MB
Happy Buddha	1,087,716	543,652	40.6 MB
Horse	96,964	48,484	1,07 MB
Igea	268,686	134,345	2.96 MB
Lucy	525,814	262,909	6.03 MB
Max Planck	98,260	49,132	1.11 MB
Thai Sculpture	1,000,000	499,999	181 MB

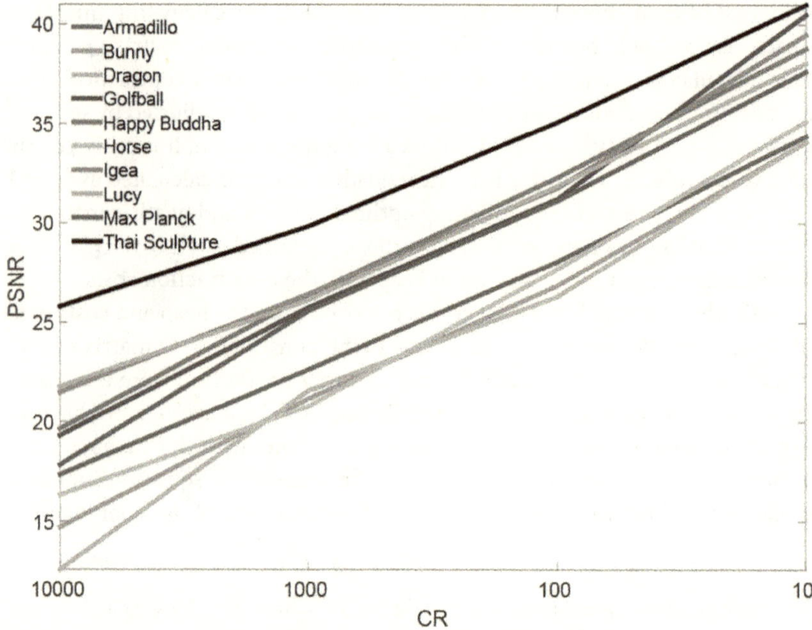

Fig. 1. The peak signal-to-noise ratio (PSNR) values of QEM based mesh simplification technique with different values of the compression ratio (CR).

Now, we have quadric matrix of any given vertex. This operation provides significant benefits: only 4×4 matrices are necessary for working with plane sets and it is enough to find summation of two matrices while computing quadric matrix of result vertex after contraction of two vertices.

IV. Results and Discussion

In this work, we used MeshLab [5] to analyze QEM based mesh simplification techniques, as QEM based mesh simplification techniques are already implemented in MeshLab. According to Table 1, Figure 1 and Figure 2, it can be seen that when number of faces and vertices increases and this implies model is getting more complex, the Peak Signal-to-Noise Ratio (PSNR) [13] results between reference and simplified models become lower than simpler models. Lower PSNR values mean that higher simplification errors between original and simplified models. For example; in Figure 2, the lower PSNR values belongs to Dragon which is the most complicated model in the study. Contrarily that, Max

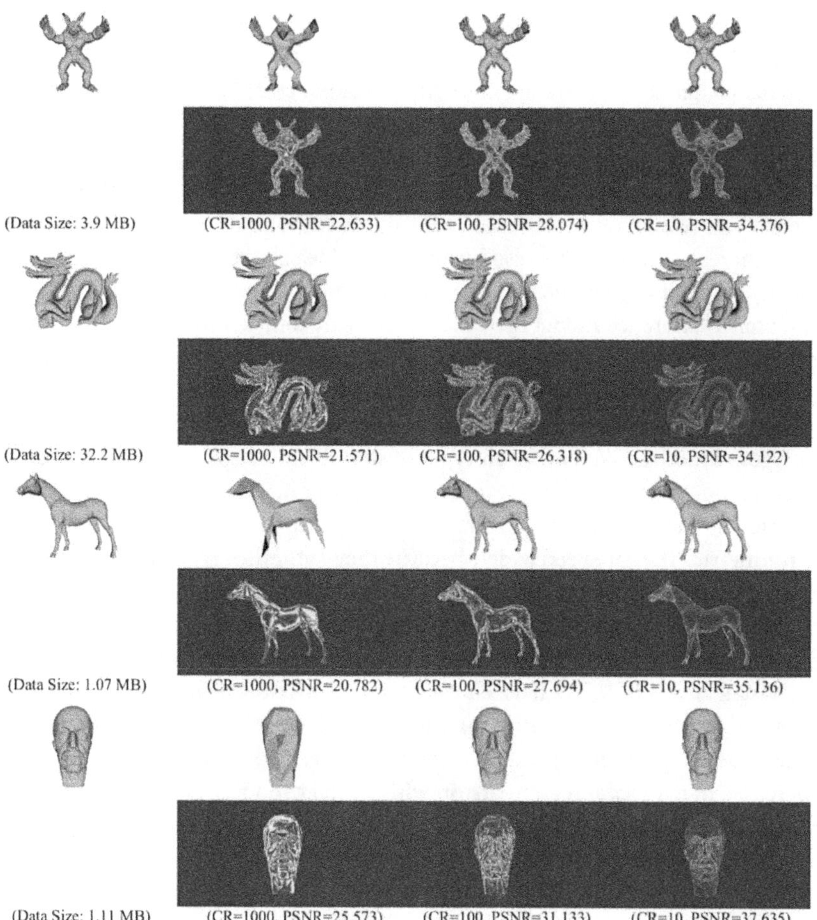

Fig. 2. A visual analysis of the QEM based mesh simplification technique on various 3D models. From top to bottom: armadillo, dragon, horse, max planck 3D objects. While the first column represents reference 3D objects, other columns represents simplified 3D objects according to various Compression Ratio (CR) parameters. Below each simplified model, we depict false-color differences between the reference 3D models and the simplified 3D models. For better comparison, false-color differences were scaled by a factor of 5. Below each simplified 3D model, we also report PSNR values (higher is better) and CR values.

Planck, which is one of the simplest models among the other models, has the highest PSNR values. Therefore, it can be understood that the more complicated and detailed model is used, the more simplification error occurs. Another finding is about relationship between compression ratio and PSNR value. In Figure 1, it is clear that when compression ratio changes, while PSNR value is affected from that situation oppositely, simplification error behaves linearly. In other words, when compression ratio goes up, PSNR value tends to go down and simplification error increases. In fact, this result supports the idea about relation between complexity of the model and PSNR value. Model simplified with compression ratio 10 is closer to reference model than the model with compression ratio 1000. Based on this, we should expect that difference (error) between reference model and simplified should be less in the simplified model with 10 than the one with 1000.

Several applications in computer graphics use simplification of complicated polygonal models. We have seen types of simplification methods such as vertex decimation, vertex clustering and iterative edge contraction using error quadric metrics. It can be said that primary advantage of contraction methods is the error metric. They are used widely because they naturally allow a multiresolution model representation. In addition to this, while vertex decimation is found to be efficient and produce good results, vertex clustering generates poor results [2]. In our study, we have performed mesh simplification on different models to see visual effects of it. We have chosen iterative edge decimation method that is supplied by MeshLab [5]. After experiments, we have realized that when model is chosen as complex, simplification error between reference and simplified models increases much more in comparison with simpler models. At the same time, if we use high compression ratio, higher simplification error is reached. Hence, it could be concluded that compression ratio affects the error linearly.

Acknowledgment

This work was supported by the Scientific and Technical Research Council of Turkey (Project No:115E203), the Scientific Research Projects Directorate of Ege University (Project No:2015/BİL/043). We thank the Stanford 3D Scanning Repository and the Visualization Virtual Services (VVS) Shape Repository for the 3D models used in this work.

References

[1] W.J. Schroeder, J.A. Zarge, and W.E. Lorensen, "Decimation of triangle meshes", *SIGGRAPH Computer Graphics*, vol. 26, no. 2, pp. 65–70, July 1992. [Online] Available: http://dl.acm.org/citation.cfm?doid=142920.134010

[2] M. Garland and P.S. Heckbert, "Surface simplification using quadric error metrics", *Proceedings of the 24th annual conference on Computer graphics and interactive techniques*, 1997, pp. 209–216.

[3] M. Garland and P.S. Heckbert, "Simplifying surfaces with color and texture using quadric error metrics", *Proceedings of the Conference on Visualization*, 1998, pp. 263–269.

[4] M. Tarini, N. Pietroni, P. Cignoni, D. Panozzo, and E. Puppo, Practical quad mesh simplification, *Computer Graphics Forum*. vol. 29, no. 2, pp. 407–418, May 2010.

[5] Visual Computing Lab ISTI-CNR, "MeshLab", 2008,. [Online] Available: http://meshlab.sourceforge.net/

[6] K.L. Low and T. Tan, "Model simplification using vertex clustering", *Proceedings of the 1997 on Interactive 3D Graphics*, 1997, pp. 75–81.

[7] Z. Hua, Z. Huang, and J. Li, Mesh simplification using vertex clustering based on principal curvature, *International Journal of Multimedia and Ubiquitous Engineering*, vol. 10, pp. 99–110, 2015.

[8] Z. Tang, S. Yan, and C. Lan, A new method of mesh simplification algorithm based on qem, *Information Technology Journal*, vol. 9, pp. 391–394, 2010.

[9] L. Yao, S. Huang, and H. Xu, Quadratic error metric mesh simplification algorithm based on discrete curvature, *Mathematical Problems in Engineering*, vol. 2015, pp. 1–7, April 2015.

[10] M Andersson, J. Gudmundsson, and C. Levcopoulos, "Restricted mesh simplification using edge contractions", *Proceedings of the European Workshop on Computational Geometry*, March 2006, pp. 121–124.

[11] H. Hoppe, "New quadric metric for simplifying meshes with appearance attributes", *Proceedings of the Conference on Visualization*, 1999, pp. 59–66.

[12] R. Ronfard, and J. Rossignac, "Full-range approximations of triangulated polyhedra", *Computer Graphics Forum*, vol. 15, no. 3, pp. 67–76, 1996.

[13] L.E. Richardson, Video Codec Design: Developing Image and Video Compression Systems, New York, USA, John Wiley & Sons, Inc., 2002.

14. Procedural City Generation Using Cellular Automata

Melek Buşra Temuçin
International Computer Institute Ege University
Izmir, Turkey busra.temucin@gmail.com

Kaya Oğuz
Department of Computer Engineering Izmir University of Economics Izmir, Turkey kaya.oguz@ieu.edu.tr

Abstract— Procedural Content Generation (PCG) algorithms are a common solution to create automatic and dynamic content for games and entertainment industry. To procedurally generate a city, its components should be handled with algorithms that are tailored to their individual characteristics. We propose a set of methods for the generation of the city layout, and work on that layout to generate and place the 3D buildings. The layout generation use cellular automata to create organic looking clusters, and the rules are repeated for each cluster to hierarchically create different levels of the city. This approach provides fractal properties which results in an organic city layout. The results are very promising; cities can be generated within seconds, and the generation can be controlled with only a few parameters.

Keywords—cellular automata; procedural 3D city generation; procedural content generation

I. Introduction

Procedural content generation (PCG) algorithms create content used in computer games to increase the re-playability value and to cut down the time needed to create the content manually. With a large number of games being published for personal computers, gaming consoles and mobile devices, PCG has become more of a necessity than a feature.

PCG algorithms have been in the arsenal of game developers since the very early years with games such as Rogue [1] and Elite [2] being the first adopters [3]. Even though the algorithms are most commonly used to generate game levels [4]–[7], there are studies about generating missions and quests [8], [9], 3D worlds [10], buildings [11] and stories [12]. A recent book on PCG and several surveys are available in the current literature [13]–[17].

Cities are heterogeneous; they are made up of different components such as terrain, layouts, buildings and roads. These components depend on and affect each other. They also require different algorithms, or different parameters for the same algorithm to be generated. This makes procedural city generation a more challenging task. One of the first approaches to tackle the problem is [18] where Kato et al. use L-systems to generate roads and genetic algorithm (GA) is to generate the buildings.

CityEngine uses geographical and sociostatistical data as input to a pipeline that generates a city [19]. The pipeline use the input data to generate the road networks using an extended L-system. The space between the roads are divided into lots that the buildings will be placed. The buildings are also procedurally generated using L-systems. Finally, the output is passed to the visualization software.

While *CityEngine* mainly depends on L-systems to generate the city, Lechner et al. make use of agents to do so [20]. The system requires a terrain descriptor where it will position the city. This terrain descriptor can also be procedurally generated, manually designed, or received from real data. The roads are generated by two agents, called extenders and connectors. The buildings are built by commercial and residential developer agents. The only difference between the building agents is that the residential developer does not prefer areas with high road density.

Instead of randomly generating the structure of the city, Greuter et al. use a different approach. They use fixed size grid to define a city, but use pseudo-random numbers to assign different numbers to different cells, so that they can pass this number to a hash function to decide the properties of buildings in that cell [21]. By passing the numbers to a hash-function, and using those numbers to generate building types, they create a seemingly infinite combinations of buildings. Since drawing all of the buildings would not be possible, they only draw the buildings in the camera's view frustum.

Kelly and McCabe have developed the *CityGen* system to interactively generate cities [22]. They divide the problem into three major parts, namely primary road generation, secondary road generation and building generation. The roads are defined as undirected graphs, and they use sampling to plot a path between a source and a destination. Even though the graph represents the roads, they generate the roads adaptive to the underlying terrain. The *CityGen* system is later extended in [23] by incorporating a building generation framework that provides more details.

Emilien et al. follow the same procedure of generating road networks first, then parcels, and finally buildings, but they focus on villages, rather than cities [24]. They generate the road networks based on interest maps that depend on properties such as how sociable and accessible a parcel is.

Lan and Ding propose the *EziCity* framework which generates scenes, not a complete city, if a 2D definition of the scene is given [25].

As also noted in [24], not all of the cities are planned in advance. This is specifically true for ancient cities, where city growth happens organically to areas that are safe and to those that provide food and water. In order to get the organic look and feel of the city, cellular automata (CA) and fractals can be used. The related work that use such approaches are described in Section II. The paper then continues to Section III where the proposed methods are discussed. The results are given in Section IV and are discussed in Section V. The paper concludes with the planned future work in Section VI.

II. Related Work

We have already discussed other methods, such as Lsystems and agents that have been used to generate cities in the previous section and we would like to focus on the studies that use cellular automata and fractals in this section.

Briefly, a cellular automaton is a set of discreet cells with a finite number of states, governed by a set of rules that are applied to each cell at discreet time intervals. The rules define the state of the cell according to some criteria, usually the state of its neighboring cells. These rules, although simple, create patterns that self organize the cells. Perhaps the most famous application of cellular automata is John Conway's "Game of Life" [26].

Cellular automata has been used for city planning before being used for generation [27]. Batty gives the basics of CA and how it can be used to generate plans for cities in [28]. The CA is made up of a grid of cells with two states, undeveloped and developed. With various simple rules, he generates patterns of organized cities. However, the paper emphasized that the real cities are not geometrically ordered, and that probability can be introduced for more variety. He defines the following rule for city development:

IF there is at least one developed cell in the cell's neighborhood
THEN the cell is developed with probability ρ

He further proposes to decrease the probability to ρ^2 the second time the cell is considered for development, and to ρ^3 the third time, which leads to ρ^n in the nth consideration.

Kato et al. use cellular automata and genetic algorithms to generate a virtual city [29]. The city is modeled as a grid of cells, where each cell represents one of the seven states that include building and road types. They present a

TABLE I: Example rules for a cellular automata that generates a virtual city from [29]

rule 1	0.01
if	Current state is vacant, and field strength of access roads is strong
then	Next state is house.
rule 2	0.01
if	Current state is vacant, and field strength of access roads is strong and field strength of shops is strong
then	Next state is shop
rule 3	0.01
if	Current state is house, and field strength of shops is strong
then	Next state is vacant

set of building-layout and road-generation rules that govern the cellular automata. The chromosomes in the GA represent the time-series pattern of the changes in the city. The order the rules are applied determine the pattern and GA searches for the proper sequence of values. This is defined by the fitness value that is made up of two sums, F_1 is the fitness of building ratio, and F_2 is the fitness of the area which is based on its size. It should be emphasized that the rules have probabilities associated with them. The example rules in the paper are listed in Table I.

Chen and Liu propose a fractal city model that use iterative and fractal processes to generate a virtual city [30]. While the iterative process finds random available areas in the space it is working in, the fractal process repeats this iterative process for every space that is generated, recursively.

III. Proposed Methods

To procedurally generate a city, its components should be handled with algorithms tailored to their individual characteristics. We propose a set of methods to procedurally generate the city layout, and work on that layout to place 3D buildings. The layout generation algorithm works on an integer matrix to generate clusters of city blocks using cellular automata, and is repeated within the clusters of that matrix, resulting in a city layout that has fractal properties.

Each run of the city layout generation is called a "stage" that consists of the following five procedures.

- Randomization
- Grouping
- Marking
- Growing
- Preparation for the next stage or Finalization

The procedures operate on a map represented by an integer matrix of size $M \times N$ at all stages, but take different sized clusters into consideration as they proceed.

The integer values of the map represent the types of the cell. The empty cells are assigned to value 0, full cells are assigned to 1 and any number less than 0 represent the space between the clusters. The full cells will contain the buildings and the space between the clusters will become the roads. As will be discussed shortly, the probability of being empty is set by the p parameter.

The number of stages are controlled by a parameter s and is known before the method begins. This variable will also be used as the calculate the space between the clusters at every stage. The first stage generates the major clusters and the space between them will be larger, representing major roads. As the following stages are executed, the distance between the generated clusters will become smaller and will be used for streets.

The first procedure randomly sets the values of the map as empty or full using the p variable. If the random value is less than p, then the cell is set as empty, and full otherwise. This randomization provides the initial set of cells for the cellular automata to work on.

The second procedure uses a cellular automaton to group the random cells together. It uses the 8-cell Moore neighbourhood. The rules are defined as follows:

IF an empty cell has four or more neighbours
THEN fill the cell
IF a full cell has less than five neighbours
THEN empty the cell

The stopping condition for the cellular automaton is defined by the number of changes of empty cells to full cells. It is observed that the number of such cells decrease as the clusters are formed. The difference limit is empirically set to $l = M \times N/1000$ and is increased 10 fold at every successive stage.

The result of this automaton is clusters of full cells. However, these clusters are very roughly shaped and are far away from each other. The following two procedures are applied to improve these conditions.

The Marking procedure uses the flood fill algorithm to mark every cluster with a different number greater than one, thus differentiating the clusters. It runs by scanning the map row by row and whenever it comes across a full cell, the flood fill algorithm is used to mark that cluster with a marking number. Once the flood fill returns, the marking number is incremented. The procedure continues until all full cells are identified.

The fourth procedure helps the clusters to grow until a constant distance is left between them. The procedure takes the distance *d* between the clusters as a parameter and by default is set to the *s* value which denotes the number of stages left to run. This procedure uses a cellular automaton for the empty cells on the edge of clusters. The rules are given below:

```
For every zero cell on the edge of a cluster
Learn the adjacent color
IF there is at least one cell in neighbourhood with a different color
THEN leave the cell empty
ELSE color the cell with adjacent color
```

The final procedure is preparing the map for the next stage or finalizing it for 3D modeling. The empty cells in this stage are marked with a negative integer to differentiate them from clusters so that they will not change in the following stages.

Once the stages are complete the resulting map is used to generate 3D buildings. For this preliminary study, the buildings are generated randomly only by considering their proximity to the major roads. The buildings are simply rectangular prisms with different heights without any texture on them. To create visual diversity the buildings closer to the major roads are higher and their material color is set to a different shade of gray.

The city layout generation is implemented using Java programming language. The resulting maps are outputted as text files that are fed into a C# script in the Unity 3D game engine (http://www.unity3d.com) that parses and generates random buildings. A sample screenshot is given in Figure 1.

IV. Results

As discussed in Section III, the proposed method works with a series of parameters. The resulting maps and the effect of these parameters are discussed in this section.

Figure 2a shows the results of the algorithm that operates on a 120×150 sized map, with the empty cell probability of $p = 0.5$, and number of stages set as $s = 1$. The map on the left shows the result of the map after a single run of the grouping procedure. If the difference limit is applied, the cellular automata at the grouping

Fig. 1: A close up of the random city from the Unity 3D application that can read the city layouts and generate random buildings accordingly. (a) The map on the left is generated with a single run of thecellular automaton in the second procedure. The map on the right uses the limit value *l* for the stopping condition of the same cell automaton. These maps have a single stage. (b) These maps have the same properties but ran for two stagesinstead of one. This figure also shows the fractal properties as the larger clusters have smaller clusters generated inside.

procedure produces the map on the right. The maps generated in Figure 2b are of the same size and probability, but the number of stages are set as $s = 2$. The effect of the p variable is tested with a map of size 120×150 and the number of stages as $s = 3$. The p variable is set to values 0.4,0.5,0.6 and 0.65 and the results are displayed in Figure 3. For values below 0.5, the full cells tend to cover the entire map. At the extreme probabilities such as 0.7 or 0.8, the generated maps are empty.

Even though the distance d between the clusters are set to the stage variable s, it is possible to set it to different values at every stage. A set of examples are given in Figure 4 with a city of size 170×250 and number of stages as $s = 3$. As expected, a large d distance creates a sparse city.

V. Discussion

We proposed a method to create random city layouts using cellular automata. The method have fractal properties since it can repeat the set of procedures within its clusters to generate a more organic city layout map. This city layout is used to generate random buildings that have different heights according to their proximity to major roads.

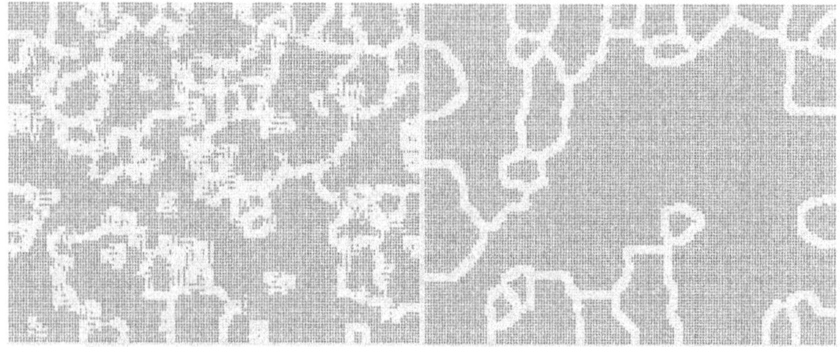

(a) The map on the left is generated with a single run of thecellular automaton in the second procedure. The map on the right uses the limit value *l* for the stopping condition of the same cell automaton. These maps have a single stage.

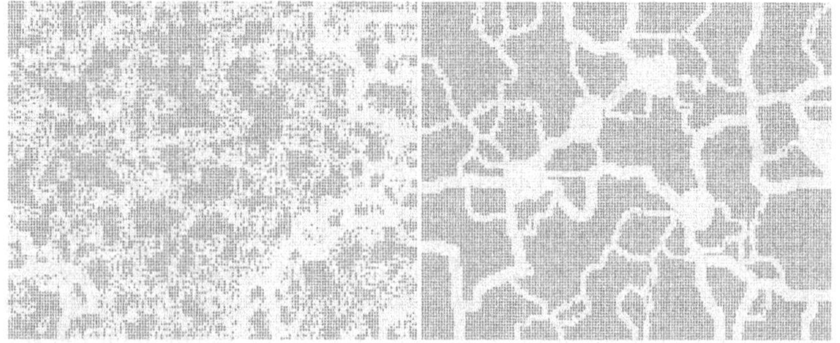

(b) These maps have the same properties but ran for two stagesinstead of one. This figure also shows the fractal properties as the larger clusters have smaller clusters generated inside.

Fig. 2: The effect of the stopping condition for the cellular automaton in the grouping procedure.

The parameters of the methods can be bound to the properties of the generated city layout. If the empty cell probability p is 0.5, the resulting city has more chaotic roads with wider clusters. A value of 0.65 generates more sparse, narrow clusters and linear roads. These values can be set differently for every stage; so a city layout with $p_{1=}$ 0.65,$p_{2=}$ 0.5 and $p_{3=}$ 0.55 for stages one to three is possible.

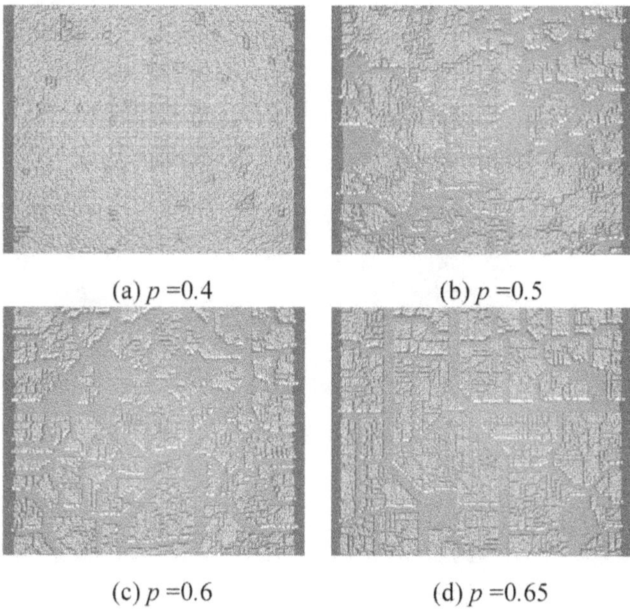

(a) p =0.4 (b) p =0.5

(c) p =0.6 (d) p =0.65

Fig. 3: The screenshots show the effect of p values to the generated city layouts. All cities are of size 120×150.

The difference limit variable l in the grouping procedure can also be used to affect the properties of the generated city. This parameter limits the number of cellular automaton steps.

The cellular automata can run repeatedly as many times as required. A low repetition count results in a fractured and chaotic city layout. Higher repetition count produces more unified clusters, however, it increases the running time of the method.

The running time of the method depends on the M and N values that define the size of the map and the difference limit l which controls the repetition of the cellular automata. The time complexity of the method is bound by the map size that the cellular automata operates on, and can be given as $O(MN)$. If $M = N$, the running time becomes $O(n^2)$. This is reflected in the running time of the city layout generation on a commodity hardware computer. Table II shows that the running time increases with $O(n^2)$ as does the size of the map.

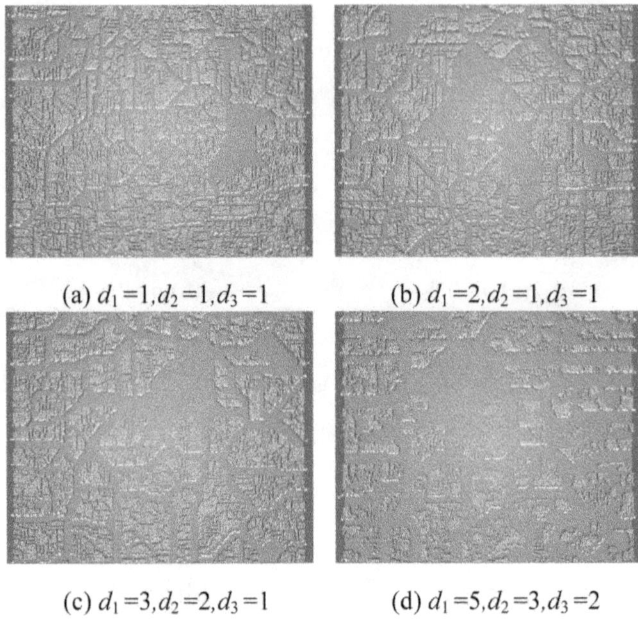

(a) $d_1=1, d_2=1, d_3=1$ (b) $d_1=2, d_2=1, d_3=1$

(c) $d_1=3, d_2=2, d_3=1$ (d) $d_1=5, d_2=3, d_3=2$

Fig. 4: Cities with different road width values. For the three stages shown in these screenshots distance is defined as d_1,d_2 and d_3.

TABLE II: Relation between city size and running time

100 x 100	less than 1 second
500 x 500	2 seconds
700 x 700	5 seconds
1000 x 1000	10 seconds
2000 x 2000	37 seconds

VI. Future Work

In this study we have mostly focused on the generation of the city layout. The resulting layouts have organic city properties that can be observed in large cities. The layout is used to generate buildings that are sensitive only to the major roads. The generated cities have a lot of potential for further improvements. It is important to state these improvements and lay the groundwork for future work.

The large spaces that can also be seen in the screenshots are defined as major roads, but they can be used as green areas such as public parks and recreation. It is also possible to set them as rivers that pass through cities. Another possibility is to set them as hills or mountains.

There are studies that can be used to generate different kinds of buildings [11]. Using the distance to major roads or parks, the building types can be defined and generated accordingly.

As future work, we plan to look into the generation of buildings, using terrain information for the city layout generation, adding road textures and green areas. For overall improvement, building placement and alignment will also be optimized.

References

[1] Michael Toy, Glenn Wichman, Ken Arnold, and Jon Lane. Rogue. Digital Game, 1980.

[2] David Braben and Ian Bell. Elite. Digital Game, 1984.

[3] Gillian Smith. An analog history of procedural content generation. In *Proceedings of the 2015 Conference on the Foundations of Digital Games (FDG 2015)*, 2015.

[4] Roland Linden, Ricardo Lopes, and Rafael Bidarra. Designing procedurally generated levels, 2013.

[5] P. Mawhorter and M. Mateas. Procedural level generation using occupancy-regulated extension. In *Proceedings of the 2010 IEEE Conference on Computational Intelligence and Games*, pages 351–358, Aug 2010.

[6] R. van der Linden, R. Lopes, and R. Bidarra. Procedural generation of dungeons. *IEEE Transactions on Computational Intelligence and AI in Games*, 6(1):78–89, March 2014.

[7] Chongyang Ma, Nicholas Vining, Sylvain Lefebvre, and Alla Sheffer. Game level layout from design specification. *Computer Graphics Forum*, 33(2):95–104, 2014.

[8] Joris Dormans. Adventures in level design: Generating missions and spaces for action adventure games. In *Proceedings of the 2010 Workshop on Procedural Content Generation in Games*, PCGames '10, pages 1:1–1:8, New York, NY, USA, 2010. ACM.

[9] Jonathon Doran and Ian Parberry. A prototype quest generator based on a structural analysis of quests from four mmorpgs. In *Proceedings of the 2Nd International Workshop on Procedural Content Generation in Games*, PCGames '11, pages 1:1–1:8, New York, NY, USA, 2011. ACM.

[10] R.M. Smelik, T. Tutenel, K.J. de Kraker, and R. Bidarra. A declarative approach to procedural modeling of virtual worlds. *Computers & Graphics*, 35(2):352–363, 2011. Virtual Reality in Brazil Visual Computing in Biology and Medicine Semantic 3D media and content Cultural Heritage.

[11] Pascal Muller, Peter Wonka, Simon Haegler, Andreas Ulmer, and Luc¨ Van Gool. Procedural modeling of buildings. *ACM Trans. Graph.*, 25(3):614–623, July 2006.

[12] B. Tearse, P. Mawhorter, M. Mateas, and N. Wardrip-Fruin. Skald: Minstrel reconstructed. *IEEE Transactions on Computational Intelligence and AI in Games*, 6(2):156–165, June 2014.

[13] Noor Shaker, Julian Togelius, and Mark J. Nelson. *Procedural Content Generation in Games: A Textbook and an Overview of Current Research.* Springer, 2016.

[14] Mark Hendrikx, Sebastiaan Meijer, Joeri Van Der Velden, and Alexandru Iosup. Procedural content generation for games: A survey. *ACM Trans. Multimedia Comput. Commun. Appl.*, 9(1):1:1–1:22, February 2013.

[15] Ruben M. Smelik, Tim Tutenel, Rafael Bidarra, and Bedrich Benes. A survey on procedural modelling for virtual worlds. *Computer Graphics Forum*, 33(6):31–50, 2014.

[16] George Kelly and Hugh McCabe. A Survey of Procedural Techniques for City Generation. *The ITB Journal*, 7(2):5, 2006.

[17] N. Brewer. Computerized dungeons and randomly generated worlds: From rogue to minecraft. *Proceedings of the IEEE*, 105(5):970–977, May 2017.

[18] N. Kato, T. Okuno, A. Okano, H. Kanoh, and S. Nishihara. An alife approach to modeling virtual cities. In *Systems, Man, and Cybernetics, 1998. 1998 IEEE International Conference on*, volume 2, pages 1168–1173 vol.2, Oct 1998.

[19] Yoav I. H. Parish and Pascal Muller. Procedural modeling of cities. In¨ *Proceedings of the 28th Annual Conference on Computer Graphics and Interactive Techniques*, SIGGRAPH '01, pages 301–308, New York, NY, USA, 2001. ACM.

[20] Thomas Lechner, Pin Ren, Ben Watson, Craig Brozefski, and Uri Wilenski. Procedural modeling of urban land use. In *ACM SIGGRAPH 2006 Research Posters*, SIGGRAPH '06, New York, NY, USA, 2006. ACM.

[21] Stefan Greuter, Jeremy Parker, Nigel Stewart, and Geoff Leach. Realtime procedural generation of 'pseudo infinite' cities. In *Proceedings of the 1st International Conference on Computer Graphics and Interactive Techniques*

in Australasia and South East Asia, GRAPHITE '03, pages 87–ff, New York, NY, USA, 2003. ACM.

[22] George Kelly and Hugh McCabe. Citygen: An interactive system for procedural city generation. In *Fifth International Conference on Game Design and Technology*, pages 8–16, 2007.

[23] Graham Whelan, George Kelly, and Hugh McCabe. Roll your own city. In *Proceedings of the 3rd International Conference on Digital Interactive Media in Entertainment and Arts*, DIMEA '08, pages 534–535, New York, NY, USA, 2008. ACM.

[24] Arnaud Emilien, Adrien Bernhardt, Adrien Peytavie, Marie-Paule Cani, and Eric Galin. Procedural generation of villages on arbitrary terrains. *The Visual Computer*, 28(6):809–818, Jun 2012.

[25] Lan Jianliang and Ding Youdong. EziCity: A Rapid Landscape Design Framework of Virtual 3d City. Atlantis Press, November 2013.

[26] Martin Gardner. Mathematical games: The fantastic combinations of John Conway's new solitaire game 'life'. *Scientific American*, 223:120–123, oct 1970.

[27] M Batty and Y Xie. From cells to cities. *Environment and Planning B: Planning and Design*, 21(7):S31–S48, 1994.

[28] Michael Batty. Cellular automata and urban form: A primer. *Journal of the American Planning Association*, 63(2):266–274, 1997.

[29] N. Kato, T. Okuno, R. Suzuki, and H. Kanoh. Modeling virtual cities based on interaction between cells. In *Systems, Man, and Cybernetics, 2000 IEEE International Conference on*, volume 1, pages 143–148 vol.1, 2000.

[30] J. Chen and H. Liu. Modeling virtual city based on fractal. In *2007 2nd International Conference on Pervasive Computing and Applications*, pages 78–83, July 2007.

15. Object Selection with New Generation Kinect Camera in 3D Environment

Ömer Faruk Çangır
Computer Engineering Department Hacettepe University Ankara, Turkey omerfarukcangir@gmail.com

Haşmet Gürçay
Mathematics Department Hacettepe University Ankara, Turkey gurcay@hacettepe.edu.tr

Abstract— Object selection is one of the most basic and important element in human computer interaction. As computer graphics environments move from two-dimensional environments to three-dimensional environments, interaction methods have begun to differentiate. This differentiation reveals various interaction problems. In this context, it is aimed to determine the most appropriate methods of selecting hand-held objects that can be applied in three-dimensional environments without the need for installation, marking wear or carrying in the work. Moreover, it is aimed to make the determined methods more efficient in terms of ease of use, speed and accuracy. For this purpose, three different experimental three-dimensional environments by using the Kinect sensor were developed in order to compare three most appropriate methods. Bubble Cursor, Depth Ray and Squad methods were compared on object sparse, object dense and object moving environments. In addition, the evaluation questionnaires filled by users after the experiment were also analyzed. As a result, although there is no significant difference between object less dense and object moving environments in terms of user preference and selection performance, in object dense environments, it was seen that users preferred to use squad method and also performed better performance by using squad method.

Keywords— human computer interaction; hands free object selection; kinect sensor

I. INTRODUCTION

With the development of virtual environments, the use of computer graphics has become widespread, and communication between people and computers has become more important. As computer user interaction processes increase, many types of interaction tools have emerged.

Whichever type of 3D computer graphic is used, selection process is one of the most basic elements used for user computer interaction [1]. For this reason, many selection methods working with 3D graphics have begun to be investigated with the aim of developing human-computer interaction with 3D graphics.

While human-computer interaction is provided, one of the most common techniques is the hand-free technique. Dam, Carvalho, Braz, Raponso and Haas [2] showed that many different operations by using the "hand free method" can be performed on selection techniques such as navigation, suppression, and holding.

In this study, object selection method is used to determine the methods that can be used to make human computer interaction more efficient. In this context, it is aimed to detect and develop methods that can be used to select objects in 3D environment more easily, faster and more precisely by using Kinect new generation camera. By examining the performance of the developed methods in different environments; the most appropriate method is identified in terms of using ease of use, speed and accuracy characteristics.

II. Related Works

One of the most common methods used to interact with 3D objects is the 'Ray Casting' method. Firstly, 'Ray Casting' method developed by Roth [3] was used for modeling and illuminating solid objects through transmitted light beams. By using this method, which appears to have effects in the object selection field, Liang and Green have developed a method named as the laser weapon [4]. In this method, the selection is carried out by means of rays transmitted from a hand-held device and the object that the beam passes through is selected. This method is also called 'Flashlight Selection' because it operates in a similar manner to the flashlight operating mechanism. The 'Flashlight Selection' method paves the way for other studies on object selection.

Another method developed by using 'Ray Casting' method was developed by Mine [5]. In this study, Mine investigated methods such as sending beams to determine the direction of movements, and examined the usages of these methods in motion, selection, modification and scaling. In addition, Mine, Brooks and Sequin [6] have proposed a method that the direction of motion by using two hands can be determined by the way of the beam. In addition, many different studies have been performed based on the 'Ray Casting' method [7] [8].

The object may intersect with more than one object in the dense environment depending on the density of the objects in the environment during the selection through the beam. Grossman and Balakrishnan [9] developed the 'Depth Ray' method based on the principle of selecting the object closest to a point on the ray to solve this problem. Another technique developed by them for selecting objects in 3D graphics environments is the 'Bubble Cursor' method [10]. They show that this method performs close to the 'Depth Ray' method.

Kopper, Bacim and Bowman [11], who specialize in object-dense environments, have developed the 'Squad' method. With this method, object selection in the dense environment has become more accurate, even though it is completed with multiple steps.

Later on, the cameras that are able to perceive the free hand motions are cheaper to produce (such as Microsoft Kinect, Asus Xtion, etc.) and the techniques that cameras are used have begun to be developed. The methods developed by using these types of tools have eliminated the need to wear any markers or hold markers on the hand. At the same time, these methods require less installation setup.

Guimbretiere and Nguyen [12], working on manual hand free interaction, have studied marked menu selection by using the first-generation Kinect sensor at close range and have proposed a selection technique applied by using two hands. Despite the significant improvement in hand free 3D interaction, this method is not preferred much because it requires the camera to be adjusted at a certain angle. Cashion [13] has proposed the 'Zoom' and 'Expand' hybrid methods to overcome the problem of selecting object in 3D dense environments. The 'Zoom' method aims to reduce the complexity of the object dense environments while the 'Expand' method facilitates the selection process by distributing the entire objects in the area to the screen.

In this study, many object selection methods mentioned by Argelageut and Andujar [14] were evaluated and it was decided that Kinect sensor is the best available to use and the most appropriate for 3D environments. The methods are considered separately and it has been determined that many methods are not appropriate for 3D environments. We conclude that 'Bubble Cursor', 'Depth Ray' and 'Squad' methods are more appropriate than other methods. It was seen that the selected methods show higher performance than the other methods compared with the related studies.

III. Experimental Game Development

Kinect new generation camera has been used as a depth sensor in 3D hand free object interaction studies, and the methods were developed to work with this camera. To ensure that users can more easily and quickly select objects in three-dimensional environments, methods developed earlier in different environments were updated to accommodate new environments. Then, in order to compare the applied methods with each other, each method was applied to

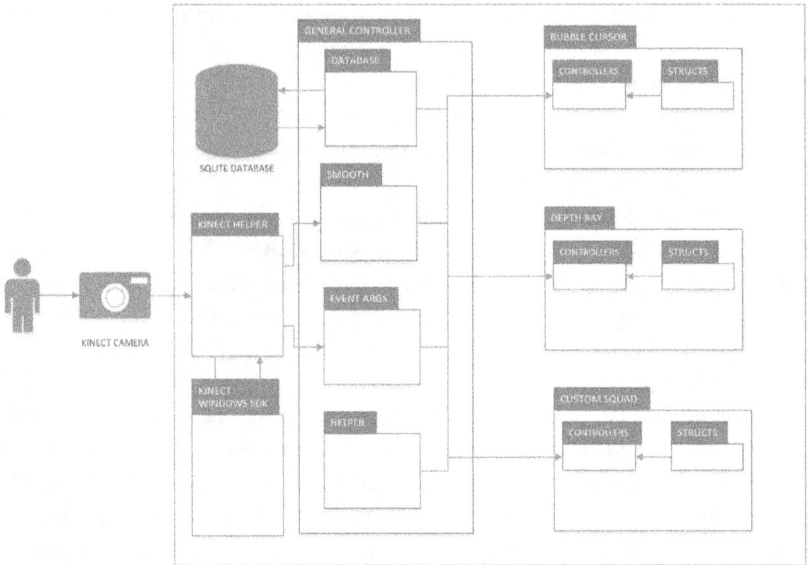

Fig. 1. 3D object selection game general system architecture

the same environments. Three different situations are discussed while the developed methods are compared with each other. In this context, object dense, object sparse and object moving environments are created. These three-dimensional environments were used to compare methods in terms of speed, ease of use and accuracy in different environments.

Object selection methods were developed on hand free methods, which are the easiest to use and do not require installation and a small game that have different levels is revealed. In this game, three different methods have developed with 3D tracking tools. These methods were updated to enable hand free interaction. In addition, many parts of the methods were modified to allow for easier, faster and more accurate selection, and these methods were compared. When comparing, the results in sparse, dense, and moving environments for each method were examined.

A. Experimental Game Design

Kinect second generation camera was preferred as a depth sensor in order to detect the user's hand movements. Software development kit was used to transfer joint positions into the application in 3D. The game was designed on the windows operating system. Unity 3D was prefereed as a game engine.

In the developed system, the joint data was extracted from the Kinect sensor by the help of software development kit and then it was used to determine the hand position and the selection gesture. As shown in Fig. 1 the joint data received via the Kinect assistant library. After that smooth packet is obtained so that movement could be obtained with less noise.

Detection of the object selection motion was performed with the aid of the event packet. The selection command sent from this package was processed in the control package of the corresponding method and the object selection was completed. In addition, the database package was used so that all the selected operations can be stored as permanent data.

B. User Interaction with Kinect in 3D Environments

The coordinates of the shoulder vertebrae intersection joint, right shoulder joint, right elbow joint and right wrist joint from the sensor were used to monitor the hand coordinate and transmit it to the screen. The left hand wrist joint coordinates were used in the method that the left hand should be used. In order to increase the ease of use performance, shoulder spinal joint was considered as the center and cursor coordinate was determined according to the position of the three-dimensional wrist joint coordinate with respect to shoulder spinal joint. In addition, the shoulder wrist distance was calculated so that the objects in the three-dimensional selection area can be easily selected wherever the position is. This distance was achieved by summing the distance between the shoulder elbow joints and the distance between the elbow wrist joints. This computed length was matched to the half of the screen width by non-linear matching, as in the 'go-go' method proposed by Poupryrev, Billinghutst, Weghorst andIchikawa [15]. This makes it easier to select the objects in the corners. In addition, the computed length was matched to the depth of the virtual environment so that the objects in the farthest back can be selected easily.

The object selection on the developed game was provided by closing the hand while it was open. In addition, the selection process was completed by using the hand moving the cursor. This makes the selection process easier.

The joint coordinates obtained with the help of the sensor must be matched to the screen coordinates after they are received from the software development kit. For this purpose, the joint data was firstly converted from the camera coordinate system to the screen coordinate system and from the screen coordinate system to the world coordinate system.

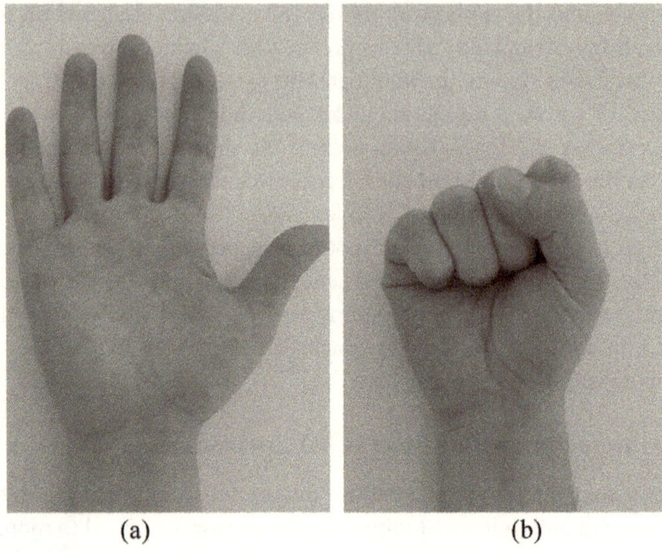

Fig. 2. Hand positions to select objects a) Hand open position is used when no selection is desired b) Hand punch position is used when selection desired.

The joint data taken from the sensor has many noise due to the distance between user and sensor. Besides, the software development kit does not work properly while trying to follow the joints. In some cases, the joints are intermingled each other. For these reasons, coordinates from the software development kit cannot be used directly. In order to avoid intermingling, filtering with arithmetic mean was used. In addition, to be able to select objects, perceiving of the selection process by punching the user's hand was preferred as shown in Fig. 2.

C. Environment Development

A level based method has been used in the experimental evaluation to ensure that different algorithms are handled in different environments. In order to examine the behavior of each method in dense, sparse and object moving environments, three different levels were created as shown in Fig. 3.

Generated levels make it possible to compare methods in depth. In sparse environment 12 selectable objects were placed and in dense environment 20 selectable objects placed. In the moving environment, 12 selectable objects in the virtual environment were allowed to move at fixed speeds between randomly specified coordinates. In this view, it was aimed to measure the effect of object

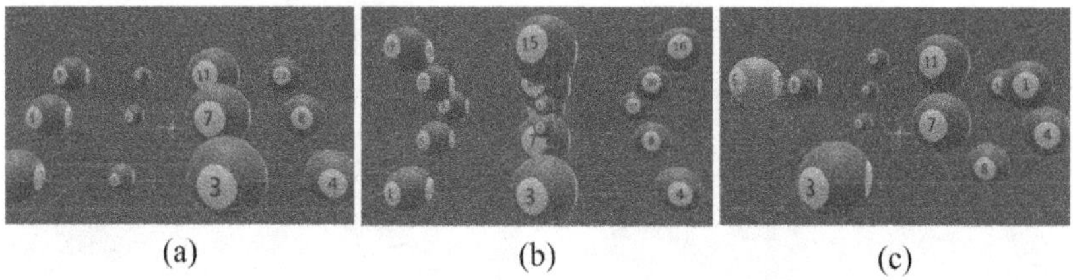

(a) (b) (c)

Fig. 3. View of different types of environments a) Sparse object environment b) Dense object environment c) Moving object environment

motion on selection performance. In addition, objects were used in different sizes in each level and it was aimed to determine the effect of object size on the object selection process. The difference in object size can be clearly seen in Fig. 4, where the object dense environment is viewed from six different angles.

While the selected methods are being applied, the object selected at that moment is displayed in blue color according to the operating principle of that method. At the same time, the object to be selected is displayed in green. In order to facilitate the selection of objects that are overlaid or behind one another, all objects on the transmitted beam and at a certain distance from the position of the cursor are set semi-transparently.

1) Modified 3D Bubble Cursor Method

In the modified method sphere around the cursor, whose dimensions are determined according to the working principle of the method, is shown in three dimensions. According to the size of this sphere, the closest object is activated and selected. An exemplary level display of the method was shown in Fig. 5.

2) Modified 3D Depth Ray Method

In the modified method, the beam was placed to the bottom of the camera. Then, the object closest to the cursor on the beam was correctly activated and ready for selection. When the 'Depth Ray' method is applied, it was preferred to extend the beam until the position of the cursor was reached, not extending to the back part of the selection area. An exemplary level display of the method was shown in Fig. 6.

3) Modified 3D Hybrid Squad Method

In the developed method, both 'Squad' method has been updated from two dimensions to three dimensions and the selection steps have been reduced from

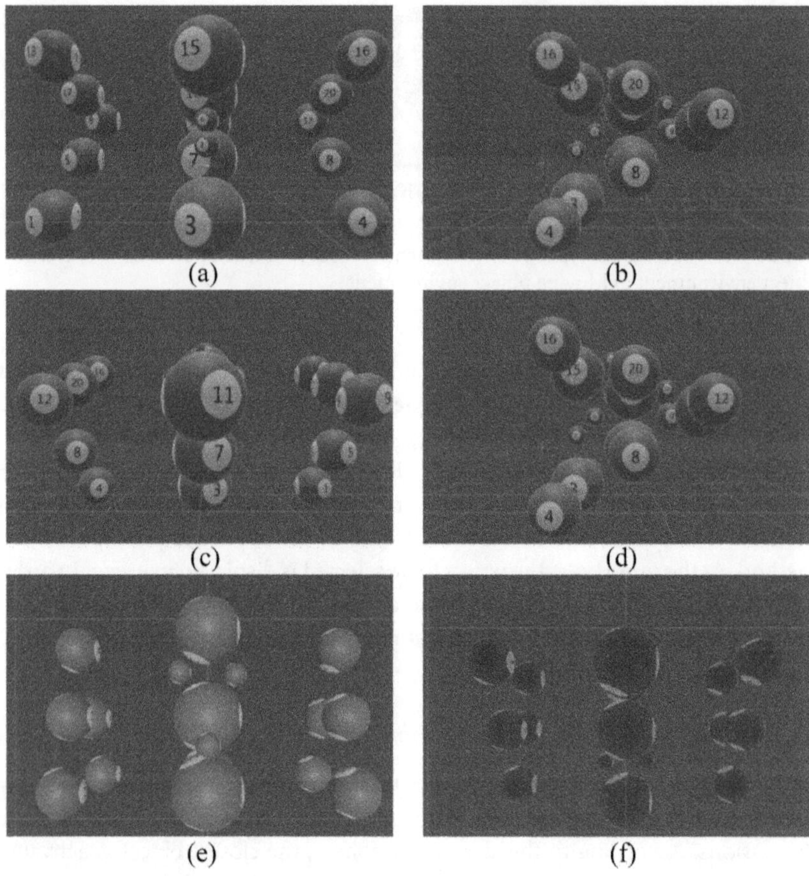

Fig. 4. Object dense environment from different views a)Front view b)Right view c) Back view d)Left view e)Top view f)Bottom view

several steps to one step. In order to realize this, a virtual ray was used as in the 'Depth Ray' method. The right hand was used to control the position of the cursor at the end of the beam, while the left hand was used to select the desired object from the selectable objects. According to the movement of the beam, all objects on the beam and at a certain distance from the cursor were listed on one side of the screen and these objects were made selectable according to the position of the left hand. Then, when the right-hand selection command was captured, the desired object was selected on the screen. An exemplary level display of the method was shown in Fig. 7.

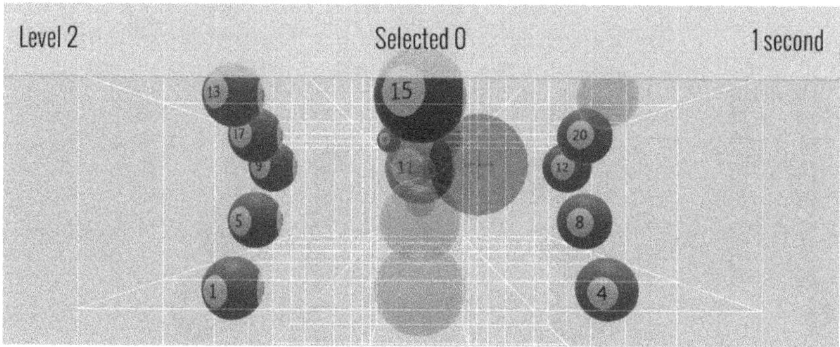

Fig. 5. Modified 3D Bubble Cursor method level screen

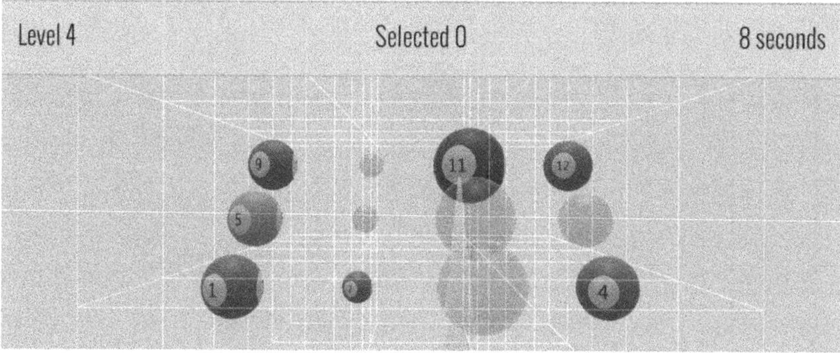

Fig. 6. Modified 3D Depth Ray method level screen

In the method screen left canvas, was used to choose the indicator value. The left hand used to select the object listed in the left canvas. An indicator value was determined according to the y-axis magnitude of the left hand shoulder vector created after left hand and shoulder positions were taken. Indicator value was classified as shown in Fig. 8.

D. Storing User Data

In order to ensure that the application runs on different platforms, sqlite database has been chosen to collect user data. In order to communicate with the sqlite database, a database package was created. Then the levels played by the users and all object selection times in the levels were recorded in the database.

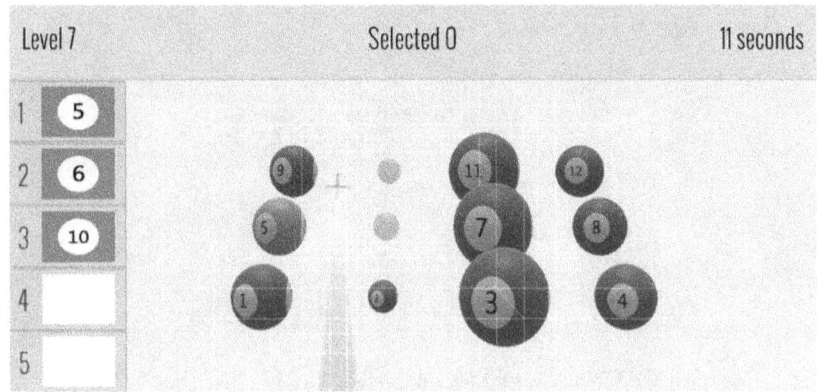

Fig. 7. Uyarlanmış 3B Hibrit Squad Yöntemi seviye ekranı

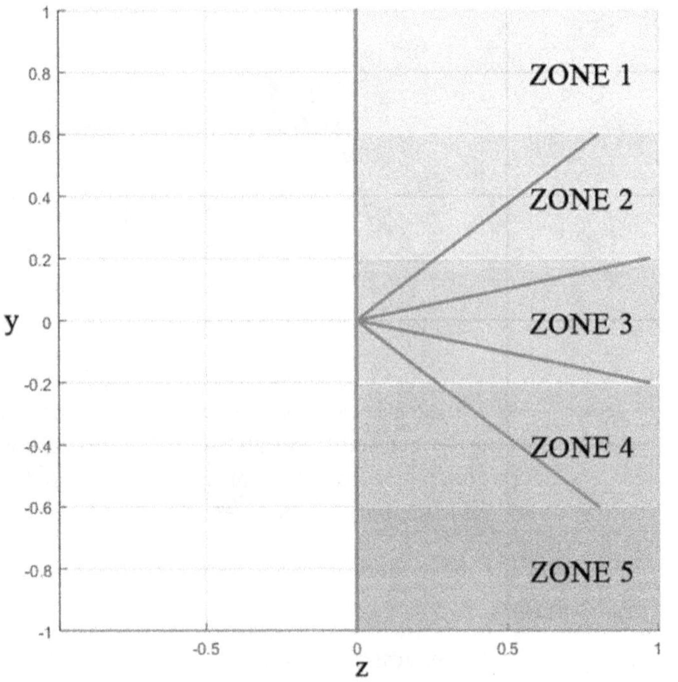

Fig. 8. Left arm selection zones

This method was intended to make level-based or object-based comparisons easier to compare methods.

IV. Application of User Tests and Analysis of the Data

A. Application of User Tests

The methods need to be tested with different users in order to compare them. Experimental 3D game developed for this purpose was tested with 17 different users (5 female, 12 male) from age range between 18 and 37 with mean age of 22. A user spent 20 minutes on average.

Before the user tests were performed, the general information was given about topics such as which objects should be selected or what the different objects on the screen express for all participants. It was aimed at the participants to be more familiar with the game.

Following the of the preparation and informing process, participants were registered to the system with the help of the developed interface. Subsequently, participants performed 72 object selection operations in 9 levels, including 8 different object selection operations at each level. All selection process times were recorded separately in the database.

The tests were carried out as shown in Fig. 9. After the end of the test session, the participants completed a questionnaire with 5 questions about their previous experiences and 8 questions about the methods they played. It was aimed to evaluate these methods for different participants in terms of ease of use, speed and accuracy.

All of the tests were completed on a 42-inch monitor that was running on an Intel Core-i5 processor-based laptop with an Intel HD 4000 graphics card with 8GB of RAM.

B. Fitts Analysis

The Fitts Law examines the relationship between the time required to complete a task, the distance between the target and the starting point, and the size of the target object [17]. Later, MacKenzie and Buxton made some improvements on this work and expanded the equality by applying it to two dimension [18]. As in (1) there was a relationship between marker moving time to target area MT, distance between start and target point D and target size W.

$$\mathrm{MT} = \mathrm{a} + \mathrm{b}log_2\left(\frac{D}{W} + 1\right) \tag{1}$$

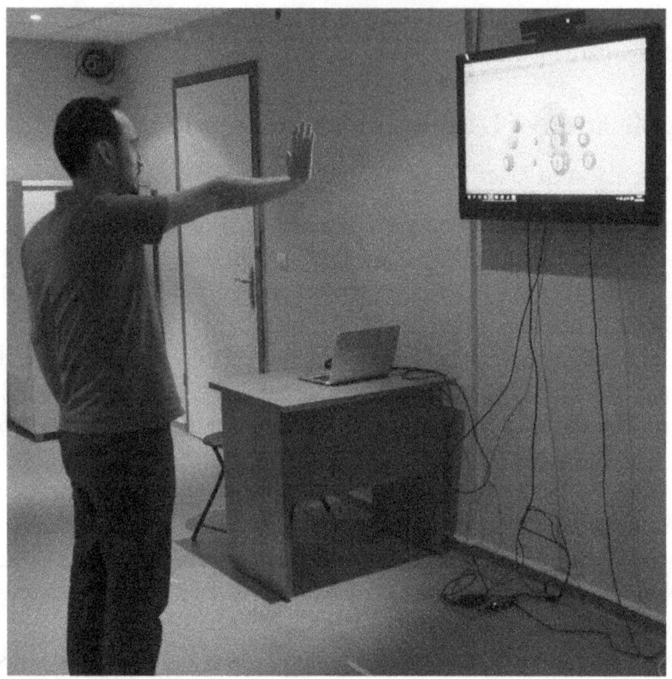

Fig. 9. Test environment where user tests are performed

In equation, phrase $(log_2 \frac{D}{W}+1)$ was treated as index of difficulty (ID). When the equation was interpreted, it can be seen that the difficulty level of the task increased when the distances between the start and end points was increased or the target size was decreased. Also, it can be deduced that there was a linear relationship between the index of difficulty and the elapsed time (a and b values are defined as environment based constants).

Test data obtained from the game were analyzed according to (1). The distance and target size were obtained from the position and radius information of the objects in the developed environment. Because of the different object numbers, locations, and selection sequences in the environments, the Fitts analysis was performed separately for dense and sparse environment. In the moving environment, this analysis was not performed because the distance value was unstable.

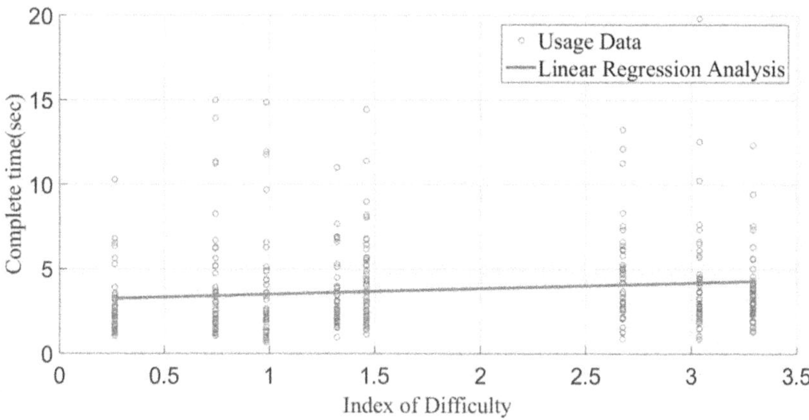

Fig. 10. Change of selection period depending on the difficulty in sparse environment

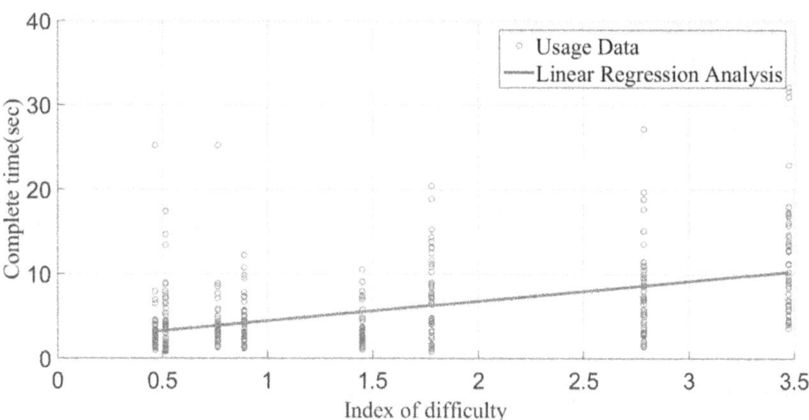

Fig. 11. Change of selection period depending on the difficulty in dense environment

The distance information of the object to be selected in equality was calculated as the Euclidean distance between the previous selected object. The distance calculation for the first selected object was made according to the origin. Also, since the objects in the environment were in a spherical shape, the target sizes used as 1, 8, 27 because of their radius values 1, 2, 3. Using all these values, index of difficulty was calculated and Fig. 10 was created for sparse environments and Fig. 11 was created for dense environment. When figures were examined, as the

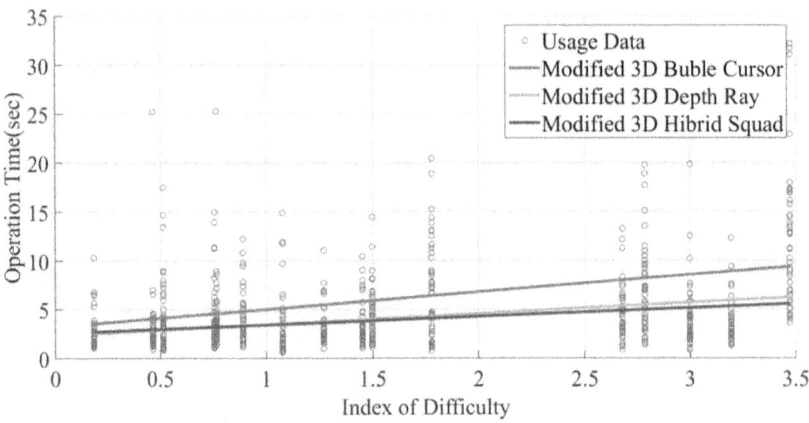

Fig. 12. Change of transaction time due to difficulty for developed methods

difficulty increases, it can be seen that the completion time increased for both environment.

While figures were created, values were obtained from raw data. For 8 different difficulty indices, the linear regression process was performed. Within 95 % confidence interval, confidence bound were [2.66, 3.71] and [0.07, 0.59] for Fig. 10 and [1.24, 2.85] and [2.93, 2.81] for Fig. 11. In addition, the linear regression coefficients a and b were 3.18 and 0.33 for Fig. 10; 2.05 and 2.37 for Fig. 11. The performance indicator value used by MacKenzie [18] in the performance evaluation of the Fitts law was evaluated by using IP = 1 / b equation. As a result of the evaluation, it can be interpreted that performance value of the object in the less dense environment (3.03) is higher than the performance value in the object dense environment (0.42).

While the methods were compared with each other, the values of each method in sparse and dense object environment were considered and compared in a single set. For the index of difficulty, a linear regression process was performed while plotting the transaction time graph as shown in Fig. 12. As a result of the regression procedure, the lower and upper confidence limits for the 'Modified 3D Bubble Cursor', 'Modified 3D Depth Ray' and 'Modified 3D Hybrid Squad' methods at 95 % confidence interval were [2.05 4.42] [1.21 2.45], [1.57 3.11] [0.70 1.51] and [1.82 3.23] [0.50 1.23]. In addition, when the linear regression coefficients were looked at, it was found that a was 3.24 and b was 1.83 for 'Modified 3D Bubble Cursor', a was 2.34 and b was 1.11 for 'Modified 3D Depth Ray' and also a was 2.53 and b was 0.87 for 'Modified 3D Hybrid Squad'. When

TABLE I. AVERAGE LEVEL COMPLETION PERIODS

Levels	Average (sec)	Data Count	Standart Deviation
Level 1	40,3197	17	13,08589
Level 2	56,8762	17	14,62668
Level 3	35,4934	17	10,12128
Level 4	24,6506	17	8,42693
Level 5	40,4128	17	13,21565
Level 6	24,9862	17	10,75386
Level 7	24,3904	17	7,76899
Level 8	38,6378	17	8,52657
Level 9	24,4620	17	7,48686
All	34,4699	153	14,78722

the performance indicator values were compared in this data, it can be seen that 'Modified 3D Hybrid Squad' method provide the highest performance with the 1.14 index of performance value.

C. Analysis of Participant Game Data

The normality test was completed before analyzing on transaction time dependent variable of the data obtained from the tests. The Kolmogorov-Smirnov normality assumption was made on all level data, and p value was less than 0.05 for all levels. In this case, normal distribution of level data was deemed appropriate. In addition, the homogeneity of the variances was tested and it was determined that the variances were homogeneous with p = 0.156 value. Table 1 shows the mean time to completion at 95 % confidence interval for the data that are normally distributed and homogeneous.

A graph of the data distribution in Table 1 according to the environment can be seen in Fig. 13.

As shown in Fig. 13, 'Modified 3D Depth Ray' and 'Modified 3D Hybrid Squad' methods can be selected in close to each other. This process can be completed in a longer time with 'Adaptive 3D Bubble Cursor' method. Although the 'Modified 3D Hybrid Squad' method does not seem to make the selection much faster than the 'Modified 3D Depth Ray' method, it needs to be analyzed whether this difference is meaningful. For this purpose, methods were grouped and ANOVA analysis was performed on the grouped data.

Kolmogorov-Smirnov normality assumption was completed on the data before ANOVA analysis and p values for all methods were found to be greater

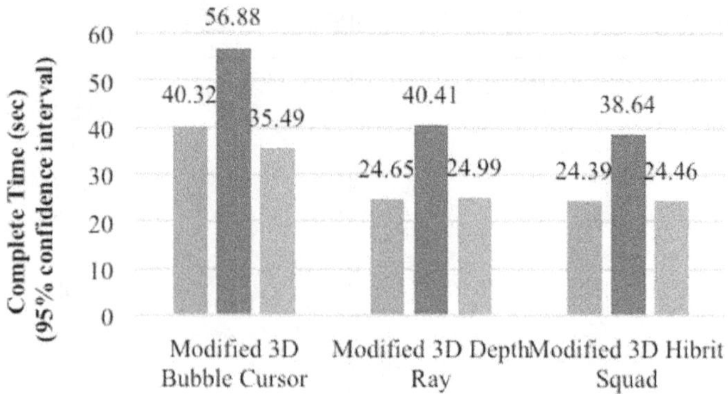

Fig. 13. Distribution of level completion time averages according to methods and environments

TABLE II. TUKEY HSD POST HOC MULTIPLE COMPARISON RESULTS THROUGH GROUPED DATA BY METHODS

(I) Method	(J) Method	Average Difference (I-J)	Standart Deviation	Sig. Level (p)
Modified 3D Bubble Cursor	Modified 3D Depth Ray	14,21323	2,60372	0,000
Modified 3D Bubble Cursor	Modified 3D Hybrid Squad	15,06634	2,60372	0,000
Modified 3D Depth Ray	Modified 3D Bubble Cursor	-14,21323	2,60372	0,000
Modified 3D Depth Ray	Modified 3D Hybrid Squad	0,85312	2,60372	0,943
Modified 3D Hybrid Squad	Modified 3D Bubble Cursor	-15,06634	2,60372	0,000
Modified 3D Hybrid Squad	Modified 3D Depth Ray	-0,85312	2,60372	0,943

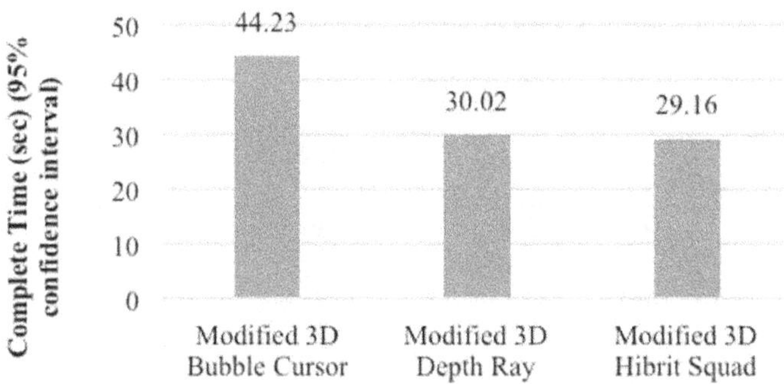

Fig. 14. Distribution of level completion time averages by methods

than 0.05. In addition, homogeneity test was performed on grouped data and it was found that the data were homogenous with p = 0.078 value. ANOVA analysis was performed on homogeneous and normally distributed data and it was seen that p was smaller than 0.05 (F = 21.130). With this data, the hypothesis showing that the methods differ in terms of completion times has been verified. The Tukey HSD post hoc multiple comparison test was performed to determine the difference between the two methods and the results are shown in Table 2.

When significance level is less than 0.05 it can be said that there is meaningful difference between methods. After the data were interpreted, it can be seen that there was no significant difference between the 'Modified 3D Depth Ray' and 'Modified 3D Hybrid Squad' method. It can be said that 'Modified 3D Bubble Cursor' method on the same data was different from other methods in terms of completion time. The graphical representation of the completion time at 95 % confidence interval of the grouped data according to the methods can be seen in Fig 14. It can be seen that the selection process using the 'Modified 3D Bubble Cursor' method was performed longer than the selection process in other environments.

In order to see the effects of environment on object selection, the data were classified as 'Object Sparse', 'Object Dense' and 'Object Moving' according to environment. Following this, Kolmogorov-Smirnov normality assumption was made on the data, and p values for all methods were found to be greater than 0.05. In addition, homogeneity test was performed on grouped data and it was found that the data were homogenous with p = 0.378 value. ANOVA analysis was performed on homogeneous and normally distributed data and it was seen that p was smaller than 0.05 (F = 28.033). With this data, the hypothesis showing that the methods differ in terms of completion times was verified. The Tukey HSD

TABLE III. TUKEY HSD POST HOC MULTIPLE COMPARISON RESULTS THROUGH GROUPED DATA BY ENVIRONMENTS

(I) Environment	(J) Environment	Average Difference (I-J)	Standart Deviation	Sig. Level (p)
Object Sparse	Object Dense	-15,52199	2,51499	0,000
Object Sparse	Object Moving	1,47304	2,51499	0,828
Object Dense	Object Sparse	15,52199	2,51499	0,000
Object Dense	Object Moving	16,99503	2,51499	0,000
Object Moving	Object Sparse	-1,47304	2,51499	0,828
Object Moving	Object Dense	-16,99503	2,51499	0,000

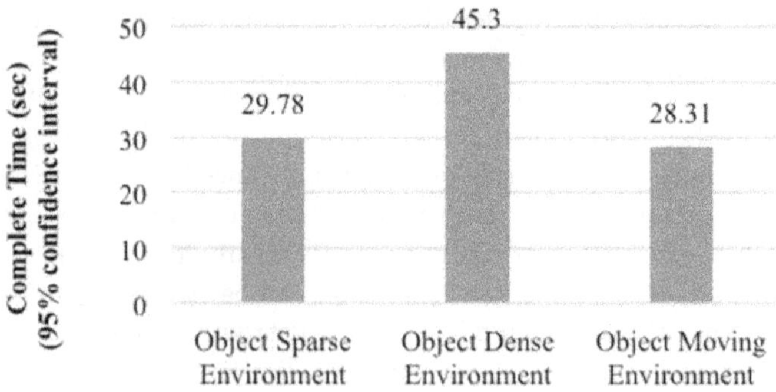

Fig. 15. Distribution of level completion time averages by environments

post hoc multiple comparison test was performed to determine the difference between the two environments and the results were shown in Table 3.

When the data were interpreted, it can be seen that there was no significant difference between sparse and moving object environments. It can be said that dense object environment on the same data was different from other environments in terms of completion time. The graphical representation of the completion time at 95 % confidence interval of the grouped data according to the environments can be seen in Fig 15.

D. Analysis of Previous Experiences of Participants

The first group of questions consisted of questions about the previous experience of the participants. When participants assessments of these questions were

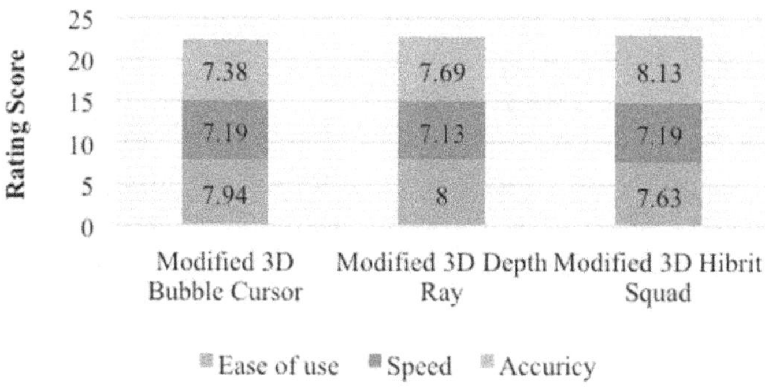

Fig. 16. Functional evaluation of methods

analyzed, it was seen that 43.75 % participants played video games at least once a week or more frequently. In addition, 75 % of the games played by participants are games that are developed for 3D environments and 12.5 % of the games played by the participants are games with parts that allow remote control. The game play rate that allows remote control with the Kinect camera was also 12.5 %. 81.3 % of participants also classified their playing skills as moderate or superior.

E. Analysis Game Evaluations of Participants

When the evaluations of first two questions in the user experience survey aiming to determine the qualification of the Kinect on the cursor movement and object selection process were analyzed, it was seen that the evaluation averages of both questions were 7 out of 10. In addition, if the evaluations lower than 5 were insufficient and higher than 5 were sufficient, the participants in both questions used 87.5 % sufficient votes.

Another group of questions in the user evaluation survey focused on the evaluation of methods in terms of ease of use, speed and accuracy. The averages of the values were compared so that these evaluations can be compared. The results of the participants scores over 10 points in terms of ease of use, speed and accuracy for each method were given in Fig. 16. It can be seen that the 'Modified 3D Hybrid Squad' method in this chart has a higher rating than the other two methods in total because it is highly evaluated for accuracy than the others.

Then normality and homogeneity test were completed by grouping data according to the methods. According to the Kolmogorov-Smirnov test, normal distribution of the data was not found to be appropriate, but the data were

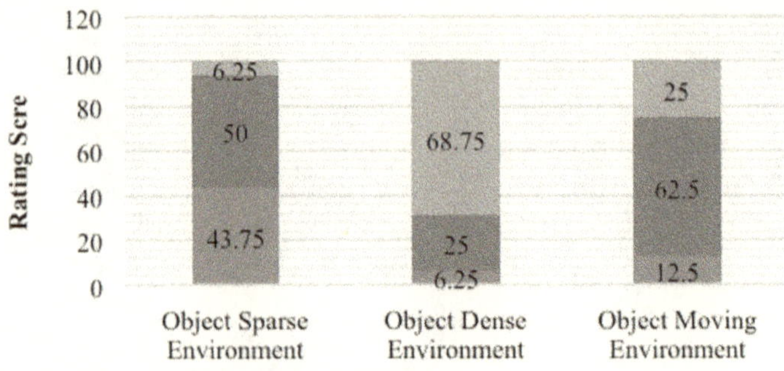

Fig. 17. Preferred environment-based methods

homogeneous. ANOVA analysis was performed after this step. (P = 0.91, F = 0.094). In this case, at 95 % confidence interval, the results were considered to be meaningless.

After evaluating the methods according to their functionality, the analysis in Fig. 17 was performed in order to determine the preferred methods in the environment. It can be seen that the 'Modified 3D Hybrid Squad' method was preferred in 'Object Dense' environments when the graphical data were analyzed, and 'Modified 3D Depth Ray' method was preferred in 'Object Dense' and 'Object Moving' environments.

The normality and homogeneity test was performed on the grouped data according to the environment. According to the Kolmogorov-Smirnov test, normal distribution of the data was not found to be appropriate, but the data were homogeneous. After this step, ANOVA analysis was completed ($p < 0.05$, F = 10,43). In this case, at 95 % confidence interval, it was achieved that the assessments differed significantly from each other. In order to determine which methods are different from each other, the Tamhane post hoc test was performed on the non-normal distribution data and the values in Table 4 were obtained. When the significance values of the table were examined, it was seen that there was no significant difference in the tests in the object moving environment and object dense environment. Also, it has been observed that the evaluations in object dense environments were different from those in other environments.

TABLE IV. POST HOC MULTIPLE COMPARISON RESULTS ON GROUPED ASSESSMENT DATA BY ENVIRONMENT

(I) Environment	(J) Environment	Average Difference (I-J)	Standart Deviation	Sig. Level (p)
Object Sparse	Object Dense	-1,000	0,219	0,000
Object Sparse	Object Moving	-0,500	0,219	0,086
Object Dense	Object Sparse	1,000	0,219	0,000
Object Dense	Object Moving	0,500	0,219	0,086
Object Moving	Object Sparse	0,500	0,219	0,086
Object Moving	Object Dense	-0,500	0,219	0,086

V. Conclusion

In the study, three methods which are most appropriate for object selection with hand free interaction were determined, and these methods were updated in accordance with the newly designed 3D virtual environment. An experimental game was created by using the designed 3D virtual environment. Kinect new generation camera was used to provide hands free interaction in the experiment. In order to observe the behaviors of the methods in different environments, sparse, dense and moving object environments were created, separately. Nine different levels were created by matching the developed methods to the designed environments.

After participants filled survey, object selection times were analyzed, and it was shown that the selection using 'Modified 3D Depth Ray' and 'Modified 3D Hybrid Squad' methods allowed selection in a shorter time than the 'Modified 3D Bubble Cursor' method. No significant difference was observed in the selection using the 'Modified 3D Depth Ray' and 'Modified 3D Hybrid Squad' methods. After object selection times were analyzed, participants' evaluations were analyzed by using a survey filled out by the participants. No significant difference was found in method evaluations in terms of ease of use, speed and accuracy. In addition, the 'Modified 3D Hybrid Squad' method was more preferred in 'object dense' environments, although there was no significant difference in between the 'object sparse' and 'object moving' environments in the analysis of preferred methods according to the environments.

While the selection process is carried out with the preferred method, sound detection capability of the Kinect camera can also be used. At the same time, both the selected method flow and the specific commands received by sound can be used to facilitate selection. For example, in the levels that the 'Modified 3D Hybrid Squad' method is used, the objects listed on the left side of the screen can be elected by picking up the number by voice. In addition, the scoring system in the heuristic methods can also be integrated into the systems.

References

[1] Bowman, D., Kruijff, E., LaViola, J., Poupyrev, I., 3D User Interfaces: Theory and Practice, Addison Wesley Longman Publishing, Redwood City,2004.

[2] Dam, P. F., Carvalho, F. G., Braz, P., Raposo, A. B., Haas, A., Hand Free Interaction Techniques for Virtual Environments, Proceedings of SBGames 2013, Sao Paulo, 100–108, 2013.

[3] Roth, S. D., Ray casting for modeling solids, Computer Graphics and Image Processing, 18, 109–144, 1982.

[4] Liang, J., Green, M., JDCAD: A highly interactive 3D modeling system, 3rd International Conference on CAD and Computer Graphics, Beijing, 18(4), 499–506, 1994.

[5] Mine, M., Virtual environments interaction techniques, University of North Carolina at Chapel Hill Chapel Hill, Chapel Hill, 1995.

[6] Mine, M. R., Brooks, F. J., Sequin, C., Moving objects in space: exploiting proprioception in virtual-environment interaction, Proceedings of the 24th annual conference on Computer graphics and interactive techniques (SIGGRAPH '97), Los Angeles, 19–26, 1997.

[7] Steed, A., Parker, C., 3d selection strategies for head tracked and non-head tracked operation of spatially immersive displays, 8th International Immersive Projection Technology Workshop, 163–170, 2004.

[8] Lee, S., Seo, J., Kim, G., Park, C. Evaluation of pointing techniques for ray casting selection in virtual environments, Proceedings of SPIE - The International Society for Optical Engineering, Hangzhou, 38–44, 2002.

[9] Grossman, T., Balakrishnan, R., The design and evaluation of selection techniques for 3D volumetric displays, Proceedings of the 19th annual ACM symposium on user interface software and technology (UIST '06), New York, 3–12, 2006.

[10] Grossman, T., Balakrishnan, R., The bubble cursor: enhancing target acquisition by dynamic resizing of the cursor's activation area, CHI 2005 Conference on Human Factors in Computing Systems (CHI '05), Portland, 281–290, 2005

[11] Kopper, R., Bacim, F., Bowman, D. A., Rapid and accurate 3D selection by progressive refinement, Proceedings of IEEE Symposium on 3D User Interfaces, Singapore, 67–74, 2011.

[12] Guimbretiere, F., Nguyen, C., Bimanual marking menu for near surface interactions, Proceedings of the 2012 ACM annual conference on human factors in computing systems (CHI '12), New York, 825–828, 2012.

[13] Cashion J., Intelligent selection techniques for virtual environments, Doktora Tezi, Engineering and Computer Science at the University of Central Florida, Florida, 2014.

[14] Argelaguet, F., Andujar, C., A survey of 3D object selection techniques for virtual environments, Computers and Graphics, 37(3), 121–136, 2013.

[15] Poupyrev, I., Billinghutst, M., Weghorst, S., Ichikawa, T., The Go-Go Interaction Technique: Non-Linear Mapping for Direct Manipulation in VR, Ninth Annual Symposium on User Interface Software and Technology (UIST '96), Seattle, 79–80, 1996.

[16] Fitts, P.M., The Information Capacity of the Human Motor System in Controlling the Amplitude of Movement, *Journal of Experimental Psychology*, 47(6), 1954.

[17] MacKenzie, S., Buxton, W., Extending Fitts' law to two-dimensional tasks, *Proceedings of the SIGCHI Conference on Human Factors in Computing Systems*, Monterey, 219–226, 1992.

[18] MacKenzie, I. S., Fitts' Law as a Performance Model in Human-Computer Interaction, Doktora Tezi, University of Toronto, Ontario, 2002.

16. 3D Fountain Modeling From Single Image

Şeyma Cengiz
Computer Engineering Department Ankara Yıldırım Beyazıt University Ankara, Turkey seyma.cengiz06@gmail.com

Abdullah Bülbül
Computer Engineering Department Ankara Yıldırım Beyazıt University Ankara, Turkey abulbul@ybu.edu.tr

Abstract— In this paper, a system is proposed to automatically generate 3D model of fountains from a single image. Although a single image is usually not sufficient for determining the 3D structure in an image; here we utilize our prior knowledge of general characteristic structures of fountains over the history such as the similarity of their top parts and their symmetrical shape. Firstly, object is detected and labelled on image using Convolutional Neural Networks (CNNs). Then, the object is segmented from background with Graph Cut technique and the contour of the object is used to estimate the shape of 3D model. Finally, 3D model of the fountain is completed by applying the texture acquired from the input image.

Keywords—Computer Graphics; 3D Modeling; Convolutional Neural Networks; Deep Learning; Graph Cut; Cultural Heritage

I. Introduction

Architectural heritage is one of the most important components of cultural heritage. Turkey is a country with a wealth and diversity at a universal level in terms of cultural heritage and ablution fountains (*şadırvan* in Turkish) form an important class of architectural heritage and they have an important place in Turkish-Islamic architecture. Through the history, various styles of fountains are built representing different eras. Figure 1 shows several important examples of Fountains.

Technological advances in ability to measure the physical world and in computer modeling capabilities have increased creation of 3D models of cultural heritage objects and environments. Generation of 3D models of these heritages on digital environment will contribute to the transfer to the generations, to the easier understanding of the architectural design. As the number and variety of 3D polygonal models increases in online repositories, there is a growing need for automatic algorithms that can obtain structural and semantic relationships from large model

Fig. 1. Fountain examples from Şehzadebaşı Mosque, Fatih Mosque, Konya Mevlana, and Hagia Sophia. Images are courtesy of Mustafa Cambaz [2].

collections [1] which led researchers to in- crease their efforts on 3D modeling. In general, reconstructing a 3D structure requires multiple views of the scene so that the coordinates of the corresponding features in different views are solved in combination which is a challenging task; in case of presence of only a single image, however, generating a 3D model becomes even more challenging. In that case, prior knowledge about the objects in the image helps revealing the 3D structure. Recent improvements in Computer Vision enables robust object detection, which makes it possible to utilize our prior knowledge of objects in an image.

In this study, our motivation is building a system for recon- structing fountains from a single image and generating a 3D mesh model of the output. Fountains have certain properties depending on the architectural style and the time period they are built in. These properties can be exploited to detect fountains and generate their 3D models. Generally, fountains have a polygonal base, a pointed dome, pool and taps on the inside and columns on side surfaces. We developed a system that generate 3D model of detected fountain objects using this information. Thus, our system contributes documenting cultural heritage by providing 3D models of historical artifacts for which we have only a single photograph.

II. Related Work

Many studies in literature have different approaches for 3D modeling of objects from images. For example, Mao and Xu developed a system [3] to detect drinking cups from single photo and reconstruct the target objects in 3D. There are three steps in that study. First step is detection part to find location of the objects. Histogram of Oriented Gradient (HOG) method as template feature matching method with sliding windows is used to detect the target object. Second step is 3D reconstruction part to find the optimal 3D cup model for detecting target object. They matched the edge maps of the target objects with parameters that specify shape and coordinates of cup. A cup model has 4 faces as outside wall, inside wall, outside bottom and inside bottom. Then, they project models on the background image and optimize their positions. Third step interactive image editing part as in [4] to adjust the parameters of the cup model.

In sweep-based modeling [5], researchers developed an interactive system to model and manipulate 3D man-made object from a single image. This method, called 3-sweep, provide the user to explicitly define the three dimensions of cylinder, cuboid or similar primitives using three sweeps. First and second sweeps are used to define the first and second dimension of a 2D profile. Third sweep is used to define the main curved axis of the primitive. This method cannot be used for modeling highly complex objects.

A group of researchers [6]–[8] used part based modeling for modeling objects having certain parts. Here, researches produce a part-based template algorithm that groups original models in clusters of models with their variations. The deformable templates are used to describe shape variations within a collection as in [9]. Each template represents a distribution of shapes with different geometric features. This algorithm is evaluated with different datasets such as collections of chairs. While this algorithm achieves higher accuracy for boxy parts, it is not very successful for shapes with complex parts.

In category-specific modeling [10], [11], researchers developed an algorithm that specifies category of models from images. The automatic object segmentation method which is 2D annotations present in computer vision datasets such as PASCAL VOC [12] is used to handle the objects from images. In their system, firstly, the objects are detected and segmented from image. Then, viewpoints and subcategories are predicted using a CNN based system. After that, the system learns the 3D shape of the model at canonical bounding box scale. The mean shape learned from the predicted subcategory is scaled by given the predicted bounding box.

Yan et al. [13] propose an algorithm to detect the object given an arbitrary 2D view using a general 3D feature model of the class. The motivation of this study is an efficient object detection system from the same class under different viewing conditions. 3D modeling of the object is considered to obtain better accuracy. In the feature modeling phase, SIFT detector is used to compute the features of 2D model views. The features computed in 2D images are mapped to the 3D model by using homographic framework.

With general 3D modeling approach in [14], a method is presented for grasping from a single view of a 3D sensor. There are two approaches to make the model as grasp planning and locating objects in camera images using the information given by the geometric clues. The model is developed for fast estimation of different kinds of symmetries and unknown objects. In this system, firstly, data is cleaned from noise. Next, a region of interest (ROI) is applied to detect the object using a variant of the RANSAC algorithm. After, estimation is made over the remaining data a contour is fitted. Finally, all fitted models are triangulated and triangulation results are merged. Rothganger et al. [15] models objects in 3D in terms of local affine-invariant descriptors of photos and video views. Geometric constraints occur due to different views of the same patches under affine projection. These constraints are combined with a normalized representation of their appearance.

The study of Jiang et al. [16], builds symmetric architecture from single image, and the aim of this study is similar to ours; however, in this study user interaction is required while our approach is fully automatic.

There are also approaches based on structure from motion (SfM) that require a vast number of views of the modeled scene [17]–[19]. These approaches find visual features, e.g, SIFT features, in input images and match corresponding features among views. Then these correspondences are used to extract camera locations of the input images as well as 3D positions of the common features among images resulting in a sparse point cloud of the scene. The advantage of these systems is their capability of generating complex scenes as long as the scene doesn't contain highly reflective and repetitive surfaces. On the other hand, for a robust reconstruction, a notable number of input images (<50) are required and the result is a sparse point cloud which is usually densified by other methods such as [20] before surface reconstruction.

III. System Overview

A system is built for image-based modeling of fountains in 3D. The system takes a single image as the input and produces a 3D mesh model as the output by applying the steps summarized in Figure 2. In the first step, our model employs deep learning for detecting the fountain object in the input image. Then the

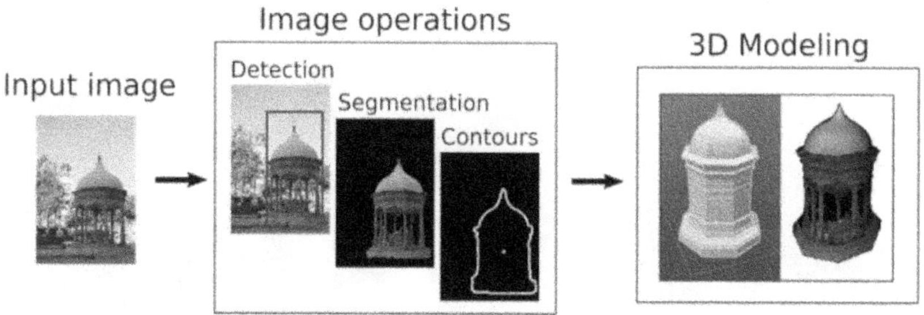

Fig. 2. Overview of the system.

background in the image is removed and boundaries of the detected object are determined. These boundaries are used for generating a 3D mesh model resembling generalized cylinders. Finally, a texture made from the input image is mapped over the 3D model.

A. Object Detection

The goal of object detection is to find an object of a predefined class in a static image or video frame. There are lots of detection algorithms [21] in computer vision including feature extraction and training steps. In this study, Convolutional Neural Networks (CNNs) that include both feature extraction and classification was used to detect fountain objects.

1) Convolutional Neural Networks (CNNs): Nowadays, CNNs are very popular in deep learning that recently have proven to be very successful at image recognition. CNNs are inspired by the visual cortex of the human brain and they are a special case of Artificial Neural Network (ANN). CNNs include feature extraction function before the normal feed-forward neural network. Computational process of operations is slow because they involve intensive processing. GPU programming with CUDA is used to speed up operations in training and testing phases.

CNNs architecture is shown on Figure 3. The input image passes to the first convolutional layer. The filters applied in the convolution layer extract relevant features from the input image to pass further. Pooling layers are then added to further reduce the number of parameters. Several convolution and pooling layers are added before the prediction is made.

Detection process is implemented with Darknet Yolo [23] which is a framework for detection and classification with CNN architecture. YOLO, abbreviated

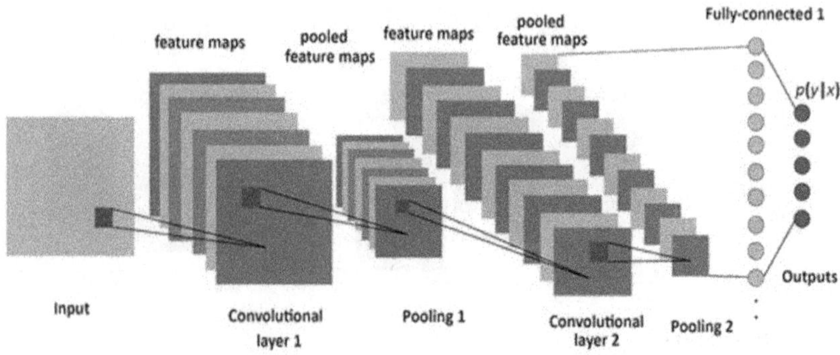

Fig. 3. CNN Architecture [22]

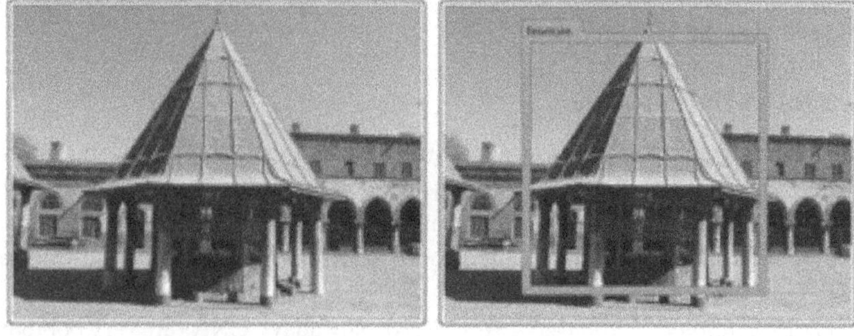

Fig. 4. Deep learning-based fountain detection Left: Test image [24] Right: Prediction image with the bounding box of the detected object.

from "You Only Look at Once", is an open source neural network framework written in C and CUDA. It is currently the state of the art on real time object detection methods. This model has several advantages over classifier-based systems. It applies a single neural network to the full image, so its predictions are informed by global context in the image. This network divides the image into regions and predicts bounding boxes and probabilities for each region. These bounding boxes are weighted by the predicted probabilities. We have trained the system with a dataset of 200 fountain images acquired from the Internet. After training the system with CNNs, the fountain objects are detected on test images as shown in Figure 4.

B. Object Segmentation

The process of separating an image into foreground and background is known as image segmentation and it is a major preprocessing step in many vision-based applications. The fountain object is segmented out from the background by using GrabCut technique [25]. GrabCut uses information encapsulated in the image as in most segmentation techniques. It makes use of both edge and region information. This information is used to create an energy function which, when minimized, produces the segmentation. Initially user draws a rectangle around the foreground region. In this study, the foreground region is determined by the bounding box obtained from the object detection step. A Min-cut/Max-Flow algorithm is used to segment the graph. This algorithm determines the minimum cost cut that will separate the Source and Sink nodes. The cost of the cut is determined by the sum of all the weights of the links that are cut. Once the Source and Sink nodes are separated, all pixel nodes connected to the Source node become part of the foreground, and the rest become part of the background. Figure 5 shows a simplified diagram of the GrabCut approach.

We have used OpenCV's implementation of GrabCut and Figure 6 shows a sample result of segmentation, eliminating background while keeping the fountain as the foreground object.

C. Generation of 3D Model

After the segmentation step, we have an isolated fountain image which needs to be converted to a 3D model. Here, we utilize the symmetric form of fountains. The general pattern of the model is formed by giving approximate radius values from top to bottom according to the symmetrical structure of all the side surfaces of the fountain. These radius values are calculated using contours of the fountain image. Contours are extracted by flood filling the foreground object and applying an edge detection method. In order to remove high frequency noise from the extracted boundaries, the image is low-pass filtered before edge detection. Contour extraction is shown in Figure 7.

Radius values of model, r_i, are measured according to distance between the x coordinate of every contour point and x coordinate of the fountain center. The important parameters are shown in Table I.

Having the radius values, a 3D model of the fountain is generated in the form of generalized cylinders as shown in Figure 8.

After the generation of the 3D model of the fountain object, texture coordinates are calculated for every vertex of the model and the input image is used as

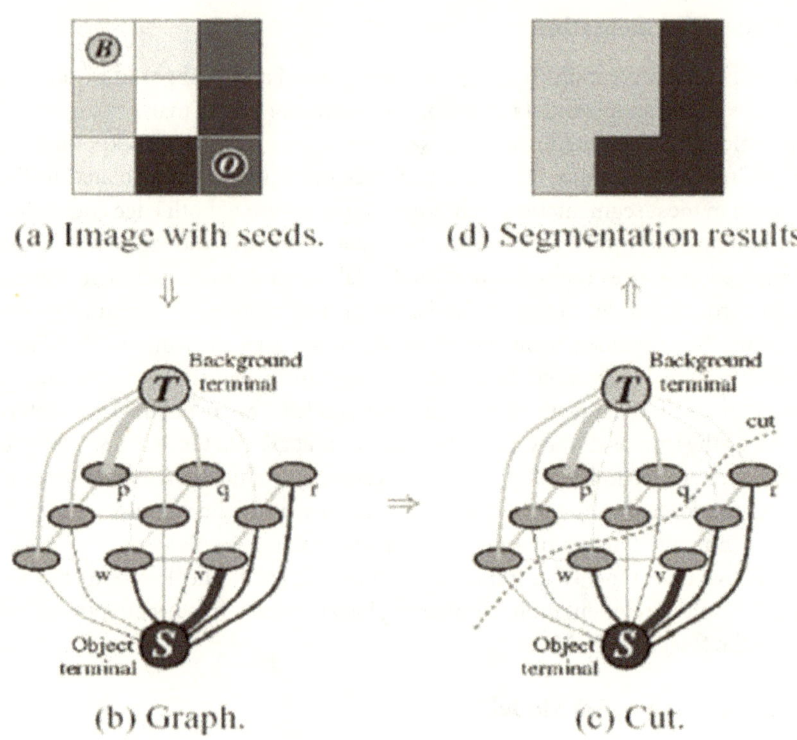

Fig. 5. Diagram of the GrabCut [26]

the texture map to increase realism of the output. Figure 9 shows texturing to 3D model of sample fountain.

IV. Results and Discussion

In this study, we have proposed a simple way for generating a 3D model of fountain object from a single image. The example outputs of project are shown on Figure 10. We create models based on the structural characteristics of the fountain objects. Firstly, to find location of the fountain on an image, detection operation is performed by taking advantage of the information that the fountains are in a hierarchical structure with a dome on top, a pool in the middle, and columns. Although they are generally in the same hierarchical structure, there are fountains with different architectural design such as differently shaped dome, hexagonal, octagonal, etc. Therefore, it would be difficult to recognize them with

Fig. 6. Left: Prediction image [27] Right: Segmented Object using GrabCut

TABLE I PARAMETERS OF A MODEL

r_i	the radius of the index-i circle face
x_0	the centers coordinate on the drawing
x	the x coordinate on the drawing
$r_i = (x_0 - x)$	where $x < x_0$

traditional machine learning methods [28]. If data show a hierarchical structure such as a cat (has whiskers, paw, etc.), deep learning has no difficulty in learning. It learns this hierarchical structure. We used CNNs and GPU acceleration from the deep learning methods to get robust results in the detection step. The results suggest that CNNs can successfully detect fountain objects despite the variety of their styles. After the detection step, 3D model is generated by taking advantage of the symmetrical structure of all the side surfaces of the fountain. To find symmetric profile of the fountain, fountain image is separated from background by applying image segmentation technique using GrabCut method. Then, contours

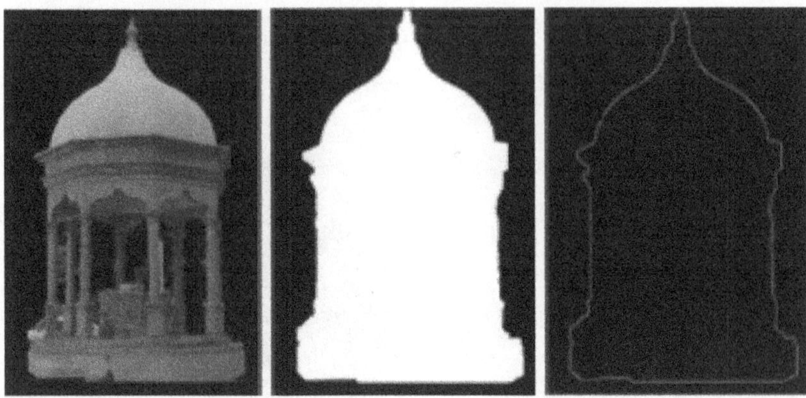

Fig. 7. Extracting contour of fountain object. Left: segmented image Middle: Foreground image by flood filling Right: contour of fountain

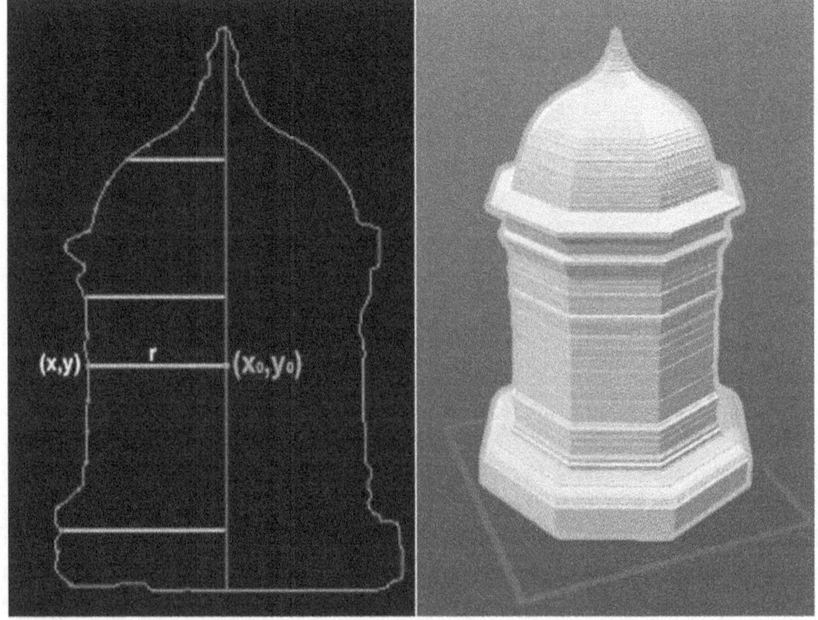

Fig. 8. Left: Determining the radii of model for each height. Right: Generation of 3D model

Fig. 9. Textured 3D model.

of the fountain image, which helps determining the radius of surface of revolution, are found.

This approach gives satisfactory outputs if fountain images are not taken from very close distances, ideal case would be images with an orthographic projection which is not possible with ordinary cameras. To increase the quality of outputs, camera position relative to the image can be estimated using vanishing points in the image. Although the system is currently used for modeling fountains, it can also be used for other structures having rotational symmetry.

Limitations. Despite the success of the proposed system to generate 3D models, it has several limitations that need further efforts. First, the current algorithm does not model every features of the fountain, but only models using symmetric surface information. Second, boundary box of some detected fountain cannot fully enclose the fountains, they occasionally extend outside the boundary boxes, causing incorrect segmentations. As stated before, determining the camera location would increase the success of 3D modeling and texture mappings steps.

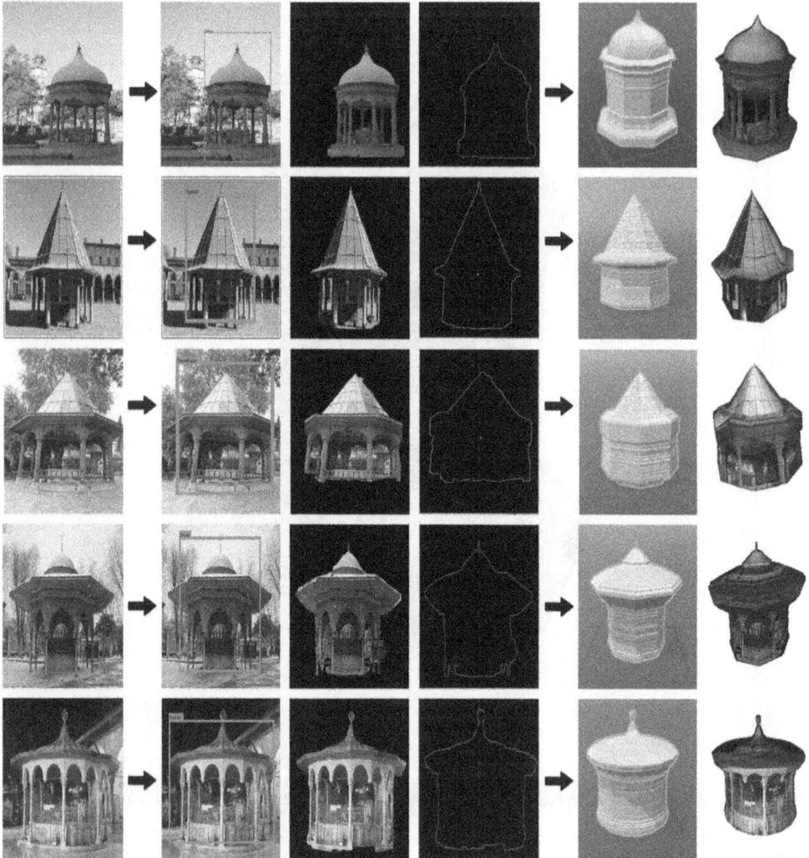

Fig. 10. Results. Left: Input images, middle: Operations to generate 3D model, right: Outputs

V. Conclusion

We have proposed a system for generating fountains from single images. The algorithm employs recent advances in machine learning to extract the fountain in an image and generates its 3D model exploiting its symmetric form. The results indicate that the system can reconstruct the general structure of the fountains in the photos. This study also helps cultural heritage by enabling easy modeling and documentation of historical architecture. In the future, the systems performance can be further increased by including more training images in object detection part. We also plan to automatically estimate the polygonal structure (how many columns) of fountains and plan to model every properties of the fountains, e.g., taps, chairs etc., using procedural modeling. Also, texturing the 3D model according to the camera viewpoint remains as a future work. We expect to develop a fully automatic system to store 3D models of historical artifacts, not only fountains, in a database. Lastly, our study can be extended to recognize the architectural style of the modeled architecture and categorize them accordingly.

Acknowledgment

One of the authors, Abdullah Bülbül, is supported by TUBITAK BIDEB 2232 program (Project number: 117C010).

References

[1] Y. Gingold, T. Igarashi, and D. Zorin, "Structured annotations for 2d-to-3d modeling," *ACM Trans. Graph.*, vol. 28, no. 5, pp. 148:1–148:9, Dec. 2009. [Online]. Available: http://doi.acm.org/10.1145/1618452.1618494

[2] M. Cambaz, http://www.mustafacambaz.com/, accessed: 2017-08-29.

[3] J. Mao and L. Xu, "Automatic 3d reconstruction via object detection and 3d transformable model matching cs 269 class project report," 2014.

[4] N. Kholgade, T. Simon, A. Efros, and Y. Sheikh, "3d object manipulation in a single photograph using stock 3d models," *ACM Trans. Graph.*, vol. 33, no. 4, pp. 127:1–127:12, Jul. 2014. [Online]. Available: http://doi.acm.org/10.1145/2601097.2601209.

[5] T. Chen, Z. Zhu, A. Shamir, S.-M. Hu, and D. Cohen-Or, "3-sweep: Extracting editable objects from a single photo," *ACM Trans. Graph.*, vol. 32, no. 6, pp. 195:1–195:10, Nov. 2013. [Online]. Available: http://doi.acm.org/10.1145/2508363.2508378

[6] V. G. Kim, W. Li, N. J. Mitra, S. Chaudhuri, S. DiVerdi, and T. Funkhouser, "Learning part-based templates from large collections of 3d shapes," *ACM Trans. Graph.*, vol. 32, no. 4, pp. 70:1–70:12, Jul. 2013. [Online]. Available: http://doi.acm.org/10.1145/2461912.2461933

[7] S. Eslami and C. Williams, "A generative model for parts-based object segmentation," in *Advances in Neural Information Processing Systems 25*, F. Pereira, C. J. C. Burges, L. Bottou, and K. Q. Weinberger, Eds. Curran Associates, Inc., 2012, pp. 100–107. [Online]. Available: http://papers.nips.cc/

paper/4774-a- generative-model-for-parts-based-object-segmentation.pdf

[8] Y. Amit and A. Trouve, “Pop: Patchwork of parts models for object recognition,” *Int. J. Comput. Vision*, vol. 75, no. 2, pp. 267–282, Nov. 2007. [Online]. Available: http://dx.doi.org/10.1007/s11263-006-0033-9

[9] A. K. Jain, Y. Zhong, and M.-P. Dubuisson-Jolly, “Deformable template models: A review,” *Signal processing*, vol. 71, no. 2, pp. 109–129, 1998.

[10] A. Kar, S. Tulsiani, J. Carreira, and J. Malik, “Category-specific object reconstruction from a single image,” in *Proceedings of the IEEE Conference on Computer Vision and Pattern Recognition*, 2015, pp. 1966–1974.

[11] R. J. Lopez-Sastre, T. Tuytelaars, and S. Savarese, “Deformable part models revisited: A performance evaluation for object category pose estimation,” in *Computer Vision Workshops (ICCV Workshops), 2011 IEEE International Conference on*. IEEE, 2011, pp. 1052–1059.

[12] “The pascal object recognition database collection,” http://host.robots.ox.ac.uk/pascal/VOC/databases.html, accessed: 2017-08-29.

[13] P. Yan, S. M. Khan, and M. Shah, “3d model-based object class detection in an arbitrary view,” in *2007 IEEE 11th International Conference on Computer Vision*, Oct 2007, pp. 1–6.

[14] Z.-C. Marton, D. Pangercic, N. Blodow, J. Kleinehellefort, and M. Beetz, “General 3d modelling of novel objects from a single view,” in *Intelligent Robots and Systems (IROS), 2010 IEEE/RSJ International Conference on*. IEEE, 2010, pp. 3700–3705.

[15] F. Rothganger, S. Lazebnik, C. Schmid, and J. Ponce, “3d object modeling and recognition from photographs and image sequences,” *Toward Category-Level Object Recognition*, pp. 105–126, 2006.

[16] N. Jiang, P. Tan, and L.-F. Cheong, “Symmetric architecture modeling with a single image,” in *ACM Transactions on Graphics (TOG)*, vol. 28, no. 5. ACM, 2009, p. 113.

[17] N. Snavely, S. M. Seitz, and R. Szeliski, “Photo tourism: Exploring photo collections in 3d,” in *SIGGRAPH Conference Proceedings*. New York, NY, USA: ACM Press, 2006, pp. 835–846.

[18] N. Snavely, S. Seitz, and R. Szeliski, “Modeling the world from internet photo collections,” *International Journal of Computer Vision*, vol. 80, no. 2, pp. 189–210, 2008. [Online]. Available: http://dx.doi.org/10.1007/s11263-007-0107-3

[19] S. Agarwal, N. Snavely, I. Simon, S. M. Seitz, and R. Szeliski, “Building rome in a day,” in *International Conference on Computer Vision*, Kyoto, Japan, 2009.

[20] Y. Furukawa and J. Ponce, "Accurate, dense, and robust multiview stereopsis," *IEEE transactions on pattern analysis and machine intelligence*, vol. 32, no. 8, pp. 1362–1376, 2010.

[21] P. F. Felzenszwalb, R. B. Girshick, D. McAllester, and D. Ramanan, "Object detection with discriminatively trained part-based models," *IEEE transactions on pattern analysis and machine intelligence*, vol. 32, no. 9, pp. 1627–1645, 2010.

[22] S. Albelwi and A. Mahmood, "A framework for designing the architectures of deep convolutional neural networks," *Entropy*, vol. 19, no. 6, p. 242, 2017.

[23] J. Redmon, S. Divvala, R. Girshick, and A. Farhadi, "You only look once: Unified, real-time object detection," in *Proceedings of the IEEE Conference on Computer Vision and Pattern Recognition*, 2016, pp. 779–788.

[24] O. Cakoglu, https://www.flickr.com/photos/ondercakoglu/4534409162/in/photostream/, accessed: 2017-08-30.

[25] C. Rother, V. Kolmogorov, and A. Blake, "Grabcut: Interactive foreground extraction using iterated graph cuts," in *ACM transactions on graphics (TOG)*, vol. 23, no. 3. ACM, 2004, pp. 309–314.

[26] M. Marsh, "Implementing the "grabcut" segmentation technique as a plugin for the gimp," http://www.cs.ru.ac.za/research/g02m1682/, ac- cessed: 2017-09-29.

[27] E. Cali, http://www.panoramio.com/photo/4983965, accessed: 2017-08- 30.

[28] M. Weber, M. Welling, and P. Perona, "Unsupervised learning of models for recognition," *Computer Vision-ECCV 2000*, pp. 18–32, 2000.

17. Real-Time Distant Light Filtering Using Gaussian Mixture Model

Özkan Anıl Töral
Department of Computer Engineering Yaşar University, Turkey ozkantoral@gmail.com

Serkan Ergun
International Computer Institute Ege University, Turkey serkan.ergun@ege.edu.tr

Aydın Öztürk
Department of Computer Engineering İzmir University, Turkey aydin.ozturk@izmir.edu.tr

Abstract— We propose a novel real-time rendering technique using GMM for environment lighting. We represent isotropic and anisotropic BRDF using sum of SG fitted by EM algorithm, which provide an accurate approximation with acceptable number of lobes. To suppress the approximation errors, we use GPU generated MIP-maps for filtering environment maps that does not require a pre-computation. MIP-mapped lookup is performed with the size of SG lobes to make filtering efficient. Based on empirical results, it is shown that both isotropic and anisotropic reflectances can be handled in real-time using our technique.

Keywords— Real-time Rendering; Precomputed Radiance Transfer; Spherical Gaussian

I. Introduction

Numerical estimation of the illumination integral can produce some approximation errors. To reduce these errors, various noise reduction techniques such as multiple importance sampling [24] have been proposed. On the other hand, similar to our technique, an interesting approach for eliminating numerical estimation would be convolving environmental maps as a pre-filtering step. Since the isotropic BRDF are radially symmetric, environment maps can easily be filtered using only the lobe direction and size, but orientation of the lobe must also be taken into account for anisotropic materials. In this work we demonstrate that GMM can provide an accurate approximation for anisotropic BRDF representation.

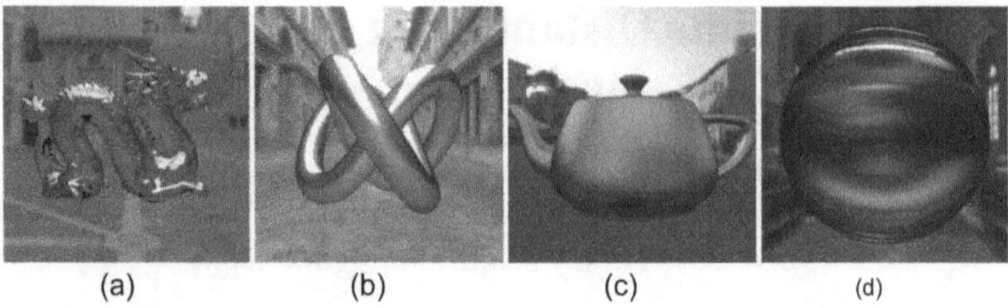

Figure 1. Our proposed method can handle a wide range of materials under different lighting conditions: (a) specular material, (b,c) glossy materials, (d) anisotropic material.

Adopting SG to represent spherical function is a widely used technique to achieve real-time rendering of reflectance, BRDF, under environment lights. Recent techniques [23, 25, 29] used SG to represent BRDF, lighting and visibility approximation.

Our main contribution in this work is to use GMM fitted by EM algorithm to represent both isotropic and anisotropic BRDF. For the pre-filtering step, we use hardware generated MIP-maps in real-time, which does not require pre-computation, to suppress the approximation errors.

Although similar approaches have been proposed [23, 25, 29, 11], our technique is more efficient than its competitors in terms of computation time of fitting GMM, and has the advantage that scene can be changed dynamically.

The rest of this paper is organized as follows. First, some of the relevant work on environment maps filtering and PRT are presented in Section 2. Background information for SG is given in Section 3. Our BRDF approximation technique, using GMM, is described in Section 4. Using GMM for distant lighting in real-time is explained in Section 5. Empirical results are presented in Section 6 and Section 7 is devoted to conclusion and future work.

II. Related Work

There has been extensive effort for evaluating the Rendering Equation [12]. A common approach for the evaluation of the underlying equation is to employ an efficient sampling of light [21, 6], or BRDF [28, 2, 14, 17]. However, outgoing radiance in Rendering Equation cannot be accurately approximated using one of these importance sampling strategies only.

The original method for PRT [20] which supports low-frequency lighting environments only, requires using SH for lighting. Frequency space approaches like SH [19] solve the pre-filtering step but they support only low-frequency lighting and BRDF remain static. Haar wavelets [16] for PRT supports all-frequency lights, but it has high memory and precomputation costs. A number of approaches have been proposed to overcome this problem. For example, factoring BRDF into separate view and light components works well for diffuse BRDF, but requires large number of terms to approximate specular BRDF [26, 15, 27, 23].

SRBF for environment light and light transport functions are used to achieve all-frequency PRT [23]. This method provides a real-time rendering with plausible image quality, but it does not support highly specular BRDF.

Similar to our approach, GMM have been applied for the PRT-based rendering routine [8, 7, 25, 29]. [8] proposed their own optimization technique to fit the Gaussian parameters which supports interpolation over view direction and vertices. This technique is limited to static scenes and its storage cost is high. [7] improved their previous method by separating reflectance and visibility which provides a light-weight visibility approximation for shadowed reflections using SH and a separate GMM for each predetermined view elevation angle. This technique does not support anisotropic BRDF and requires expensive precomputation and storage even for isotropic BRDF. [25] fit Gaussians for the anisotropic BRDF using L-BFGS-B solver [30], and the view directions are approximated by SG through their own spherical warp technique. The anisotropic BRDF require too many SG for the approximation and errors occur in grazing angles. [29] reduce these errors and the lobe count for the approximation by using ASG. [11] represents the convolution of the BVNDF and the small-scale BRDF as SG. This technique supports only isotropic BRDF as small-scale materials.

[4] proposed a GPU-based filtered importance sampling algorithm for image-based lighting. This method supports dynamic scenes but it can only work with moderately anisotropic BRDF. [13] proposed a pre-filtering technique of anisotropic environment maps using Banks model [1]. In this method, Lambert's term is discarded to decrease the dimensionality which causes reflectance problems, and pre-computing environment is required for every BRDF parameters. [22] use GMM for multiresolution reflectance of the geometric structure within each pixel. In this technique GMM are aligned with their own modified EM algorithm. [9] proposed a similar approach by using GMM for normal map filtering.

III. Spherical Gaussian

Von Mises-Fisher distribution is a probability distribution on the (p-1)-dimensional sphere in R^p. On a unit sphere in 3 dimensions (p=3), the probability distribution function can be written as:

$$g(x;\mu,\kappa)=\kappa/4\pi\sinh\kappa\ e^{(\kappa\mu^T x)} \quad (1)$$

where $\mu \in S^2$ is the lobe axis and $\kappa \in (0,+\infty)$ is the lobe sharpness. The direction vector $x \in S^2$ is the spherical parameter of the resulting function.

It is shown in [9] that von Mises-Fischer distribution is equivalent to Gaussian distribution with $\sigma^2=1/2\kappa$ that is

$$g(x;\mu,\kappa) \cong \kappa/2\pi\ e^{\kappa(x\cdot\mu-1)}. \quad (2)$$

Spherical Gaussian distributions, or von Mises-Fischer distributions are symmetric around their μ axis. Therefore, it is trivial to rotate these distributions just by rotating the μ axis.

IV. Brdf Approximation with Mixture of Sg

We used GMM to approximate BRDF. For a set of fixed viewing directions $V = \{\mathbf{o}_j\}$, specular component of the BRDF multiplied by the cosine term is modeled as:

$$\rho\left(\mathbf{i},\mathbf{o}_j\right)\left(\mathbf{n}\cdot\mathbf{i}\right) \cong \sum_{k=1}^{N} \alpha_{k,j} g\left(\mathbf{i};\mu_{k,j},\kappa_{k,j}\right) \quad (3)$$

where, ρ is the BRDF function, **i** is the incoming light direction, $\mathbf{o}_j$ is a fixed viewing direction, **n** is the surface normal, and *g* is a SG distribution with parameters $\mu_{k,j}$ and $\kappa_{k,j}$. Once these parameters are estimated, they can be stored in a texture (parameterized by k and j) to be used later in the rendering process.

Parameters of a probability model can be estimated using maximum likelihood estimation method. Unfortunately due to the summation terms in GMM, no closed-form solution to the maximization problem is possible and the parameters must be estimated numerically using optimization methods. In this work, we used the EM algorithm [22, 5, 3]. It is an iterative nonlinear optimization method for estimating parameters of a probability model that depends on unobserved latent variables.

Using samples $X = \{\mathbf{x}_i; 1 \leq i \leq M\}$ drawn from the original distribution, the parameter vector Θ of a GMM can be estimated by maximizing the likelihood.

$$\begin{aligned}\Theta &= argmax P(X|\Theta) \\ &= argmax \prod_{i=1}^{M} \left(\sum_{k=1}^{N} \alpha_k g\left(\mathbf{x}_i;\mu_k,\kappa_k\right)\right).\end{aligned} \quad (4)$$

While estimating the parameters of this model using EM algorithm, the latent variables $Z = \{\mathbf{z}_i; 1 \leq i \leq M\}$, determine the component from which the observation originates.

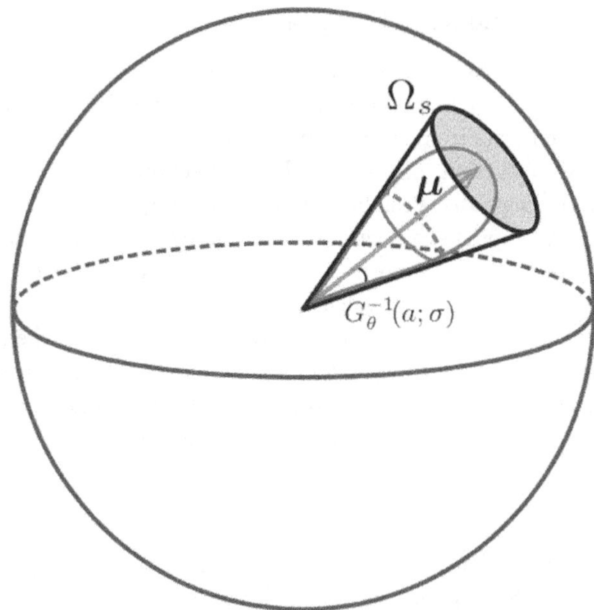

Figure 2: Spherical Gaussian and the solid angle subtended by it

$$X_i | (Z_i = k) \sim G(\mu_k, \kappa_k)$$
$$\textit{where} \quad P(Z_i = k) = \alpha_k \quad \textit{and} \quad \sum \alpha_k = 1 \tag{5}$$

EM algorithm iteratively alternates between two steps. In the first step named expectation step, the EM algorithm finds the expected value of log-likelihood with respect to the unknown data Z given the observed data X and the current parameter estimates. In the second step, that is the maximization step, algorithm estimates parameters that maximizes the expectation computed in the first step. These alternating steps guarantee convergence to a local maxima. However, local maxima may not be the optimum result. To ensure good results, initial guess for the parameters must be carefully used.

In order to provide a better guess for the initial parameters, first we calculate the spherical coordinates $\theta \quad and \quad \phi$ of the samples. Next, we sort the samples along the axis having the highest variance. Then, we cluster the sorted samples into N equally sized sets. Finally we assume each set originates from a different SG distribution and we estimate the parameters of each SG distribution from the samples in the corresponding set.

V. Real Time Rendering Using Sg

Since SG are radially symmetric, they can easily be used for real-time distant light filtering. The mean direction vector, $\boldsymbol{\mu}$, of a SG can be used as a parameter to look up light intensities from an environment map. And the variance parameter, κ, defines how much of the pixels around this direction should be averaged. As in [4], this averaging process can be simplified by using the mipmaping capabilities of GPU. The mipmap level can be calculated by using the number of environment map pixels covered mostly by the SG distribution, thus the solid angle subtended by the lobe should be calculated first.

We first rotate the distribution to align the direction $\boldsymbol{\mu}$ with the unit-z axis. This procedure simplifies the calculations and it does not change the solid angle, Ω_s, subtended by the distribution. We then calculate the cumulative distribution of the θ marginal of the distribution where θ is the azimuth angle in spherical coordinates.

$$G_\theta(\theta;\kappa) \cong 1 - e^{\kappa(\cos\theta - 1)} \quad (6)$$

In order to find the θ span of the distribution, which is shown in Figure 2, we can use the inverse of this cumulative distribution function.

$$G_\theta^{-1}(a;\kappa) = \arccos\left(\frac{\log(1-a)}{\kappa} + 1\right) \quad (7)$$

where a is the percentage of the coverage.

The solid angle subtended by the distribution can then be found by:

$$\Omega_s = 2\pi \int_0^{G_\theta^{-1}(a;\kappa)} \sin\theta d\theta = -\frac{2\pi\log(1-a)}{\kappa}. \quad (8)$$

In order to calculate the number of environment map pixels we also need to find the solid angle subtended by a single pixel of the highest detail mipmap of the environment map, Ω_p, as in [4]:

$$\Omega_p = \frac{d(\mathbf{i})}{w \cdot h} \quad (9)$$

where d(i) is the distortion factor and w,h are the environment map width and height in pixels consecutively.

Finally the mip level can be calculated by the logarithm of the ratio of the solid angle subtended by the distribution and the solid angle subtended by a single pixel:

$$I = \frac{1}{2}\log_2\left(\frac{\Omega_s}{\Omega_p}\right)$$

$$I = \frac{1}{2}(\log_2(-2\pi\log(1-a)) - \log_2\kappa) + \frac{1}{2}(\log_2(wh) - \log_2 d(i) \tag{10}$$

The first term in Eq. 10 depends only on a, which controls the smoothness. The second term only depends on the SG parameters and the third term depends only on environment map size. All these three terms can be precomputed with minimal computation cost. The final term depends on the mapping distortion factor, and since we use dual paraboloid environment maps [10], it can simply be calculated as:

$$d(\mathbf{i}) = 4b^2(|\mathbf{i}_z| + 1)^2 \tag{11}$$

where b is a scaling parameter to allow each paraboloid to have information from the opposite direction as in [4].

To summarize our method; for each $\mathbf{o}_j$ the EM algorithm is used for fitting 8 SG, the $\alpha_{k,j}$, $\kappa_{k,j}$ parameters and spherical coordinates of $\mu_{k,j}$ are stored in a texture as a precomputation step. For rendering, first the out-going direction vector in tangent space is calculated. Then for each SG, the parameters of the distribution are looked up from the precomputed texture. A rotation is performed to transform direction into world-space and calculate the texture mapping coordinates for the environment map. The mipmap lod index is calculated in a similar way as described above. Finally, the contribution of this SG is looked up from the environment map using the calculated lod index.

VI. Results

We have implemented three methods for comparison: our proposed method, Colbert and Křivánek's filtered importance sampling [4] method and the all-frequency rendering method proposed by Wang et al. [25]. We also prepared reference images using PBRT offline renderer [18]. We used a simple sphere model lit by a distant light as a test scene. Instead of measured BRDF data, we have used generated data from the Ward BRDF model [28] for our test purposes.

Empirical results have shown that the Filtered Importance sampling method is the slowest of these three methods. It requires large samples (about 40 samples) in order to achieve similar quality. On the other hand, All-frequency method requires only a single sample, making it the fastest of these methods. Our proposed method uses 8 SG lobes, which makes it slower than all-frequncy

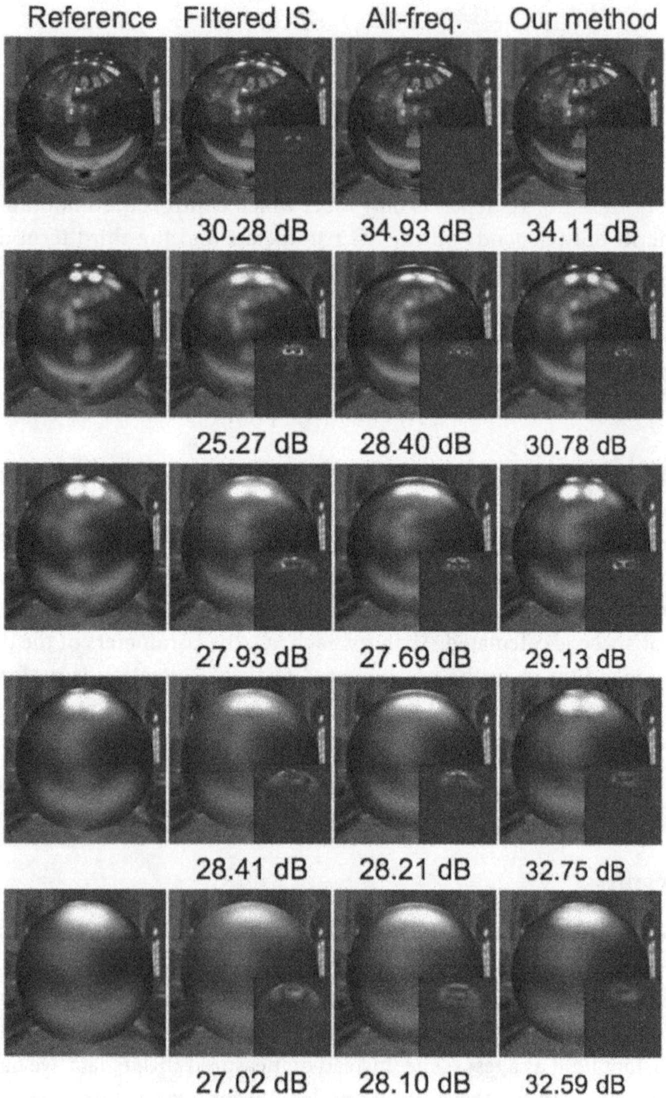

Figure 3. Rendered spheres based on different methods and specularity levels. Columns left to right: reference image, filtered importance sampling, all-frequency rendering, and our method. Rows top to bottom: images obtained using the specularity levels (= 0:01; 0:05; 0:1; 0:15; 0:2). The insets for each image represents the difference image between the images produced by the corresponding method and the reference images.

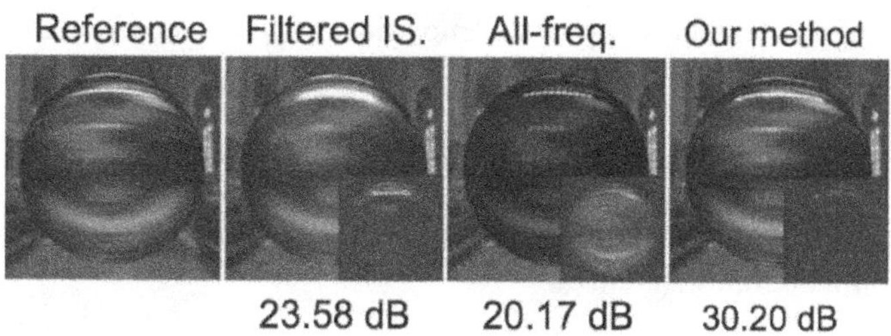

Figure 4: Spheres with an anisotropic material $(\boldsymbol{\alpha_x = 0.01, \alpha_y = 0.2})$ are rendered using a) Filtered importance sampling, b) All-frequency rendering and c) Our method. Reference image is shown in the first column and difference images are shown in insets. PSNR values are given below each image.

method but faster than filtered importance sampling. On our test machine, (Intel Core i7-920, NVIDIA GeForce GTX 480, 12 GB ram), all-frequency rendering method runs at 5200 frames per second (fps). Filtered Importance sampling method runs at 490 fps and our method runs at 990 fps.

We have also compared the quality of rendered images using five specularity levels, that is $\alpha = \{0.01, 0.05, 0.1, 0.15, 0.2\}$ Rendering results are presented in Figure 1. Reference images are shown in the first column of this figure. Rendered images using filtered importance sampling, all-frequency rendering, and our method are presented in the following columns. The insets for each image represents the difference image between the images produced by the corresponding method and the reference images. The Peak Signal-to-Noise Ratio (PSNR) values are also given for each image. For both specular and glossy materials, our method produces competitive quality images. It can also be seen from Figure 2 that our method can handle anisotropic materials as well.

VII. Conclusion and Future Work

We have proposed a real-time method to render scenes illuminated with distant light. Although it is slightly slower, our proposed algorithm produces better quality images as compared to its competitors. It can also handle both isotropic and anisotropic materials.

Using fixed viewing directions and fitting them independently produces blocking artifacts. These artifacts can be avoided by linearly interpolating the SG parameters. Since the fitting is done independently for each viewing direction, the resulting SG parameters cannot be interpolated. This can be handled by considering the neighboring view directions in the fitting procedure. Developing an algorithm for an efficient parameter interpolation is left as a future work.

Acknowledgement

The authors would like to thank the anonymous reviewers and Murat Kurt for their comments and suggestions.

References

[1] David C. Banks. Illumination in diverse codimensions. In Proceedings of the 21st Annual Conference on Computer Graphics and Interactive Techniques, SIGGRAPH '94, pages 327–334, New York, NY, USA, 1994. ACM.

[2] Ahmet Bilgili, Aydın Öztürk, and Murat Kurt. A general brdf representation based on tensor decomposition. Computer Graphics Forum, 30(8):2427–2439, December 2011.

[3] Jeff Bilmes. A gentle tutorial of the em algorithm and its application to parameter estimation for gaussian mixture and hidden markov models. Technical report, 1998.

[4] Mark Colbert and Jaroslav Křivánek. Real-time shading with filtered importance sampling. In ACM SIGGRAPH 2007 sketches, SIGGRAPH '07, New York, NY, USA, 2007. ACM.

[5] A. P. Dempster, N. M. Laird, and D. B. Rubin. Maximum likelihood from incomplete data via the em algorithm. JOURNAL OF THE ROYAL STATISTICAL SOCIETY, SERIES B, 39(1):1–38, 1977.

[6] Serkan Ergun, Murat Kurt, and Aydın Öztürk. Real-time kd-tree based importance sampling of environment maps. In Proceedings of the 28th Spring Conference on Computer Graphics, SCCG '12, pages 77–84, New York, NY, USA, 2012. ACM.

[7] Paul Green, Jan Kautz, and Frédo Durand. Efficient reflectance and visibility approximations for environment map rendering. Computer Graphics Forum (Proc. EUROGRAPHICS), 26(3):495–502, 2007.

[8] Paul Green, Jan Kautz, Wojciech Matusik, and Frédo Durand. View-dependent precomputed light transport using nonlinear gaussian function approximations. In Proceedings of the 2006 Symposium on Interactive 3D Graphics and Games, I3D '06, pages 7–14, New York, NY, USA, 2006. ACM.

[9] Charles Han, Bo Sun, Ravi Ramamoorthi, and Eitan Grinspun. Frequency domain normal map filtering. ACM Trans. Graph., 26(3), July 2007.

[10] Wolfgang Heidrich and Hans-Peter Seidel. View-independent environment maps. In Proceedings of the ACM SIGGRAPH/EUROGRAPHICS

Workshop on Graphics Hardware, HWWS '98, pages 39–ff., New York, NY, USA, 1998. ACM.

[11] Kei Iwasaki, Yoshinori Dobashi, and Tomoyuki Nishita. Interactive bi-scale editing of highly glossy materials. ACM Trans. Graph., 31(6):144:1–144:7, November 2012.

[12] James T. Kajiya. The rendering equation. Computer Graphics, 20(4):143–150, 1986. (Proc. SIGGRAPH '86).

[13] Jan Kautz, Pere-Pau Vázquez, Wolfgang Heidrich, and Hans-Peter Seidel. Unified approach to prefiltered environment maps. In Proceedings of the Eurographics Workshop on Rendering Techniques 2000, pages 185–196, London, UK, UK, 2000. Springer-Verlag.

[14] Jason Lawrence, Szymon Rusinkiewicz, and Ravi Ramamoorthi. Efficient BRDF importance sampling using a factored representation. ACM Transactions on Graphics, 23(3):496–505, 2004. (Proc. SIGGRAPH '04).

[15] Xinguo Liu, Peter-Pike Sloan, Heung-Yeung Shum, and John Snyder. All-frequency precomputed radiance transfer for glossy objects. In Proceedings of the Fifteenth Eurographics Conference on Rendering Techniques, EGSR'04, pages 337–344, Aire-la-Ville, Switzerland, Switzerland, 2004. Eurographics Association.

[16] Ren Ng, Ravi Ramamoorthi, and Pat Hanrahan. All-frequency shadows using non-linear wavelet lighting approximation. ACM Trans. Graph., 22(3):376–381, July 2003.

[17] R. Pacanowski, Oliver Salazar Celis, C. Schlick, X. Granier, P. Poulin, and A. Cuyt. Rational brdf. IEEE Transactions on Visualization and Computer Graphics, 18(11):1824–1835, 2012.

[18] Matt Pharr and Greg Humphreys. Physically Based Rendering, Second Edition: From Theory To Implementation. Morgan Kaufmann Publishers Inc., San Francisco, CA, USA, 2nd edition, 2010.

[19] Ravi Ramamoorthi and Pat Hanrahan. An efficient representation for irradiance environment maps. In Proceedings of the 28th Annual Conference on Computer Graphics and Interactive Techniques, SIGGRAPH '01, pages 497–500, New York, NY, USA, 2001. ACM.

[20] Peter-Pike Sloan, Jan Kautz, and John Snyder. Precomputed radiance transfer for real-time rendering in dynamic, low-frequency lighting environments. ACM Trans. Graph., 21(3):527–536, July 2002.

[21] László Szécsi, László Szirmay-Kalos, Murat Kurt, and Balázs Csébfalvi. Adaptive sampling for environment mapping. In Proceedings of the 26th Spring Conference on Computer Graphics, SCCG '10, pages 69–76, New York, NY, USA, 2010. ACM.

[22] Ping Tan, Stephen Lin, Long Quan, Baining Guo, and Heung-Yeung Shum. Multiresolution reflectance filtering. In Proceedings of the Sixteenth Eurographics Conference on Rendering Techniques, EGSR'05, pages 111–116, Aire-la-Ville, Switzerland, Switzerland, 2005. Eurographics Association.

[23] Yu-Ting Tsai and Zen-Chung Shih. All-frequency precomputed radiance transfer using spherical radial basis functions and clustered tensor approximation. ACM Trans. Graph., 25(3):967–976, July 2006.

[24] Eric Veach. Robust Monte Carlo Methods for Light Transport Simulation. PhD thesis, Stanford, CA, USA, 1998. AAI9837162.

[25] Jiaping Wang, Peiran Ren, Minmin Gong, John Snyder, and Baining Guo. All-frequency rendering of dynamic, spatially-varying reflectance. ACM Trans. Graph., 28(5):133:1–133:10, December 2009.

[26] Rui Wang, John Tran, and David Luebke. All-frequency relighting of non-diffuse objects using separable brdf approximation. In Proceedings of the Fifteenth Eurographics Conference on Rendering Techniques, EGSR'04, pages 345–354, Aire-la-Ville, Switzerland, Switzerland, 2004. Eurographics Association.

[27] Rui Wang, John Tran, and David Luebke. All-frequency relighting of glossy objects. ACM Trans. Graph., 25(2):293–318, April 2006.

[28] Gregory J. Ward. Measuring and modeling anisotropic reflection. Computer Graphics, 26(2):265–272, 1992. (Proc. SIGGRAPH '92).

[29] Kun Xu, Wei-Lun Sun, Zhao Dong, Dan-Yong Zhao, Run-Dong Wu, and Shi-Min Hu. Anisotropic spherical gaussians. ACM Trans. Graph., 32(6):209:1–209:11, November 2013.

[30] Ciyou Zhu, Richard H. Byrd, Peihuang Lu, and Jorge Nocedal. Algorithm 778: L-bfgs-b: Fortran subroutines for large-scale bound-constrained optimization. ACM Trans. Math. Softw., 23(4):550–560, December 1997.

18. An Efficient Plugin for Representing Heterogeneous Translucent Materials

Sermet Önel
Department of Computer Engineering Yaşar University sermet.onel@yasar.edu.tr

Murat Kurt
International Computer Institute Ege University murat.kurt@ege.edu.tr

Aydın Öztürk
Department of Computer Engineering İzmir University aydin.ozturk@izmir.edu.tr

Abstract— This paper presents a plugin that adds an efficient representation of heterogeneous translucent materials to the Blender 3D modeling tool. Algorithm of the plugin is based on Singular Value Decomposition (SVD) method and Mitsuba renderer is the default rendering software used by the proposed plugin. We validate the efficiency of the proposed plugin by using a set of measured heterogeneous subsurface scattering data sets.

Keywords—BSSRDF; Subsurface Scattering Model; Factorization; Heterogeneous Subsurface Scattering; Mitsuba Renderer; Blender 3D Modeling Tool

I. INTRODUCTION

Efficient representation of heterogeneous translucent materials in computer graphics is a common problem. A number of efficient methods have been proposed to represent the Bidirectional Scattering Surface Reflectance Distribution Function (BSSRDF) for homogeneous translucent materials[1][2]. However, none of these methods could be generalized to provide proper outputs for heterogeneous translucent materials. The characteristics of having structural deficiencies, impurities and composite structures inside the object volume of heterogeneous translucent materials require approaches with a different view point[3][4][5][6][7]. On the other hand, high storage needs and computational costs of these algorithms remain to be a major problem to be resolved.

In this study, we use the Singular Value Decomposition (SVD) method for the BSSRDF representation of heterogeneous translucent materials. The proposed approach was implemented in C++ and included in the source codes of Mitsuba renderer project [8]. The integration plugin which has already been available for use by [9], was modified and imported into the three dimensional (3D) Blender modeling tool[10].

Our plugin helps to render heterogeneous translucent materials accurately and efficiently. As it can be seen in Figure 7 and Figure 8, the rendering output of the plugin gives heterogeneous subsurface scattering effects visually plausibly.

II. Related Work

The problem of representing BSSRDF for heterogeneous translucent materials has an extensive literature. Major efforts have been devoted to the development of some approximation models. The underlying approaches broadly can be classified into two groups. The first group includes the techniques that extend the Jensen's *Dipole Diffusion Approximation Model* [1] and the second group consists of the techniques that are based on development of new material models.

Jensen's Dipole Diffusion Approximation model reduces computation time of eight dimensional BSSRDF[11] to acceptable rates. This approach is effective on homogeneous translucent materials and extended by many researchers [12],[13],[14],[15],[16]. Nevertheless, the main observation of light being isotropic and modeling homogeneous translucent materials make this model inappropriate for heterogeneous BSSRDF representation.

Jakob et al. [12] extended the Dipole Diffusion Approximation model by using anisotropic approach. This model improved the Jensen et al.'s model, however, the output of this model does not give visually plausible heterogeneous subsurface scattering effects. Mertens et al. [13] modeled human skin by using an interactive method to achieve local subsurface scattering. Donner and Jensen's [14] study is based on using multiple dipoles and they represented their study on paper and human skin. Another study was presented by Jimenez et al. [15] in which human skin was represented. Considering the psychological states of human face, Jimenez et al. [16] also modeled facial appearance for different regions on the face.

The second group of techniques includes Goesele et al.'s [5] compact model depending on underlying geometry, Tong et al.'s [6] model of quasi-homogeneous materials and Song et al.'s [7] SubEdit representation which allows interactive editing and rendering of translucent materials. Although these techniques classified in this group have made some improvements on heterogeneous BSSRDF representation, their design issues still prevent them to provide efficient solution for the problem.

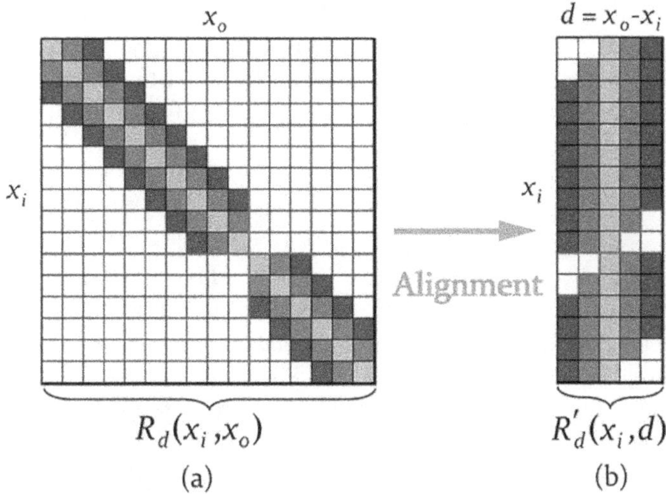

Fig. 1. The diffuse BSSRDF matrix (a) is transformed by first aligning the diagonal by a change of variables to $R'_d\,(x_i, d)$ (b) [3].

The Peers et al.s' work [17] was an important step for the solution of the problem. In their study, they employed the Non-Negative Matrix Factorization (NMF) algorithm. Replacing the Tucker-based factorization on tensor decomposition with the NMF algorithm, the same algorithm was also used by Kurt et al. [3]. Kurt [18] also used Singular Value Decomposition method on tensor decomposition and reported that the underlying algorithm has improved the computational efficiency for the data with two dimensional (2D) matrices [18]. This was the main motivation in our work and following his work, his corresponding algorithm was imported through our plugin and an efficient heterogeneous subsurface scattering representation was put into service for 3D modeling tools.

III. Subsurface Scattering with Svd Method

A. Preparation of the Test Data

Heterogeneous translucent materials are represented by BSSRDF [11],[1].

$$L_o\left(x_o,\vec{\omega}_o\right)=\int\limits_{A}\int\limits_{\Omega+}L_i\left(x_i,\vec{\omega}_i\right)S\left(x_i,\vec{\omega}_i;x_o,\vec{\omega}_o\right)\left(\vec{\omega}_i\cdot\vec{n}\right)d\vec{\omega}_i dx_i. \tag{1}$$

This function relates to the outgoing radiance at one point to the incident flux at another. Eq. (1) can be separated into a local and a global component where the local component represents the reflected light and the global component represents the scattering light in the material volume [3]. The global component is represented by the diffuse BSSRDF [17],[3]

$$S_d\left(x_i,\vec{\omega}_i;x_o,\vec{\omega}_o\right)=\frac{1}{\pi}F_i\left(x_i,\vec{\omega}_i\right)R_d\left(x_i,x_o\right)F_o\left(x_o,\vec{\omega}_o\right). \tag{2}$$

Eq. (2) represents the diffuse BSSRDF S_d with a four dimensional (4D) spatial subsurface scattering component R_d and the directionally dependent components F_i and F_o. The directionally dependent components are ignored and R_d is focused in the modeling process [17], [3], [5], [7].

For the factorization operation, 4D R_d value is transformed into 2D matrix. As it can be seen in Figure 1, Kurt et al. [3] reorganized this matrix by changing the variable $d = x_0 - x_i$. With this reorganization, R'_d matrix is found which is a more compact matrix for representing measured heterogeneous translucent materials.

B. Factorization

The subsurface scattering data is factorized by Singular Value Decomposition (SVD) as explained below. This method is a subset of tensor decomposition methods used in the representation of subsurface scattering effects. The study of Peers et al. [17] based on NMF algorithm and the study of Kurt et al. [3] based on Tucker-based factorization are examples of other tensor decomposition methods.

The subsurface scattering data has multi-dimensional features. It can be represented through a tensor model. For example, a 2D matrix can be considered as a second degree tensor. Thus, the subsurface scattering data in the form of a matrix provides a convenient form of data set for tensor decomposition operation.

In SVD operation an $M \times N$ matrix is defined as the product of an U matrix with dimensions $M \times K$ and a matrix V with dimensions $K \times N$ and a core tensor with dimensions $K \times K$. This is a similar decomposition operation in Kurt et al.'s study [3], however, if the value of K is chosen to be 1 that is a scalar then the dimensions of U and V matrices become $M \times 1$ and $1 \times N$, respectively, and the core tensor becomes a scalar [18].

In this operation, R'_d is taken into consideration as it is the most compact data. Another consideration in the factorization operation is making the data stay in the positive values which leads to physically correct results. This is achieved by another transformation that is [18]:

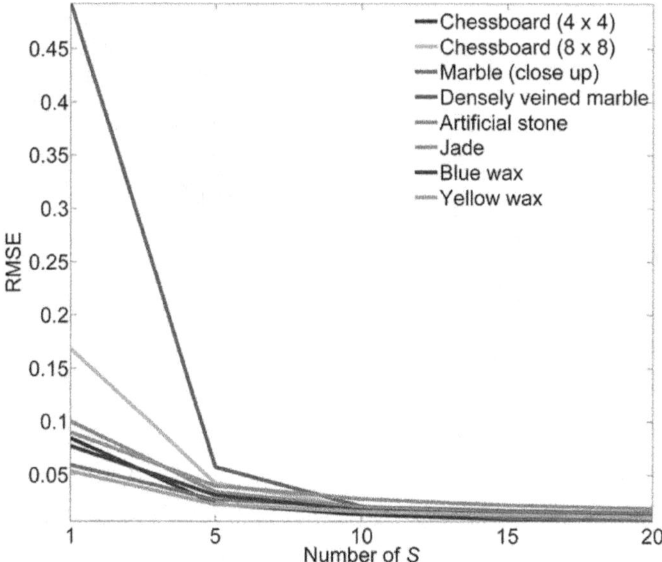

Fig. 2. The root-mean-square error (RMSE) values of SVD-based subsurface scattering model with different *S* parameter [18].

$$R_d^m(x_i, d) = \ln \frac{R_d'(x_i, d)}{A} + B. \tag{3}$$

By choosing the most appropriate values for *A* and *B* to minimize the error values, $R^m{}_d$ matrix is factorized using SVD and error terms for each color channel is modeled using Bilgili et al.'s approach [19]. This procedure is repeated *S* times to improve the accuracy of the approximation. Accordingly, $R^m{}_d$ becomes [18]:

$$R_d^m(x_i, d) \approx \sum_{j=1}^{S} f_j(x_i)\, h_j(d). \tag{4}$$

More details about SVD-based subsurface scattering representation can be found in Kurt's [18] work.

C. Analysis

SVD-based subsurface scattering representation depends only on a single parameter, *S* that is the number of terms in the factorization operation.

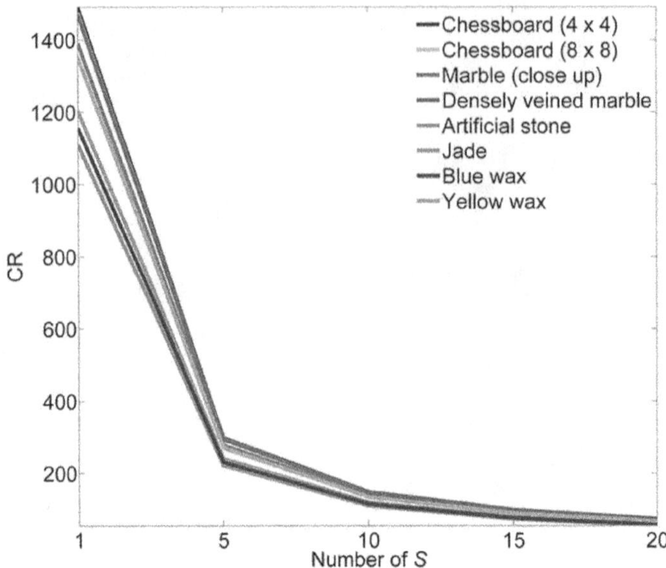

Fig. 3. The compression rates (CR) of SVD-based subsurface scattering model with different *S* parameter [18].

The work of Kurt [18] shows that SVD-based subsurface scattering representation gives convincing results for representing heterogeneous translucent materials. In this work, it is also emphasized that the approximation results are visually acceptable for some test materials even for small number of iteration $(S \leq 5)$. The effect of the number of iterations on the model errors for different materials is illustrated in Figure 2. This efficiency was the main motivation in this study, which leaded us to use SVD-based subsurface scattering representation in our plugin instead of other factorization methods such as Tucker-based factorization model [3].

Furthermore, the *S* value should be chosen carefully since the factorization is applied for each channel, separately and the compression rate can be a problem for obtaining close approximations. The effect of the number of iterations on the compression ratio for different materials is illustrated in Figure 3.

IV. The Integration Plugin

Constructing images through modeling and representing the BSSRDF for heterogeneous translucent materials is a complicated procedure. Developing a

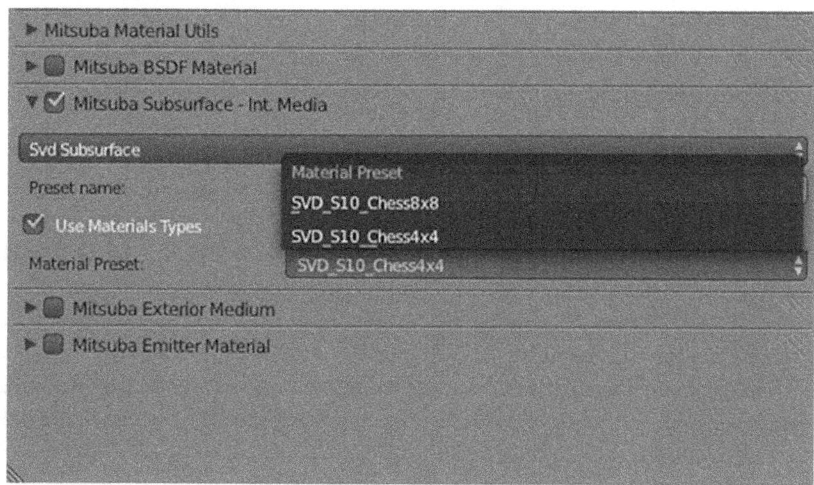

Fig. 4. The graphical user interface of our integration plugin in the Blender 3D Modeling Tool [10].

plugin to support the heterogeneous subsurface scattering effects of the underlying materials will provide a useful tool for many applications. In this work, we aim to develop a plugin for rendering purposes. Such a plugin must be an interface between the renderer and the software modeling tool. Because, the renderer which has the capability to represent the heterogeneous translucent materials is not the only task to be carried out. Details of lighting information, shading, texture mapping and others in the scene are defined by the 3D modeling tool and these details are also important properties that should be considered in the representation. Therefore, the plugin should send the details available on the scene to the renderer, and then the SVD-based subsurface scattering representation can be a solution for the common problem.

Our procedure for developing the plugin has two steps: the first step consists of the implementation of SVD-based subsurface scattering representation in the renderer, and the second step is to create a script that is able to send the details in the scene to the SVD-based subsurface scattering representation for a visually plausible rendering.

The 3D modeling tools are powerful software to model the details of the scene. Light sources, materials and their details can easily be defined using these tools. Our main choice was the Blender project [10] which is an open source developing platform and uses Python [20] as its plugin scripting language. These features were effective in our choice since Python has the ability to call C++ functions without the need of any software. It is also effective in the sense that an open source application would be useful for the documentation and faster development.

We chose Mitsuba renderer [8] as the default renderer because it is highly optimized and it has good performance, scalability, easy usability and robustness are the other features of this renderer includes. As the project is in C++ and Blender supports Python, the implementation of our plugin could become easier. The version of the Blender used in this work is 2.69 which is compatible with Mitsuba version 0.5. We modified the source codes of these versions to finalize our study successfully.

The integrator class in Python script is defined as "class mitsuba_sss_svd". This class has control parameters for using different material types which are the possible parameters for the heterogeneous translucent materials. We tested our plugin on chessboard (4×4) and chessboard (8×8) heterogeneous translucent materials, which were measured by Peers et al. [17].

The integration plugin renders the material according to the material type choice and sends the data to the Mitsuba renderer.exe file, which consists of the necessary operations. These parameters are updated a function called "params. update" which is defined in the Python script. The data is kept on the object of the material chosen, by using the default constructor self of Python language.

Graphical User Interface (GUI) of the integration plugin is shown in Figure 4. The representation of heterogeneous translucent materials is classified under the section of Subsurface Scattering of the already available integration plugin [9] and the details of the material type is chosen using the GUI of the plugin. As the Blender project [10] supports for modeling different types of materials and other details in the scene, the effect of the chosen operation is completed after the rendering operation. A general overview of the Blender 3D modeling tool and our integration plugin can be seen in Figure 5.

V. Experimental Results

The proposed plugin was developed and tested by using various objects of some sample materials on which SVD-based heterogeneous subsurface representation were applied. Empirical results have shown that the rendering operation based on the underlying plugin simulates the heterogeneous subsurface scattering effects successfully. Furthermore, two critical issues that are the processing time the plugin requires to complete the rendering process and the storage requirements were checked to see if there is any preclusion.

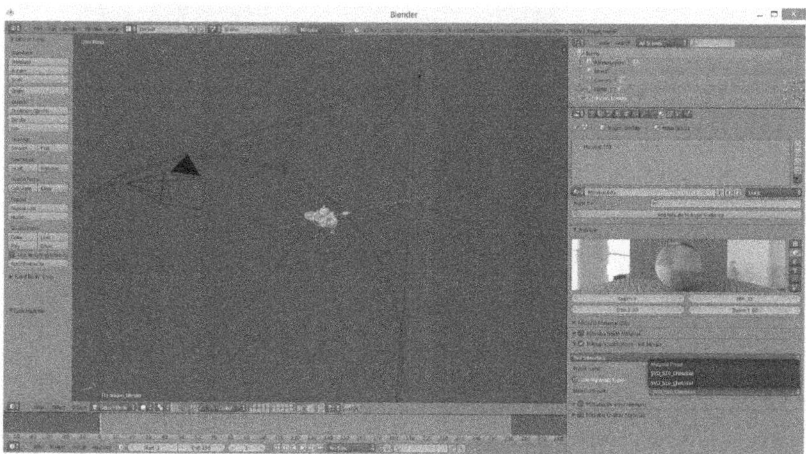

Fig. 5. A general overview of the Blender 3D Modeling Tool [10] and our integration plugin.

The materials chosen for the test were applied on a dragon and a kitten objects. We checked whether the plugin could perform subsurface scattering effects on the objects as it does directly with Mitsuba renderer. It's important that SVD-based subsurface scattering representation is a texture-space based model. The texture coordinates on the objects were carefully determined by a 3D modeling tool and these texture coordinates were sent to Mitsuba renderer. Thus the details of the scene were exported to the renderer successfully.

Figure 6 shows the preview of subsurface scattering effects on a sphere solid object. A similar effect can be rendered on an object chosen in the scene by the artist. As it is seen in Figure 7(a), our plugin helps to render heterogeneous translucent materials correctly and it took 20 minutes to take this rendering output on a computer with i7-3630QM processor with 8 GB RAM and NVIDIA GTX660M/2GB GDDR5. The storage of the output was 36.1 MB. Figure 7(b) illustrates another tested material, chessboard (8×8) on a dragon object. The output needed 38.42 MB on the same computer and the rendering process took only 20.9 minutes.

As it is seen in Figure 8, the last rendering test was done on a kitten object and on the same PC. Chessboard (4×4) and chessboard (8×8) materials were tested on the kitten object. It took 16.157 and 17.36 minutes to render chessboard (4×4) and chessboard (8×8) materials, respectively. The storage space needed

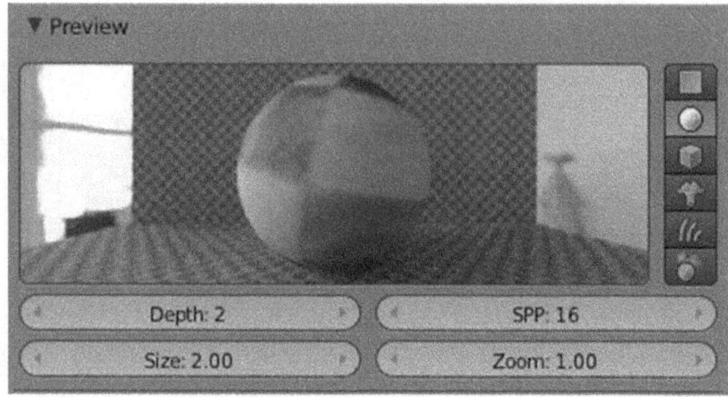

Fig. 6. A preview scene of our integration plugin in the Blender 3D Modeling Tool. Chessboard (4×4) material was represented in this rendering.

(a) (b)

Fig. 7. A dragon under spot lighting was rendered using our integration plugin. (a) chessboard (4×4) material. (b) chessboard (8×8) material.

32.86 MB and 34.72 MB for rendering chessboard (4×4) and chessboard (8×8) materials, respectively.

VI. Conclusion

In this paper, we presented an integration plugin for rendering heterogeneous translucent materials and empirically demonstrated that it enables to render visually plausible scenes. Our plugin relies on SVD-based subsurface scattering model which was proposed by Kurt [18]. In this paper, we also showed that

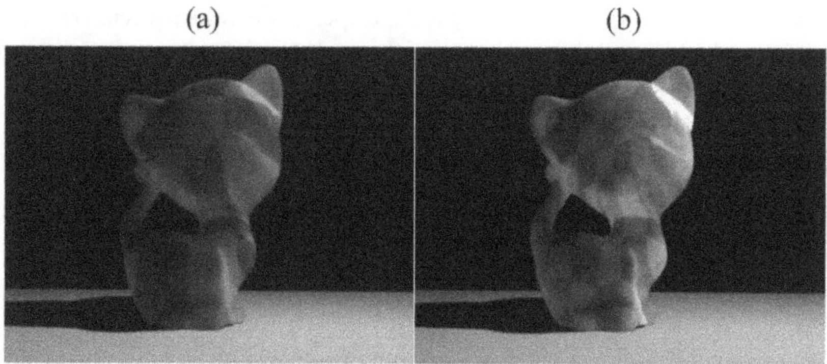

Fig. 8. A kitten under spot lighting was rendered using our integration plugin. (a) chessboard (4×4) material. (b) chessboard (8×8) material

the plugin works correctly by communicating with Mitsuba renderer and 3D modeling tool.

The plugin may be developed by adding different subsurface scattering models for rendering heterogeneous translucent materials.

VII. Future Works

This study enables the use of SVD-based subsurface scattering model for rendering heterogeneous translucent materials on a 3D modeling tool. As the proposed plugin is compatible with Blender project [10] and Mitsuba renderer [8], the functionality is strictly related with the correct versions of these tools.

As a future development, the availability of other factorization based models such as Tucker-based factorization model can be supported in the integration plugin. There is also a lack of the availability of different subsurface scattering data which will be supported in the future versions.

It is also important that there are lots of different tools for 3D modeling which brings the need for extending this plugin for different platforms. The compatibility of plugin for other platforms is also considered as a future study.

Acknowledgment

This study was supported by the Scientific and Technical Research Council of Turkey (Project No: 111E208). We would like to thank Pieter Peers for sharing his measured heterogeneous subsurface scattering data set.

We would also like to thank for the developer of the Mitsuba renderer project Wenzel Jakob and the developers of the exporter plugin Fransesc Juhe and Bartosz Styperek for their support in the development process.

References

[1] H. W. Jensen, S. R. Marschner, M. Levoy, and P. Hanrahan, "A practical model for subsurface light transport," in Proc. SIGGRAPH '01, 2001, pp. 511–518.

[2] H. W. Jensen, "Global illumination using photon maps," in Proceedings of the Eurographics Workshop on Rendering Techniques '96. London, UK, UK: Springer-Verlag, 1996, pp. 21–30.

[3] M. Kurt, A. Öztürk, and P. Peers, "A compact tucker-based factorization model for heterogeneous subsurface scattering," in Proceedings of the 11th Theory and Practice of Computer Graphics, ser. TPCG '13. Bath, United Kingdom: Eurographics Association, 2013, pp. 85–92.

[4] E. d'Eon and G. Irving, "A quantized-diffusion model for rendering translucent materials," ACM Transactions on Graphics, vol. 30, no. 4, pp. 56:1–56:14, Jul. 2011, (Proc. SIGGRAPH '11).

[5] M. Goesele, H. P. A. Lensch, J. Lang, C. Fuchs, and H.-P. Seidel, "DISCO: acquisition of translucent objects," ACM Transactions on Graphics, vol. 23, no. 3, pp. 835–844, Aug. 2004, (Proc. SIGGRAPH '04).

[6] X. Tong, J. Wang, S. Lin, B. Guo, and H.-Y. Shum, "Modeling and rendering of quasi-homogeneous materials," ACM Transactions on Graphics, vol. 24, no. 3, pp. 1054–1061, Jul. 2005, (Proc. SIGGRAPH '05).

[7] Y. Song, X. Tong, F. Pellacini, and P. Peers, "Subedit: a representation for editing measured heterogeneous subsurface scattering," ACM Transactions on Graphics, vol. 28, no. 3, pp. 31:1–31:10, Jul. 2009, (Proc. SIGGRAPH '09).

[8] W. Jakob, "Mitsuba renderer," 2013, http://www.mitsuba-renderer.org.

[9] B. Styperek and F. Juhe, "The blender plugin," 2011, https://www.mitsuba-renderer.org/plugins.html.

[10] Blender, "Blender foundation," 2003, http://www.blender.org/.

[11] F. E. Nicodemus, J. C. Richmond, J. J. Hsia, I. W. Ginsberg, and T. Limperis, "Geometrical considerations and nomenclature for reflectance," National Bureau of Standards (US), Monograph, Oct. 1977.

[12] W. Jakob, A. Arbree, J. T. Moon, K. Bala, and S. Marschner, "A radiative transfer framework for rendering materials with anisotropic structure," ACM Transactions on Graphics, vol. 29, no. 4, pp. 53:1–53:13, Jul. 2010, (Proc. SIGGRAPH '10).

[13] T. Mertens, J. Kautz, P. Bekaert, F. V. Reeth, and H.-P. Seidel, "Efficient rendering of local subsurface scattering," Computer Graphics Forum, vol. 24, no. 1, pp. 41–49, 2005.

[14] C. Donner and H. W. Jensen, "Light diffusion in multi-layered translucent materials," ACM Transactions on Graphics, vol. 24, no. 3, pp. 1032–1039, Jul. 2005, (Proc. SIGGRAPH '05).

[15] J. Jimenez, D. Whelan, V. Sundstedt, and D. Gutierrez, "Real-time realistic skin translucency," IEEE Computer Graphics and Applications, vol. 30, no. 4, pp. 32–41, 2010.

[16] J. Jimenez, T. Scully, N. Barbosa, C. Donner, X. Alvarez, T. Vieira, P. Matts, V. Orvalho, D. Gutierrez, and T. Weyrich, "A practical appearance model for dynamic facial color," ACM Transactions on Graphics, vol. 29, no. 6, pp. 141:1–141:10, Dec. 2010, (Proc. SIGGRAPH Asia '10).

[17] P. Peers, K. vom Berge, W. Matusik, R. Ramamoorthi, J. Lawrence, S. Rusinkiewicz, and P. Dutre, "A compact factored representation of heterogeneous subsurface scattering," ACM Transactions on Graphics, vol. 25, no. 3, pp. 746–753, Jul. 2006, (Proc. SIGGRAPH '06).

[18] M. Kurt, "An efficient model for subsurface scattering in translucent materials," Ph.D. dissertation, International Computer Institute, Ege University, Izmir, Turkey, January 2014, 122 pages.

[19] A. Bilgili, A. Öztürk, and M. Kurt, "A general brdf representation based on tensor decomposition," Computer Graphics Forum, vol. 30, no. 8, pp. 2427–2439, December 2011.

[20] G. V. Rossum, "Python," 1989, https://www.python.org/

19. Fast Data Parallel Radix Sort Implementation in Directx 11 Compute Shader to Accelerate Ray Tracing Algorithms

Arturo García
Visual and Parallel computing group Intel Corporation arturo.garcia@intel.com

Omar Alvizo
Universidad de Guadalajara Guadalajara, México omar.alvizo@alumno.udg.mx

Ulises Olivares-Pinto
Universidad Nacional Autónoma de México, uolivares@unam.mx

Félix Ramos
Centro de Investigación y de Estudios Avanzados del Instituto Politécnico Nacional Unidad Guadalajara framos@gdl.cinvestav.mx

Abstract— In this paper we present a fast data parallel implementation of the Radix Sort on the Direct Compute software development kit (SDK). We also discuss in detail the various optimization strategies that were used to increase the performance of our Radix Sort, and we show how these strategies can be generalized for any video card that supports the Direct Compute model. The insights that we share in this paper should be of use to any General-Purpose Graphics Processing Unit (GPGPU) programmer regardless of the video card being used. Finally, we discuss how Radix Sort can be used to accelerate ray tracing and we present some results in this area as well.

Keywords: Ray tracing; radix sort; algorithms

I. Introduction

Sorting algorithms is one of the most heavily researched topics in the history of computer science. These algorithms have multiple applications in various areas since they are a necessary step for fast information retrieval. For instance, specialized sorting algorithms exist for external storage (databases) due to their much slower access time.

On the other hand, sorting algorithms that operate on RAM have been extensively studied for decades. The aim of this research is to get greater performance.

For instance, in [1], Batcher investigates the use of sorting networks to increase the performance of the Bitonic Sort. Later, Zagha and Blelloch studied the problem of parallelizing the Radix Sort algorithm [2].

Nowadays, it is possible for the masses to have access to parallel computing architectures like multi-core processors, General-Purpose GPU (GPGPU) and Many Integrated Core architectures [3]. In the area of sorting algorithms, the Compute Unified Device Architecture (CUDA) data-parallel primitive (CUDPP) library implementation included in the Thrust productivity library [4] and presented by Merrill et al. in [5] is currently one of the fastest Radix Sort algorithms on GPGPU architectures.

One of the most eye-catching uses for the sorting algorithms on the GPGPU is the implementation of real-time ray tracing. Ray tracing is a technique that simulates how light is transported in a 3D scene and thus can achieve photo-realistic images. However, it requires the computation of several intersections between light, rays and objects. Therefore, in order to achieve real-time frame rates using ray tracing, it is necessary to sort the geometry first. In this paper we will exemplify the use of the GPGPU Radix Sort to implement a real-time ray tracing engine.

As previously mentioned, the CUDPP library has the fastest Radix Sort implementation for GPGPU [4]. This implementation has several optimizations that are CUDA-specific and therefore, it is hard to implement this algorithm in other video cards while maintaining the same performance. This paper focuses on these optimizations and how they can be generalized to remove their dependencies from the CUDA SDK, therefore making it possible to write competitive Radix Sort algorithms in a multitude of video cards.

The implementation presented in this paper was performed on Direct Compute, so it runs on AMD Radeon, Intel and NVIDIA video cards. However, the discussion of the optimizations should allow the implementation of a fast Radix Sort using other GPGPU programming interfaces like Open Computing Language (OpenCL), and they should also prove to be useful for the fine-tuning of other GPGPU algorithms.

II. Related Work

Owens et al. presented a survey in [6] of a mapping of various algorithms into the GPGPU architectures. This is also a good introductory read to the area of GPU computing as it presents several basic concepts and a multitude of algorithms including sorts.

Hillis et al. [7] described a data parallel algorithm to sum an array of *n* numbers that can be computed in time *O(log n)* by organizing the addends at the leaves of a binary tree and performing the sums at each level of the tree in parallel. This algorithm is the base for the implementation of an efficient scan algorithm, which in turn is a building block necessary for a fast Radix Sort algorithm.

Blelloch [8] presented an early specification of the Radix Sort and scan algorithms for generic parallel architectures. The current Radix Sort implementations for data parallel computing architectures are based on the scan algorithms presented in this book.

Harris et al. discussed an efficient implementation of the prefix sum in [9] for CUDA. The prefix sum is an algorithm that can be used to increase the performance of the Radix Sort by pre-computing the offsets where the elements will be stored. In this way, when elements are swapped during the sort, the algorithm already knows a lower index for each element. Further code and details are provided in [10] where segmented scan, intra-block scan, intra-warp scan and global scan algorithms are presented, again for the CUDA SDK.

In [11], Satish et al. described the design of a high- performance parallel Radix Sort for GPGPUs. Their discussion was centered around the CUDA SDK. At the time the paper was published, it was the fastest implementation of the Radix Sort. The core of their optimization relies on reducing the global communication between different threads to a minimum by breaking tasks into sizes that are compatible with the underlying hardware; minimizing the number of scatters to global memory and maximizing the coherence of scatters. The last point is achieved by using an on-chip shared memory to locally sort data blocks.

Merrill et al. in [5] superseded the Satish work by presenting a high performance and scalable Radix Sort implementation where it applies a Radix Sort strategy based on the fusion of kernels, multi-scan and thread-block serialization reducing the aggregate memory workload and thread synchronizations.

In [12], Eric Young discusses several optimization strategies for the Direct Compute SDK; but it also relies on the CUDA architecture for his discussion. In that presentation are discussed techniques such as coalesced access, proper thread group, shared memory usage, memory bank conflicts, maximization of hardware occupancy and proper decomposition of tasks to avoid global synchronization and other subjects are discussed.

It can be seen that most material on GPGPU optimization is centered on the CUDA SDK. In this paper, we take several points from this material and extend the discussion to other hardware architectures.

III. Data Parallel Computing

General-Purpose computing on graphics hardware can provide a significant advantage to implement efficient data parallel algorithms. Today, modern GPGPUs provide high data throughput, memory bandwidth and high-level programming languages to abstract the programmable units. The programing environments abstract these hardware programmable units to allow efficient data parallel processing, data transfer, matching the parallel processing resources and memory system available on the GPU.

While this programmability was first used to accelerate graphic applications, it has transformed the GPGPUs into powerful platforms for high performance computing with a wide variety of applications such as Sorting Algorithms, Matrix Operations, Ray Tracing, Collision Detection, Linear Algebra, among others.

A. GPGPU Architectures

This section presents the HW architecture of three different GPU architectures that can be used for data parallel computing implemented in the DirectX 11 Compute Shader:

1) AMD Radeon HD 7970 Graphics
2) NVIDIA Geforce GTX780
3) Intel HD Graphics 4000

It is not the purpose of this section to compare the performance between these three architectures; instead, we will focus on providing the HW and SW details to implement a portable fast Radix Sort implementation across these architectures.

1) AMD Radeon HD 7970 Graphics: The AMD Radeon HD 7970 is based on the AMD's Graphics Core Next (GCN) architecture [13]. This video card has 32 compute units (CUs). Each CU has 4 SIMD units for vector processing and each SIMD unit is assigned its own 40-bit program counter and instruction buffer for 10 wavefronts and it can execute up to 40 wavefronts. A wavefront groups 64 threads running in parallel. Thus, the AMD Radeon HD 7970 video card can issue up to 81,920 work items at a time in 2048 stream processors. Figure 1 shows a block diagram of the Radeon HD 7970 architecture.

2) NVIDIA Geforce GTX780: The NVIDIA Geforce GTX780 video card is based on NVIDIA's Kepler architecture [14] and consists of twelve next-generation Streaming Multiprocessors (SMX) with 192 CUDA cores per SMX. Thus the GeForce GTX 780 implementation has 2304 CUDA cores.

Figure 2 shows a block diagram of the GeForce GTX 780 architecture [14].

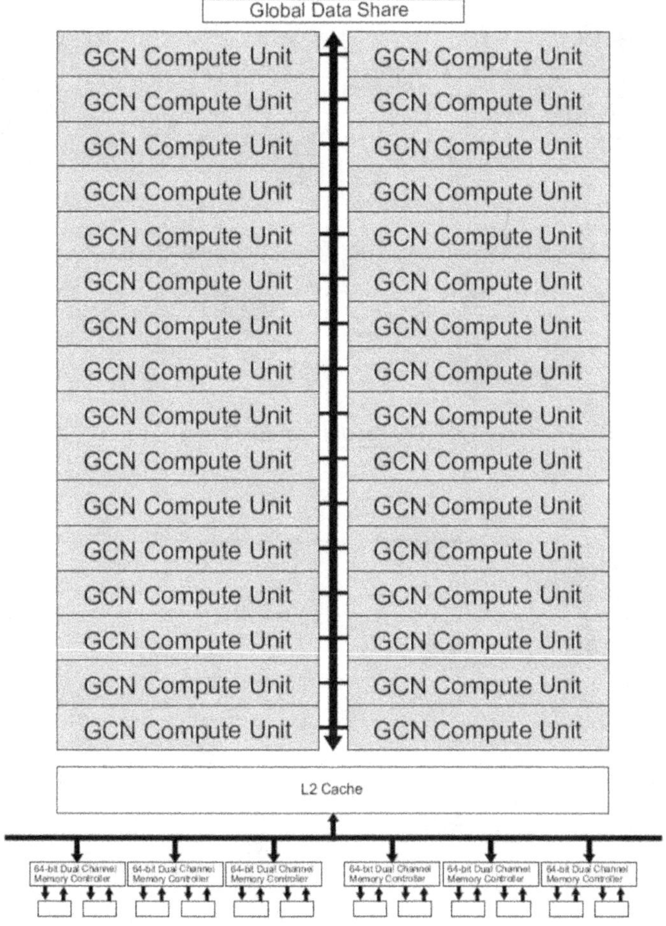

Figure 1. AMD Radeon HD 7970 Block Diagram [13].

The SMX schedules threads in groups of 32 parallel threads called warps. Each SMX features four warp schedulers and eight instruction dispatch units, allowing four warps to be issued and executed concurrently and two independent instructions per warp can be dispatched on each cycle.

3) Intel HD Graphics 4000: Intel HD Graphics 4000 is a GPU that is integrated in the 3rd generation Intel Core processor [15]. This GPU contains 16 execution units (EU) with 8 threads/EU. Figure 3 shows a block diagram of the Intel Processor Graphics architecture.

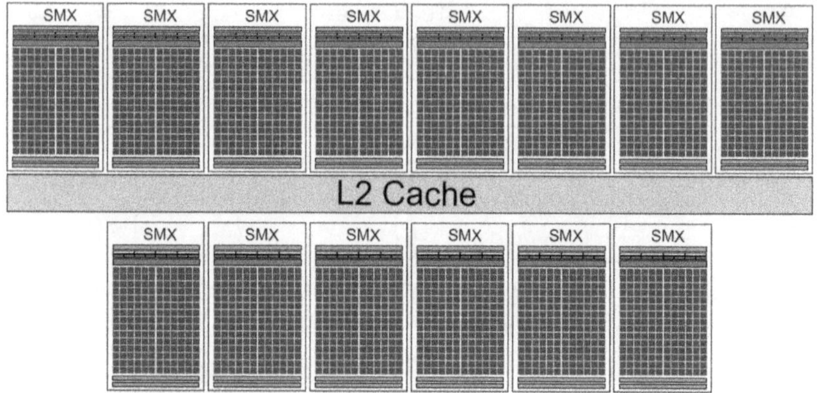

Figure 2. NVIDIA GeForce GTX 780 Block Diagram [14].

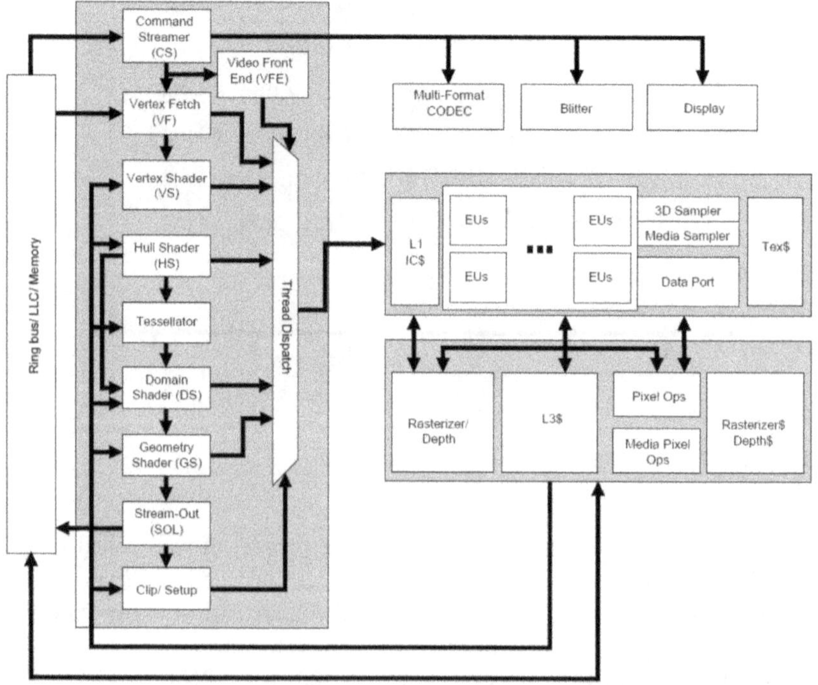

Figure 3. Overview of the Intel Processor Graphics [15].

B. General-Purpose Programing in GPUs

The main application programing interface (API) environments used to develop general-purpose applications on GPUs are the NVIDIA's CUDA C [16], Microsoft DirectX 11 Direct Compute [17] and OpenCL [18]. These APIs provide a C-like syntax programing environment with tools to easily build and debug complex applications [19].

Unlike DirectCompute and OpenCL, CUDA-enabled GPUs are only available from NVIDIA architecture. We will use DirectCompute for our Radix Sort implementation in order to enable the execution across different hardware architectures. The DirectCompute is a programmable shader stage that expands Microsoft's Direct3D 11 beyond graphics programming and enables the GPU processing units for general purpose programing. This programmable shader is designed and implemented with HLSL [20]. HLSL abstracts the capabilities of the underlying GPU HW architecture and allows the programmer to write GPU programs in a HW-agnostic way with a more familiar C-like programming language [21].

1) *GPGPU programing Model:* The programing model used for general purpose computing in GPUs is Single Program Multiple Data (SPMD) which means that all threads execute the same code. Today, GPU's can process hundreds of threads in parallel form, a thread is also called "work item" or element. Threads are grouped into blocks that can be seen as arrays of threads. The blocks are grouped to form a grid of threads in which, each thread has an ID that can be used to calculate its position within a group or the grid by using a 3D vector (x, y, z).
2) *GPGPU Memory Model:* The understanding of the cost and bottlenecks of the memory access can help improve the performance of a program running in a GPU [22]. The threads in a GPU application perform to the following memory requests:

    - Private Memory
        - Fastest access
        - Visible per thread
        - Thread lifetime
    - Shared Memory
        - Shared across threads within a Group
        - Very low access latency
        - Block lifetime
    - Global Memory

  - Accessible by all threads as well as host (CPU)
  - High access latency and finite bandwidth
  - Program lifetime
- Constant Memory
  - Short access latency
  - Read only
  - Kernel/Dispatch execution lifetime

Based on the memory access characteristics, we can design a strategy to maximize the memory bandwidth utilization, minimize the access latency and improve the performance in memory request such as, in the case of scatter and gather which are one of the most frequent operations in the implementation of sorting algorithms and suffers from low utilization of the memory bandwidth and consequently long memory latency.

IV. Radix Sort

Radix Sort is a sorting algorithm that rearranges individual components of the elements to be sorted (called keys) represented in a base-R notation. In high data-parallel computing architectures, it is very efficient to implement Radix Sort algorithms using integers that can be decomposed in keys represented as binary numbers $R = 2$ with or a power of two $R = 2^s$. It also helps if the keys are aligned with integer boundaries that match the size of the GPU registers. However, this algorithm is not limited to sorting integers, it can be used for sorting any kind of keys including string and floating-point values.

There are two different approaches to implement a Radix Sort algorithm:

1) Most significant digit (MSD) Radix Sorts
2) Least significant digit (LSD) Radix Sorts

The first group examines the digits of the keys in a left- to-right order, working with the most significant digits first. MSD Radix Sorts process the minimum amount of information necessary to get a sorting job done. The LSD Radix Sort examines the digits in a right-to-left order, working with the least significant digits first. This approach could spend processing time on digits that cannot affect the result, but it is easy to mitigate this problem and the latter group is the method of choice for most of the sorting applications.

To illustrate how an LSD radix algorithm with $R = 10$ works, consider the input array of 2-digit values shown in Figure 4.

The sorting algorithm consists of 2 passes extracting and rearranging the i-th digits of each key in order from least to most significant digit. Each pass uses

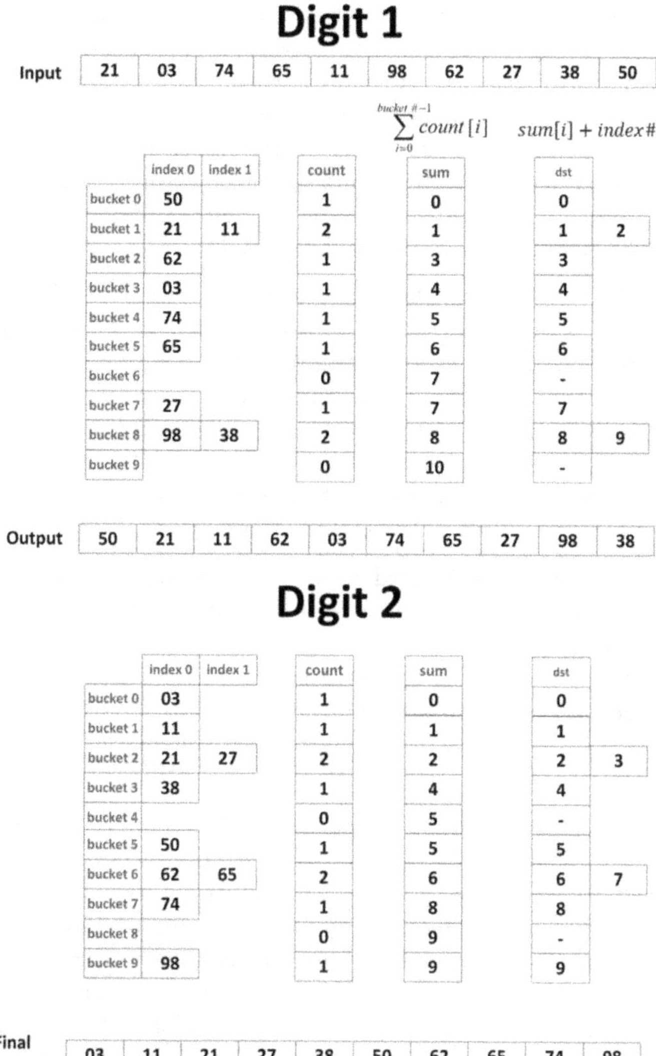

Figure 4. LSD radix R=10 sort diagram.

R counting buckets to store the digits based on the individual values from 0 to $R - 1$ to compute the destination index at which the key should be written. The destination index is calculated by counting the number of elements in the lower counting buckets plus the index of the element in the current bucket. Having computed the destination index of each element, the elements are scattered into the output array in the location determined by their destination index.

A. Radix Sort implementation in DirectX 11

The radix algorithm implementation in the DirectX 11 Compute Shader sorts an input array of 32-bit numbers de- composed in 4-bit keys of integers in a radix base of $R = 2^4$. This requires 16 counting buckets in each pass. In order to parallelize the algorithm in the GPGPU, the input array is divided into blocks that can be processed independently in a processing core. This approach was described in [11] and [23].

The algorithm sorts an input array of 32-bit numbers for the least significant 4-bit digit in 8 cycles. Each cycle uses three dispatches:

1) Each block sorts in local shared memory the 4-bit keys according to the i-th bit, using the split primitive described in [8] and [11] and compute offsets for the 16 buckets
2) Perform a prefix sum over the global buckets
3) Compute the destination index into the global array

After the 8 cycles, the output array is transferred to external memory back to the host system.

1) Local Sort Dispatch: The local sort dispatch performs a scan operation in an input array of 4-bit keys. The array is divided into blocks. Each block uses 512 threads and each thread processes one 4-bit key in LSD order. The process uses a local scan in shared memory. At a high level, the local sort dispatch is a radix $R = 2$ sort algorithm that is performed in each block, scanning one i-th bit from each key from least to most significant bit, then the keys are split placing all the keys with a in that 0 digit before all keys with a 1 in that digit. This process is repeated s times to sort the input list of N keys with respect to the i-th bit extracted in each cycle, in order from least to most significant as shown in Figure 5.

In the local sort step, each thread processes one element of the input array, extracting a bit of the key to be passed as the argument *"pred"* to the split operation. The scanBlock function performs a scan operation to obtain the total number of true predicates and the number of threads with lower groupIndex

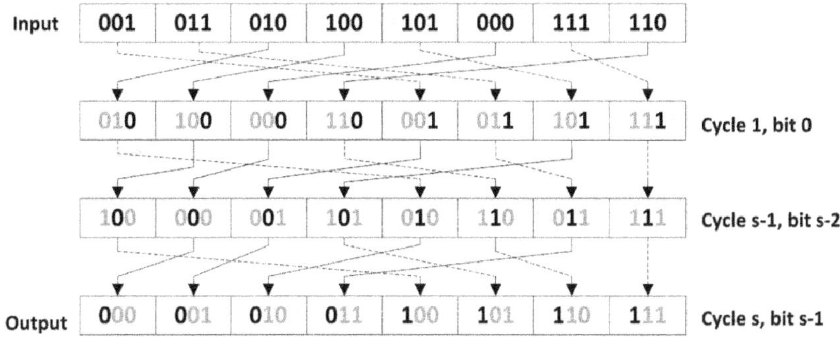

Figure 5. Radix Sort $R = 2^3$.

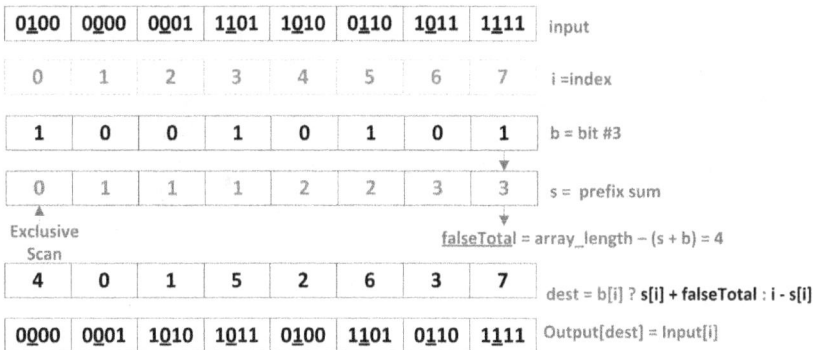

0100	0000	0001	1101	1010	0110	1011	1111	input
0	1	2	3	4	5	6	7	i =index
1	0	0	1	0	1	0	1	b = bit #3
0	1	1	1	2	2	3	3	s = prefix sum
Exclusive Scan								falseTotal = array_length – (s + b) = 4
4	0	1	5	2	6	3	7	dest = b[i] ? s[i] + falseTotal : i - s[i]
0000	0001	1010	1011	0100	1101	0110	1111	Output[dest] = Input[i]

Figure 6. Split primitive block diagram.

that have a true value in *pred*. With these values we can calculate the output position of the key for the current i-th bit-digit value (see Figure 6). The input array of 4-bit keys will be sorted after four successive split and scatter operations in local shared memory.

The scan is a common operation for data-parallel computation, it is best known as the "all prefix sum" operation. Blelloch [8] describes the scan as a primitive operation that can be executed in a fixed unit of time. The exclusive scan primitive takes a binary operator $\oplus$ with identity *i*, and an ordered array $[a_0, \; a_1, \ldots, \; a_{n-1}]$ of *n* elements, and returns the ordered array $[i, \; a0, \; (a_0 \;\; a_1), \ldots, (a_0 \;\; a_1 \; \ldots \;\; a_{n-2})]$. An "inclusive scan" version of this algorithm also returns an ordered array $[a_0, \; (a_0 \; \oplus \; a_1), \ldots, \; (a_0 \; \oplus \; a_1 \; \oplus \ldots \; \oplus \; a_{n-1})]$.

For example, applying the operator ⊕ as an addition to the following array:

$$A = [3 \quad 11 \quad 21 \quad 27 \quad 38 \quad 50 \quad 62]$$

The exclusive scan would return:

$$scan(A, +, excl) = [0\ 3\ 14\ 35\ 62\ 100\ 150]$$

And the inclusive scan output would be:

$$scan(A, +, incl) = [3\ 14\ 35\ 62\ 100\ 150\ 212]$$

In the case of the exclusive scan, the first value is the identity element for the operator that in this case is 0.

Sengupta et al. [10] presented an efficient parallel scan algorithm optimized to take advantage for the CUDA programming model and wrap granularity of the NVIDIA hardware architecture in order to maximize the execution efficiency. By using this approach, the threads executed within a block can be organized into groups to shared memory and take advantage of the synchronous execution of threads in a warp to eliminate the need for barrier synchronization.

The algorithm implements the parallel scan algorithm presented by Hillis in [7] to perform a scan operation with segmented warps/wavefronts of *k* threads. Figure 7 depicts the memory access pattern to implement this method. This function has the ability to select either an exclusive or inclusive scan operation via the scanType parameter. This algorithm applies the operator ⊕, across a block of size $b = (THREADS\ PER\ BLOCK)$, $O(k\ log2\ k)$ times for a warp/wavefront of size k.

Using this Scan primitive, we can build a function to scan all the elements in a block arranging all the threads in groups of k threads. The Scan Block function implementation is shown in the Block diagram of Figure 8. This ScanBlock function does $O(b \log_2 k)$ work for a block of size *b* and a multiple of *k* containing $\frac{b}{k}$ warps/wavefronts, and each thread processes one value of the block. Then, the LocalSort dispatch for a global array of size n will have n blocks with a work complexity of $O(n\ log_2\ k)$.

The offsets function stores the bucket offsets corresponding to the initial location in a block for each radix digit and computes the number of keys in each of the 2*s* buckets by counting the sorted digits for each block based on the individual values from 0 to $2^s - 1$. This counting is easily obtained by calculating the difference between adjacent bucket offsets. These values are written into a global buckets matrix that stores the counting values per block.

2) Prefix Sum Over the Global Buckets Matrix: The bucket sizes are obtained by performing a prefix sum dispatch on the buckets matrix. The prefix sum

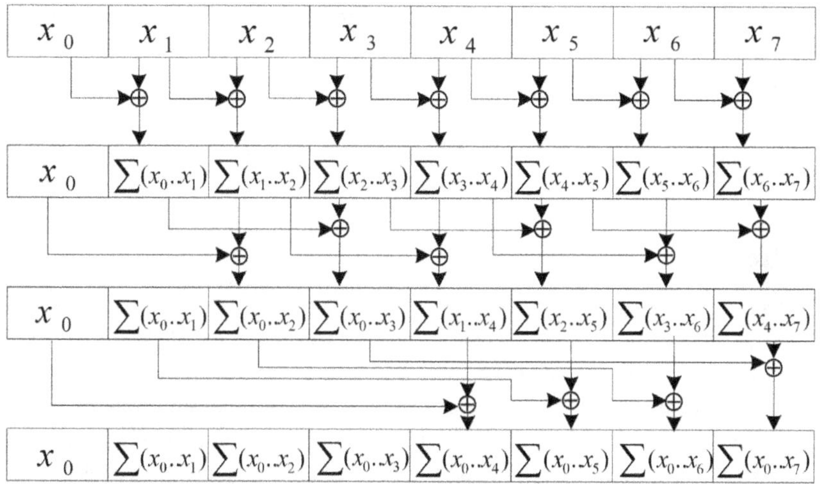

Figure 7. Memory Access Pattern of the Parallel Scan Operation.

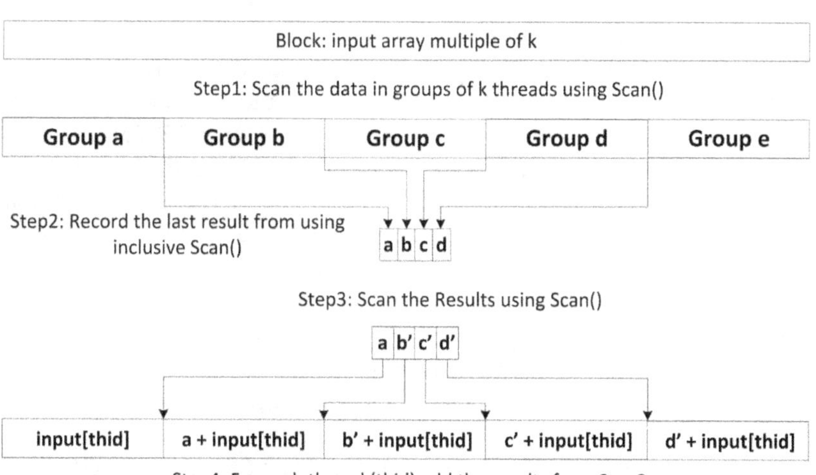

Figure 8. Constructing a ScanBlock() function.

can be done by applying the scan function explained in the LocalSort dispatch section.

3) Compute the Global Destination index: The global scatter dispatch computes the global position of the number locally sorted at its block level with respect to the 4-bit key being processed by adding its local offset in the block to the offset in its bucket.

V. Results

This section analyzes the performance of the Radix Sort implementation in the DirectX 11 Compute Shader running our algorithm in an AMD Radeon HD 7970 and NVIDIA Geforce GTX780 GPU; and compares the execution times to the Radix Sort implementation in the Thrust CUDA library running on the NVIDIA Geforce GTX780. The results obtained in the Intel HD Graphics 4000 GPU are not included in the comparison because it is integrated in a mobile platform. The NVIDA Geforce and AMD Radeon GPUs used are add-in cards for Desktop and workstation platforms where physical space and power consumption are not a constrain allowing higher graphics performance.

The timings were performed with varying number of inputs ranging from 65000 to 2 million elements. The running time measured in milliseconds is averaged over 100 runs, with the inputs redistributed randomly for each run. The running times do not include the data transfer time between CPU and GPU.

As seen in the Figure 9, the Radix Sort implementation in the Thrust CUDA library gives the best performance and the compute shader implementations running in AMD Radeon HD 7970 gives competitive performance for 2 million elements. With bigger arrays the Thrust CUDA implementation is 1.7*X* to 2*X* faster.

It is clear that the Radix Sort implementation in the Thrust library [4] is highly optimized for CUDA and also applies dynamic optimizations to improve the sorting performance and increase the speed by 2*X*, which is consistent with the results showed.

Figure 10 shows the execution times on the Intel HD Graphics 4000 GPU with the intention of demonstrating that the algorithm can be executed on any GPU with support for DirectX 11 Compute Shader.

VI. Radix Sort Applied in Ray Tracing

Ray tracing is an advanced illumination technique that simulates the effects of light by tracing rays through the screen on an image plane [24]. However, the

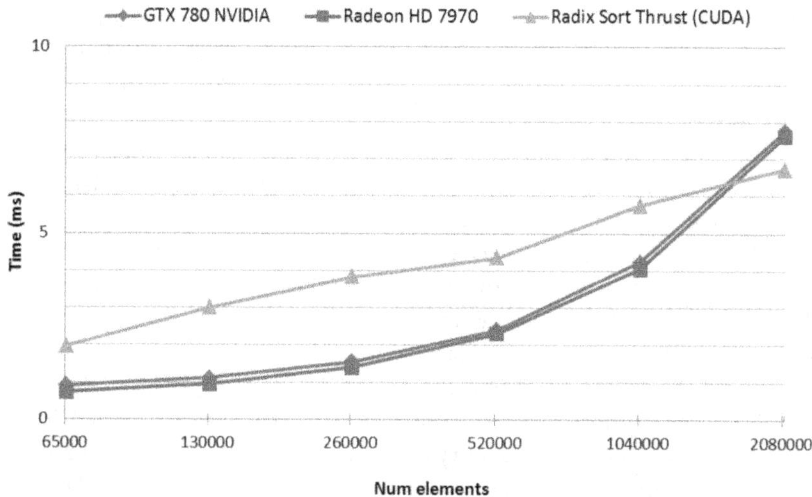

Figure 9. DirectX 11 Compute Shader Radix Sort Performance in AMD Radeon HD 7970 and NVIDIA Geforce GTX780 GPU's compared with Thrust CUDA Radix Sort running in the NVIDIA Geforce GTX780 GPU.

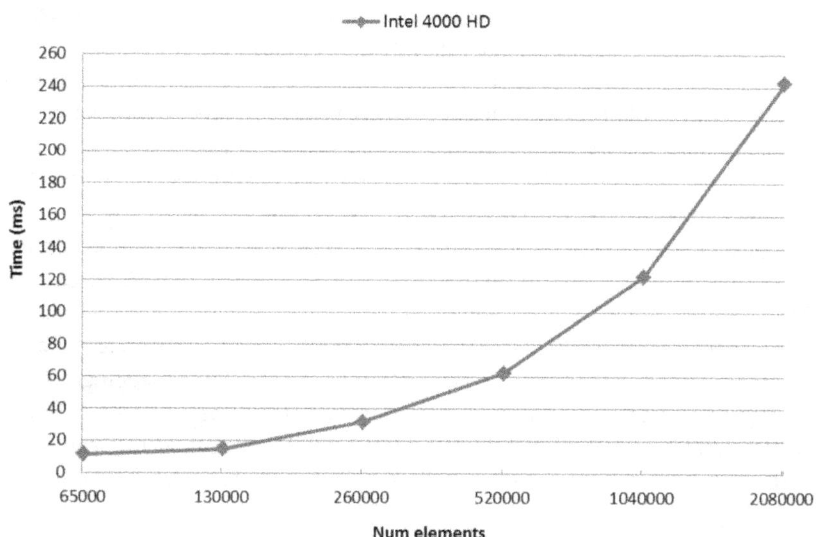

Figure 10. DirectX 11 Compute Shader Radix Sort Performance in the Intel HD Graphics 4000 GPU.

computational cost is so high that ray tracing has mostly been used for offline rendering [25]. Nevertheless, real-time ray tracing applications are available today due to constant hardware improvements and algorithmic advances, which yield more efficient algorithms.

Ray tracing is traditionally known to be a trivially-parallel implementation problem given that the color computation for each pixel is independent from its neighbors. On the other hand, both CPUs and GPUs are becoming more powerful year after year; but they are intrinsically different architectures, thus, it is also necessary to have a proper understanding of the interaction between algorithms and hardware in order to achieve a good performance. In any case, memory-handling remains a constant bottleneck [26], and it is one of the key points needed to achieve high frame rates.

To solve the memory bottleneck, efficient data structures are needed. These structures should allow fast memory access and fast ray-geometry intersection discovery. Currently, it is common to use acceleration structures to avoid unnecessary calculations during ray traversal. The most used acceleration structures are kd-trees, Bounding Volume Hierarchies (BVHs) and grids, and their variants [27]. At the core of the acceleration structures based on trees lies the implementation of a sorting algorithm in order to perform a fast tree construction. For dynamic scenes it is usual to rebuild the acceleration structures from scratch in each frame, therefore a fast sorting algorithm is of paramount importance. In most cases, the sorting algorithm used is the Radix Sort.

A stackless Linear BVH (SLBVH) implementation presented in [28] uses a bitonic algorithm to sort the primitives of the models in order to accelerate the construction of the tree structure. That implementation allows the building of models as big as 2^{18} (262,144) primitives. The use of a Radix Sort will permit the building of bigger models from scratch on every frame.

Figure 11 shows a ray tracing application [29] executing the Stanford Bunny (69,451 primitives), Stanford Dragon (871,414 primitives) and Welsh Dragon (2,097,152 primitives) models rendered in real time using the SLBVH in an AMD Radeon HD 7970 GPU. Our Radix Sort was used to accelerate the construction of the SLBVH tree. The application used using gloss mapping, Phong shading, one ray for shadows with one light source. Eight cameras with different positions and directions were set-up to measure the frame rate of the scenes.

The whole tree hierarchy was rebuilt in each frame, thus, allowing real-time ray tracing with dynamic scenes.

Table I presents the construction times of the SLBVH using our Radix Sort algorithm compared with a BVH-SAH CPU- based implementation based on the BVH of the PBRT frame- work [30]. The BVH construction was executed

Figure 11. Bunny (69K primitives, 77 FPS), Stanford Dragon (871K primitives, 23 FPS) and Welsh Dragon (2M primitives, 18 FPS) ray traced in real time with full tree reconstruction on each frame using our Radix Sort algorithm.

TABLE I. ACCELERATION STRUCTURE CONSTRUCTION TIMES

Model	Structure	Construction
Bunny(69K)	BVH CPU	0.142 sec
	SLBVH GPU	0.013 sec
Dragon(871K)	BVH CPU	2.091 sec
	SLBVH GPU	0.043 sec
Welsh-Dragon(2.2M)	BVH CPU	5.161 sec
	SLBVH GPU	0.083 sec

on a 3.19GHz Intel Core i7 CPU with 6Gb of DDR3 RAM compiled as a 32-bit application. The SLBVH GPU construction is using our Radix Sort implementation in AMD Radeon HD 7970 GPU. The results showed a very high advantage of the data parallel construction in GPU, this is a key factor in the reconstruction of acceleration structures for real-time ray tracing.

VII. Conclusion

A fast data parallel Radix Sort Implementation in the DirectX 11 Compute Shader was presented. As far as the authors know, at the time of writing this work, it is the fastest implementation published for the Compute Shader. The algorithm implemented several optimization techniques to take advantage of the HW architecture for the AMD, NVDIA and Intel GPUs, such as: taking advantage of kernel fusion strategy, the synchronous execution of threads in a warp/waveform to eliminate the need for barrier synchronization, using shared memory across threads within a group, management of bank conflicts, eliminate divergence by avoiding branch conditions and complete unrolling of loops, use of adequate group/thread dimensions to increase HW occupancy and application of highly data-parallel algorithms to accelerate the scan operations.

Although the results showed that our Radix Sort implementation gives competitive performance in AMD Radeon HD7970 and NVIDIA Geforce GTX780 GPUs sorting 2 million elements as fast as the Thrust CUDA implementation, our code still has room for improvement like the implementation of dynamic optimizations described in [5] and [4] and the optimization of the gather and scatter operations extensively used in the scan operations of the local sort dispatch and the global scatter dispatch. In [31], He et al. presented a probabilistic model to estimate the memory performance of scatter and gather operations and proposes optimization techniques to improve the bandwidth utilization of these two operations by improving the memory locality in data accesses yielding $2 - 4\ X$ improvement on the GPU bandwidth utilization and 30–50 % improvement in response time. Once these optimizations are in place, we expect to have an even better performance. However, this is the first implementation (that the authors know of) that is competitive against the Thrust CUDA library; with the additional advantage that it can run on any Compute Shader- compliant computing platform.

Finally, the algorithm was also used in a ray tracing application demonstrating its efficiency in the reconstruction of acceleration structures for real-time ray tracing, allowing the rendering of dynamic scenes at interactive frame-rates.

Acknowledgments

The authors would like to thank the Stanford Computer Graphics Laboratory for the Happy Buddha, and the Stanford Bunny models [32]; and the Bangor University for the Welsh Dragon model [33].

References

[1] K. E. Batcher, "Sorting networks and their applications." in *AFIPS Spring Joint Computing Conference*, ser. AFIPS Conference Proceed- ings, vol. 32. Thomson Book Company, Washington D.C., 1968, pp. 307–314.

[2] M. Zagha and G. E. Blelloch, "Radix Sort for vector multiprocessors," in *In Proceedings Supercomputing '91*, 1991, pp. 712–721.

[3] Intel. (2012) Product brief the intel xeon phi product family. [Online]. Available: http://www.intel.com/content/dam/www/public/us/en/documents/product-briefs/high-performance-xeon-phi-coprocessor-brief.pdf

[4] N. Bell and J. Hoberock, *A Productivity-Oriented Library for CUDA*. Morgan Kaufmann, 2012, ch. 26.

[5] D. Merrill and A. S. Grimshaw, "High performance and scalable Radix Sorting: a case study of implementing dynamic parallelism for gpu computing." *Parallel Processing Letters*, vol. 21, pp. 245–272, 2011.

[6] J. D. Owens, D. Luebke, N. Govindaraju, M. Harris, J. Krger, A. Lefohn, and T. J. Purcell, "A survey of general-purpose computation on graphics hardware," *Computer Graphics Forum*, vol. 26, no. 1, pp. 80–113, 2007.

[7] W. D. Hillis and G. L. Steele, Jr., "Data parallel algorithms," *Commun. ACM*, vol. 29, no. 12, pp. 1170–1183, 1986. [Online]. Available: http://doi.acm.org/10.1145/7902.7903

[8] G. E. Blelloch, *Vector models for data-parallel computing*. Cambridge, MA, USA: MIT Press, 1990.

[9] M. Harris, S. Sengupta, and J. D. Owens, *Par- allel Prefix Sum (Scan) with CUDA*. Addison Wes- ley, 2007, ch. 19, pp. 851–876. [Online]. Available: http://http.developer.nvidia.com/GPUGems3/gpugems3ch39.html

[10] S. Sengupta, M. Harris, M. Garland, and J. D. Owens, *Efficient Parallel Scan Algorithms for Many-core GPUs*. Taylor & Francis, 2011, ch. 19, pp. 413–442. [Online]. Available: http://www.taylorandfrancis.com/books/details/9781439825365/

[11] N. Satish, M. Harris, and M. Garland, "Designing efficient sorting algorithms for manycore gpus," in *Proceedings of the 2009 IEEE International Symposium on Parallel&Distributed Processing*, ser. IPDPS '09. Washington, DC, USA: IEEE Computer Society, 2009, pp. 1–10. [Online]. Available: http://dx.doi.org/10.1109/IPDPS.2009.5161005

[12] E. Young. (2010) Directcompute optimizations and best practices. [Online]. Available: http://www.nvidia.com/content/GTC- 2010/pdfs/2260 GTC2010.pdf

[13] AMD. (2012) AMD graphics cores next (gcn) architecture. [Online]. Available: http://www.amd.com/us/Documents/GCN Architecture whitepaper.pdf

[14] NVIDA. (2014) Nvidia geforce gtx 780 specifications. [Online]. Available: http://www.geforce.com/hardware/desktop-gpus/geforce-gtx-780/specifications

[15] Intel. (2012) Directx developers guide for for intel processor graphics. [Online]. Available: http://software.intel.com/m/d/4/1/d/8/Ivy Bridge Guide2.pdf

[16] NVIDA. (2013) Cuda programming guide. [Online]. Available: http://docs.nvidia.com/cuda/cuda-c-programming-guide/

[17] Microsoft. (2013) Compute shader overview. [Online]. Available: http://msdn.microsoft.com/en-us/library/windows/desktop/ff476331(v=vs.85).aspx

[18] Khronos. (2013) The open standard for parallel programming of heterogeneous systems. [Online]. Available: https://www.khronos.org/opencl/

[19] N. Kinayman, "Parallel programming with gpus," *Microwave Magazine*, vol. 14, no. 4, pp. 102–115, 2013.

[20] Microsoft. (2013) Hlsl. [On-line]. Available:http://msdn.microsoft.com/en-us/library/windows/desktop/bb509561(v=vs.85).aspx

[21] D. Feinstein, *Hlsl Development Cookbook*. Packt Publishing, Limited, 2013. [Online]. Available: http://books.google.com/books?id=yhC1mgEACAAJ

[22] S. Hong and H. Kim, "An analytical model for a gpu architecture with memory-level and thread-level parallelism awareness," in *Proceedings of the 36th Annual International Symposium on Computer Architecture*, ser. ISCA '09. New York, NY, USA: ACM, 2009, pp. 152–163. [Online]. Available: http://doi.acm.org/10.1145/1555754.1555775

[23] L. Ha, J. Krueger, and C. Silva, "Fast 4-way parallel Radix Sorting on gpus," 2009. [Online]. Available: http://www.sci.utah.edu/publications/ha09/Ha CGF2009.pdf

[24] I. Wald and P. Slusallek, "State-of-the-Art in Interactive Ray Tracing," in *Eurographics State of the Art Reports*, 2001, pp. 21–42.

[25] I. Wald, P. Slusallek, C. Benthin, and M. Wagner, "Interactive rendering with coherent ray tracing," in *Computer Graphics Forum*, 2001, pp. 153–164.

[26] I. Wald, W. R. Mark, J. Günther, S. Boulos, T. Ize, W. Hunt, S. G. Parker, and P. Shirley, "State of the art in ray tracing animated scenes," in *Computer Graphics Forum*, 2009.

[27] N. Thrane and L. Simonsen, *A Comparison of Acceler- ation Structures for GPU Assisted Ray Tracing*. Aarhus Universitet, Datalogisk Institut, 2005. [Online]. Available: http://books.google.com/books?id=lKEIcgAACAAJ

[28] S. Murguia, F. Avila, L. Reyes, and A. Garcia, *Bit-trail Traversal for Stackless LBVH on DirectCompute*, 1st ed. CRC Press, 2013, pp. 319–335. [Online]. Available: http://www.amazon.com/GPU-Pro-Advanced-Rendering-Techniques/dp/1466567430#reader 1466567430

[29] A. Garcia, F. Avila, S. Murguia, and L. Reyes, *Interactive Ray Tracing Using DirectX11 on the Compute Shader*, 1st ed. A K Peters/CRC Press, 2012, pp. 353–376. [Online]. Available: http://www.amazon.com/GPU-PRO-Advanced-Rendering-Techniques/dp/1439887829/ref=sr 1 1?ie=UTF8&qid=1336734285&sr=8-1

[30] M. Pharr and G. Humphreys, *Physically Based Rendering: From Theory to Implementation*. San Francisco, CA, USA: Morgan Kaufmann Publishers Inc., 2004.

[31] B. He, N. K. Govindaraju, Q. Luo, and B. Smith, "Efficient gather and scatter operations on graphics processors," in *Proceedings of the 2007 ACM/IEEE conference on Supercomputing*, ser. SC '07. New York, NY, USA: ACM, 2007, pp. 46:1–46:12. [Online]. Available: http://doi.acm.org/10.1145/1362622.1362684

[32] S. University, "The stanford 3d scanning repository," 2013. [Online]. Available: http://www.graphics.stanford.edu/data/3Dscanrep/

[33] B. University, "Eg 2011 welsh dragon," 2011. [Online]. Available: http://eg2011.bangor.ac.uk/dragon/Welsh-Dragon.html

20. Bottlenecks in Distributed Real-Time Visualization of Huge Data on Heterogeneous Systems

Gokce Yıldırım Kalkan
Simsoft Bilg. Tekn. Ltd. Sti. Ankara, Turkey
gokce@simsoft.com.tr

Veysi İşler
Dept. of Computer Engineering Middle East Technical University, Ankara, Turkey isler@ceng.metu.edu.tr

Abstract— In computer graphics, generating high-quality im-ages at high frame rates for rendering complex scenes from huge data is a challenging task. A practical solution to tackling this problem is distributing the rendering load among heterogeneous computers or graphical processing units. In this article, we investigate the bottlenecks in a single processing unit or the connection between the several processing units that need to be overcome in such approaches. We provide simulation results for these bottlenecks and outline guidelines that can be useful for other researchers working on distributed real-time visualization of huge data.

Keywords - visualization; heterogeneous systems; bottlenecks

I. Introduction

Rendering high-resolution and large-data involving com-plex objects in real-time applications such as training simu-lators is an extant challenge in Computer Graphics. Despite the advances in graphics hardware, because of the memory constraints and bandwidth bottlenecks, visualization of huge data is very difficult or impossible on a single graphics system. At this point, scalability of graphics systems and so parallel rendering on distributed systems plays an important role for improving the performance of computer graphics software. In other words, the rendering workload should be distributed among many processing units that are connected either in a single computer (cluster) or a computer network.

There exist many studies for distributed rendering of huge data, as reviewed in Section I-A. However, these studies either (i) use specialized hardware or computer networks (e.g., optic network [1]) that are not affordable in small and cheap visualization systems, or (ii) employ algorithmic solutions for distributing the primitives-to-be-rendered, which are mostly unrealistic [2], e.g. many body problem [3], or not interactive

[4]. In this article, we investigate the bottlenecks in real-time visualization of complex data with standard computers connected with a standard network.

A. Related Studies

As outlined in Figure 1, for distributed rendering, there are two main approaches: inter-frame and intra-frame based distribution. In inter-frame rendering, individual frames are generated by the processing units and the generated frames are combined to a single display unit. On the other hand, in intra-frame rendering, parts of a frame or a scene are distributed.

According to what is distributed, intra-frame rendering can be of three types: sort-first, sort-middle and sort-last classes [5]. There are various APIs which facilitate one of these approaches. Some of these are WireGL [6], Chromium [7], Multi OpenGL Multipipe SDK (MPK) [8] and Equalizer [9]. Among these APIs, Equalizer provides a much more scalable, flexible and compatible interface with less implementation overhead [10].

In the sort-first approach, a camera view is divided into several subviews, or partitions, and each subview is assign to a processing unit [11], [12], [13] - see also Figure 1. The outputs of the processing units are easily tiled together by a server for a final display device [5]. This approach is dependent on the size of the input dataset while it is irrespective of image resolution. Therefore, it is best to use this approach for applications whose final frame rate decreases with the increasing image resolution.

In the sort-last approach, the scene is partitioned into sets of objects or entities and these sets are shared among the processing units [11], [14] - see also Figure 1. The outputs of the processing units are only snapshots of only a part of the scene, and they are combined by a server possibly with a post-processing stage, where the depths of the individual objects are taken into account for a coherent snapshot of the scene [5]. This approach is dependent on image resolution while it is irrespective of input dataset. Therefore, this approach is well suited for applications whose final frame rate decreases with the increasing 3D input database.

In the sort-middle approach, a hybrid of sort-last and sort-first is performed, where parts of the view as well as the objects in sub-views are distributed [11], [15].

B. This Study

Different from the real-time interactive distributed visu-alization systems which are dependent on very fast network infrastructures, our work focuses on

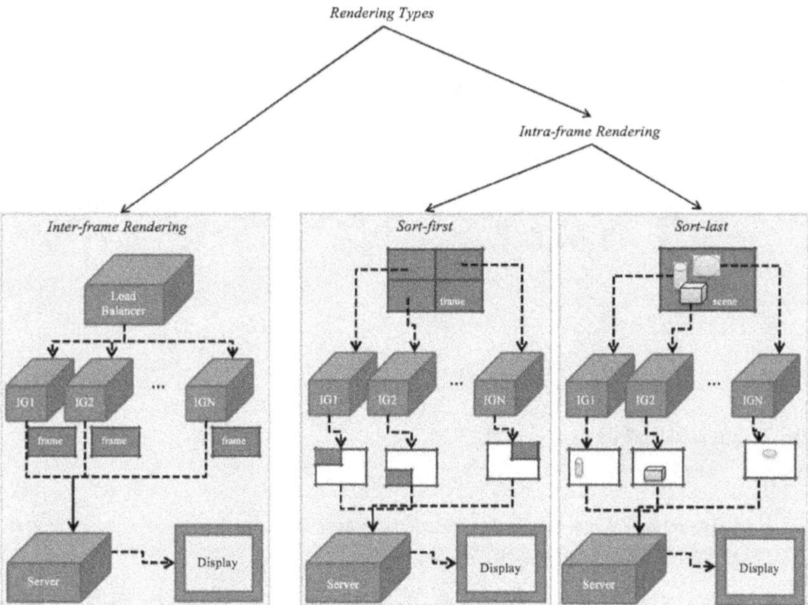

Fig. 1. Overview of the different approaches to distributed rendering.

real-time interactive dis-tributed visualization systems on standard 1 Gbit network infrastructures and proposes design guidelines for different load balancing strategies for such architectures.

We implement and compare two load balancing methods against no load balancing: The first method distributes load among a set of computers whereas the second method dis-tributes the load on the same computer among many GPUs.

II. Our Load Balancing Strategies

In this study, to share the rendering load and achieve higher refresh rates, we propose load balancing strategies for dis-tributed network environment and local distribution strategies and compare their results with scenarios which do not use load balancing strategies.

Network Sharing Method (NSM) is based on sharing rendering load on among distributed computers whereas GPU Sharing Method (GSM) is based on sharing rendering load in the same computer locally.

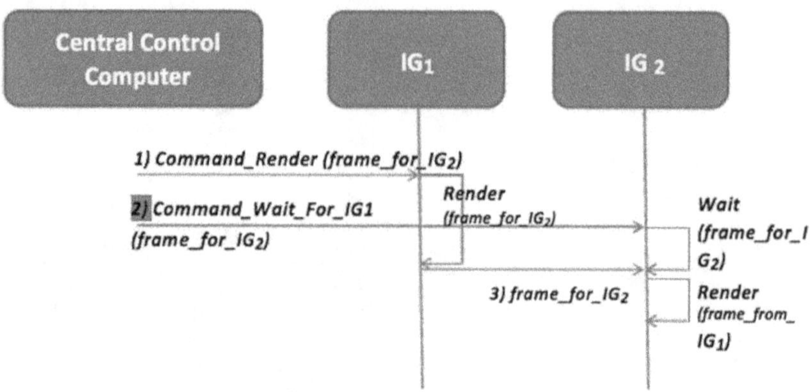

Fig. 2. Flowchart of the scenario of NSM.

A. Rendering with a network of computers (Network Sharing Method - NSM)

In this method, load is distributed among a network of computers. The distributed environment consists of the Central Control Computer and a number of Image Generators (IGs). An IG is an image generator computer which renders the scene. The Central Control Computer is the master computer which listens to the states of the IGs in terms of refresh rate and decides to give IGs commands to help other IGs rendering their frames if their refresh rate is under a threshold. The flow is processed as shown in Figure 2. The Central Control Computer commands IG_1 to render the frame for IG_2. When IG_1 completes IG_2's frame, it sends the frame over fast network to IG_2. IG_2 receives the frame and renders. The decision for distributing load among IGs is given by central control computer according to the current refresh rates.

B. Rendering locally among GPUs (GPU Sharing Method - GSM)

In this method, the load is distributed on the same computer among different Graphical Processing Units (GPUs). The IG application is responsible for rendering the scene. In this method, the flow is processed as shown in Figure 3. The IG application commands helper GPU ($GP\,U_2$) to render the frame which includes post-processing effects such as particles for the main GPU ($GP\,U_1$). $GP\,U_1$ is the GPU which is responsible for rendering all the scene. When $GP\,U_2$ completes the frame, it shares the frame with $GP\,U_1$ via shared memory of the IG. In the meantime, $GP\,U_1$ renders the main scene and blends the frame

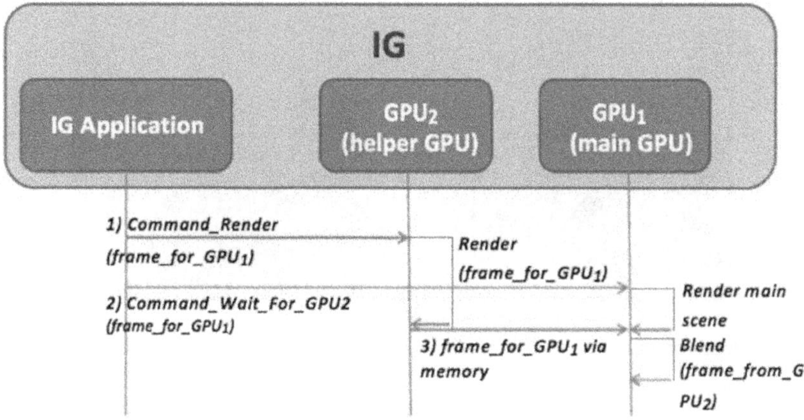

Fig. 3. Flowchart of the scenario of GSM.

which includes post-processing effects such as particles rendered by $GP\ U_2$. In this method, the decision for load balancing is made locally in the IG by the IG application according to the state of $GP\ U_1$ in terms of refresh rate. When the refresh rate is under a threshold, IG application decides that the main GPU distribute its load to the helper GPU. The output of the main GPU is used as the IG output. The output of the helper GPU is only used as in input for the main GPU.

III. Experiments and Results

The experiments were performed on a network of com-puters each of which contains two GTX 680 graphic cards (2GB RAM, 256 Bit), Intel i7 2700K processors and 16GB main memory - see Figure 5 for an outline of the experimental environment. The scene was selected as a large terrain in which some dust is scattered as particles in some regions of the terrain (see Figure 4 for a detailed snapshot). Each IG controls a window which has a camera view from the same viewpoint with contiguous field of views so that we can have a large field of view for the scene.

A. Network Sharing Method

In the network, there is significant latency due to transmis-sion of a packet, which can lead to incoherence in displayed frames. Two solutions to this problem

Fig. 4. A few snapshots from the rendered scenes.

are: (i) Block the IG receiving help until the next frame in sequence arrives. (ii) Send the frames to the slow IG latest at t - ΔtN, where ΔtN is the delay due to network transmission and the related processing. In this section, we will have a look at both. In any case, for load distribution to be worth the effort for an IG (IG_1) that needs help, the time for rendering one frame should be costing at least the time for a packet to travel on network and the rendering time for the helper IG (IG_2):

$$\Delta tIG^1 > \Delta tIG^2 + \Delta tN, \tag{1}$$

where Δt_{IG1} and Δt_{IG2} are frame rendering times for IG_1 and IG_2, respectively. The decision for the Central Control Computer should be based on the frame rendering times of IG_1 and IG_2 according to the criteria in Equation 1.

In addition, for an application that needs to achieve 25 fps, the time for the helper IG (IG_2) to render one frame for the other IG should be smaller than the time for rendering its own frame (40ms in the case of 25 fps) minus the time it takes helper IG to render the other IG's frame:

$$\Delta tIG + \Delta tOF < 40ms, \tag{2}$$

where ΔtIG shows the time of rendering its own frame for a helper IG, and ΔtOF is the time it takes a helper IG to render the other IG's frame. Based on this result, we can conclude that the decision for the Central Control Computer should also be based on the frame render times of helper IGs according to the criteria in Equation 2.

In Figure 6, we analyze the effect of the received help on the refresh rate of the IG receiving help. In these results, the helping GPU sends a frame every 50ms. In the blocking case, we see that, if the IG is fast enough, the received

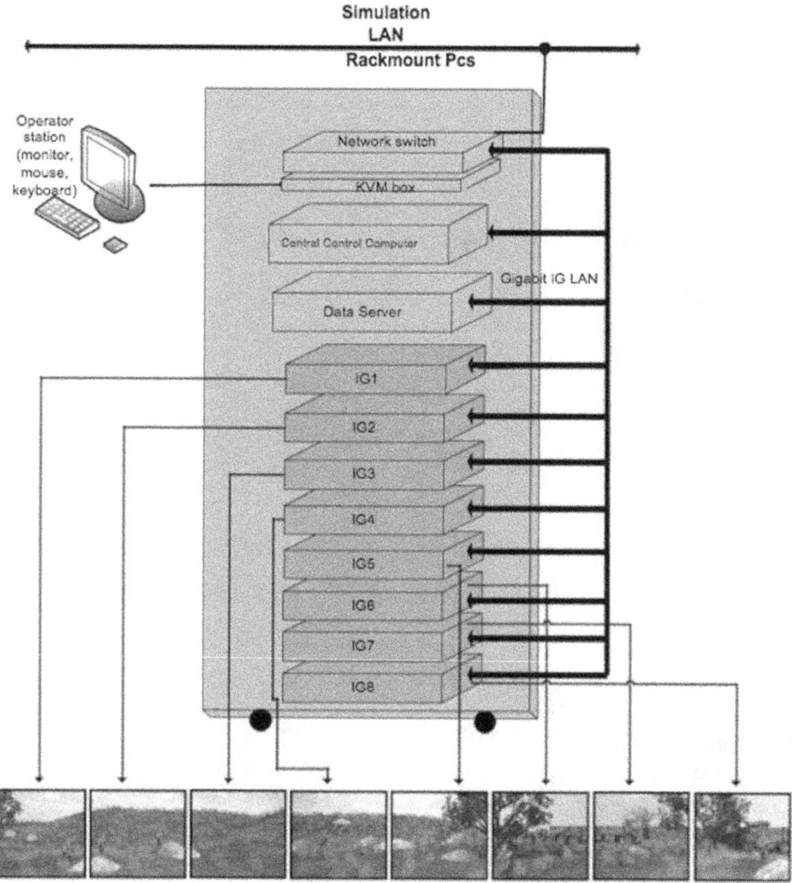

Fig. 5. Experiment Environment.

help cannot increase its refresh rate because getting help means waiting for a packet from the network, which costs more than rendering the frame itself. In the non-blocking case, however, whatever how fast the IG is, receiving help increases its refresh rate. However, the amount of increase decreases when the speed increases.

In Figure 7, we see the effect of the Network Sharing Method on the refresh rate of the helping IG. We see that the helping IG is only slightly affected. This is due to the fact that the IG prepares and sends frames over the network in a separate thread than the one rendering the scene.

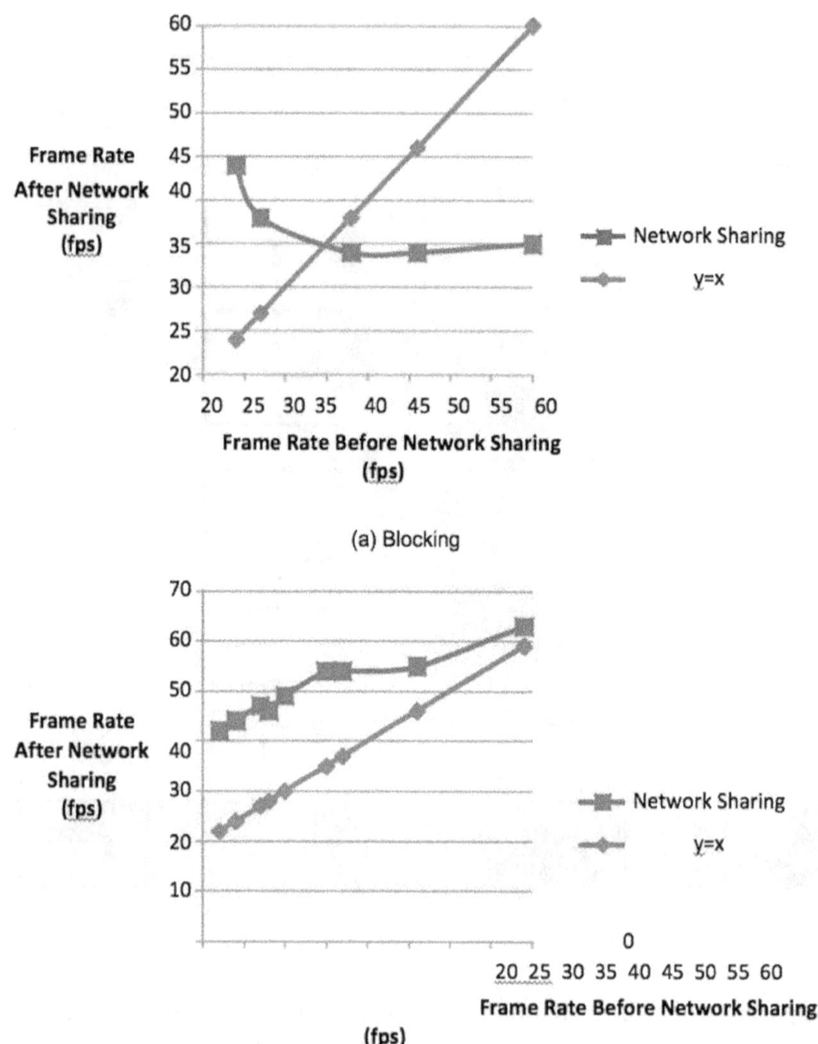

Fig. 6. The effect of Network Sharing Method on the IG receiving help. (a) With blocking the IG receiving help, (b) Non-blocking the IG receiving help.

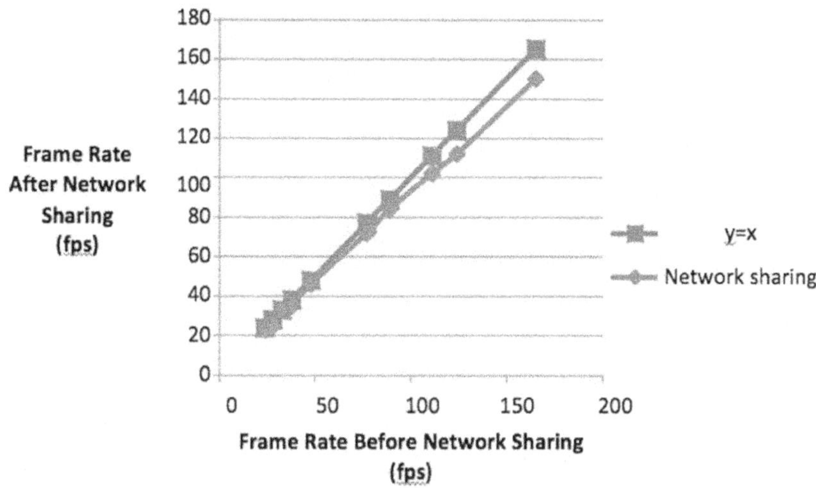

Fig. 7. The effect of Network Sharing Method on the IG giving help.

B. GPU Sharing Method

In Figure 8, we analyze the effect of the received help on the refresh rate of the GPU receiving help. The helping GPU sends every frame to the helpee GPU. We observe that, whatever the slow GPU's frame rate is, the help from the fast GPU leads to approximately similar frame rates, both in the blocking and the non-blocking cases. This is due to the fact that shared memory allows very fast transfer of frame to the helpee GPU and the helpee GPU does not get much chance to render a frame itself. Therefore, Figure 8 shows us that GPU-GPU sharing has a limit, no matter the original speed of the helpee GPU.

In Figure 9, we see the effect of the GPU Sharing Method on the refresh rate of the helping GPU. We see that the helping GPU is very much affected. This is due to the fact that the time spent for the helping GPU to copy the rendered data to the shared memory is roughly three times than rendering a scenes.

C. Guidelines

Based on the results provided in this section, we provide the following guidelines:

1) Although the distributed rendering literature has mostly focused on intra-frame load distribution strate-gies, inter-frame rendering is still plausible despite the network latency.

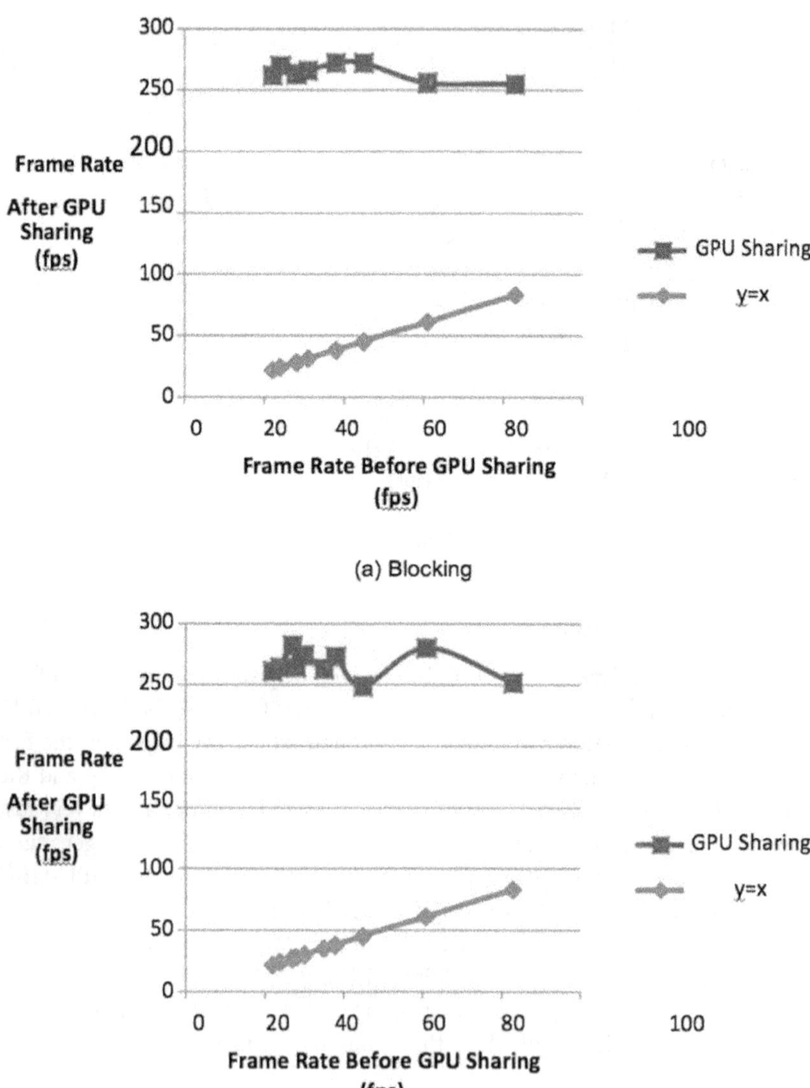

Fig. 8. The effect of GPU Sharing Method on the GPU receiving help. (a) With blocking the GPU receiving help, (b) Non-blocking the GPU receiving help.

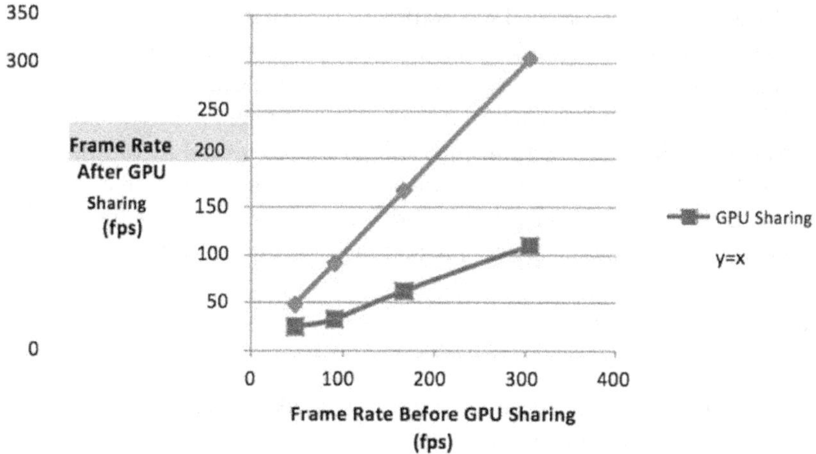

Fig. 9. The effect of GPU Sharing Method on the GPU giving help.

2) For overcoming the network latency problem, several strategies can be adopted based on the demands of the rendering problem. One is blocking the rendering node while waiting for the frame from another IG, and the other approach is non-blocking the IG re-ceiving help, and taking care of the frame coherence issue by making sure that the frames are sent at least t_N before the IG finishes rendering its own frame.

GPU-GPU sharing is much faster than IG-IG sharing. The latency problem, though less severe, exists for GPU-GPU communication as well. The same solu-tions proposed for IG-IG communication apply to GPU-GPU latency.

4) For enhanced distributed rendering, IG-IG and GPU-GPU distributed rendering should both be utilized.

IV. Conclusion

With the desire to visualize huge data or simulate complex scenes in high resolution, it has become a necessity to use parallel and distributed rendering techniques or architectures for fast, real-time, interactive simulation systems. Existing approaches either share individual frames (inter-frame meth-ods) or parts of a single frame (sort-first, sort-last and sort-middle methods), and generally, they use advanced hardware connected together in very expensive ultra fast networks.

The article has investigated the bottlenecks in parallel and distributed rendering systems with simulations. It has shown that in a locally distributed rendering system, the transfer from one GPU to the other needs to go over the CPU and the memory, which is a limiting factor. Moreover, for distributed rendering using a network of computers, the network speed is a bottleneck. We argue that, under these bottlenecks, rendering can be distributed provided that the rendering speed of a processing unit is slow enough to compensate for the time delay for the data transfer, either in the computer or in the network.

Acknowledgment

This work is partially funded by the Ministry of Science under project number SANTEZ 00906.STZ.2011-1. Moreover, we would like to thank Simsoft for providing us the software and the hardware environment for testing algorithms developed by this work.

References

[1] R. E. De Grande and A. Boukerche, "A dynamic, distributed, hierarchical load balancing for hla-based simulations on large-scale environments," in *Euro-Par 2010-Parallel Processing*. Springer, 2010, pp. 242–253.

[2] S. Marchesin, C. Mongenet, and J.-M. Dischler, "Dynamic load balanc- ing for parallel volume rendering," *Eurographics Symposium on Parallel Graphics and Visualization*, 2006.

[3] R. Hagan and Y. Cao, "Multi-gpu load balancing for in-situ visu- alization," in *International Conference on Parallel and Distributed Processing Techniques and Applications*, 2011.

[4] R. E. De Grande and A. Boukerche, "Predictive dynamic load balancing for large-scale hla-based simulations," in *Proceedings of the 2011 IEEE/ACM 15th International Symposium on Distributed Simulation and Real Time Applications*. IEEE Computer Society, 2011, pp. 4–11.

[5] P. Yin, X. Jiang, J. Shi, and R. Zhou, "Multi-screen tiled displayed, parallel rendering system for a large terrain dataset," *International Journal of Virtual Reality*, vol. 5, no. 4, pp. 47–54, 2006.

[6] G. Humphreys, M. Eldridge, I. Buck, G. Stoll, M. Everett, and P. Han- rahan, "Wiregl: a scalable graphics system for clusters," in *SIGGRAPH*, vol. 1, 2001, pp. 129–140.

[7] G. Humphreys, M. Houston, R. Ng, R. Frank, S. Ahern, P. D. Kirchner, and J. T. Klosowski, "Chromium: a stream-processing framework for interactive rendering on clusters," *ACM Transactions on Graphics (TOG)*, vol. 21, no. 3, pp. 693–702, 2002.

[8] OpenGL-Multipipe, "Opengl multipipeTM sdk white paper, document number: 007-4516-003," Sgi Techpubs Library, 2004.

[9] S. Eilemann, M. Makhinya, and R. Pajarola, "Equalizer: A scalable parallel rendering framework," *IEEE Transactions on Visualization and Computer Graphics*, vol. 15, no. 3, pp. 436–452, 2009.

[10] U. Gun, "Interactive editing of complex terrains on parallel graphics architectures," M.Sc. Thesis, Middle East Technical University, Department of Computer Engineering, 2009.

[11] S. Molnar, M. Cox, D. Ellsworth, and H. Fuchs, "A sorting classification of parallel rendering," *IEEE Computer Graphics and Applications*, vol. 14, no. 4, pp. 23–32, 1994.

[12] B. Moloney, M. Ament, D. Weiskopf, and T. Moller, "Sort-first parallel volume rendering," *IEEE Transactions on Visualization and Computer Graphics*, vol. 17, no. 8, pp. 1164–1177, 2011.

[13] E. Bethel, G. Humphreys, B. Paul, and J. D. Brederson, "Sort-first, distributed memory parallel visualization and rendering," in *Proceedings of the 2003 IEEE symposium on parallel and large-data visualization and graphics*. IEEE Computer Society, 2003, p. 7.

[14] T. Fogal, H. Childs, S. Shankar, J. Krüger, R. D. Bergeron, and P. Hatcher, "Large data visualization on distributed memory multi- gpu clusters," in *Proceedings of the Conference on High Performance Graphics*. Eurographics Association, 2010, pp. 57–66.

[15] R. Samanta, T. Funkhouser, K. Li, and J. P. Singh, "Hybrid sort-first and sort-last parallel rendering with a cluster of pcs," in *Proceedings of the ACM SIGGRAPH/EUROGRAPHICS workshop on Graphics hardware*. ACM, 2000, pp. 97–108.

21. Transfer Function Refinement for Exploring Volume Data

Shengzhou Luo
Graphics Vision and Visualisation Group Trinity College Dublin, Ireland luos@tcd.ie

John Dingliana
Graphics Vision and Visualisation Group Trinity College Dublin, Ireland John.Dingliana@tcd.ie

Abstract – Volume visualization is a powerful technique for depicting layered structures in 3D volume data sets. However, it is a major challenge to obtain clear visualizations of a volume with layers clearly revealed. In particular, the specification of the transfer function is frequently a time-consuming and unintuitive task in volume rendering. In this paper, we describe a global optimization and two user-driven refinement methods for modulating transfer functions in order to assist the exploration of volume data. This optimization is dependent on the distribution of scalar values of the volume data set and is designed to reduce general occlusion and improve the clarity of layers of structures in the resulting images. The user can explore a volume by interactively specifying different priority intensity ranges and observe which layers of structures are revealed. In addition, we show how the technique can be applied for time-varying volume data sets by adaptively refining the transfer function based on the histogram of each time-step. Experimental results on various data sets are presented to demonstrate the effectiveness of our method.

Keywords – visualization; transfer functions; volume data

I. Introduction

Transfer functions play an essential role in volume rendering. Through the assignment of visual properties, including color and opacity, to the volume data being visualized, transfer functions impact the final rendering of the data set and thus affect which structures will be visible to the user. However, obtaining an effective transfer function is a non-trivial task, which often entails time-consuming tweaking until a desired aesthetic quality is achieved in the resulting rendering. Although a number of automatic or semi-automatic approaches have been developed, transfer function design remains a challenging problem [1] [2].

For end users, such as physicians, who may not have much experience in volume rendering and transfer function design, a user-friendly approach that allows them to intuitively explore volume data sets is very desirable. As fully automatic approaches cannot currently guarantee satisfactory results in every case, exploratory approaches with simple and efficient interaction are highly desirable.

In this paper, we propose a novel approach to refine the transfer function based on the distribution of the scalar values of the volume data set. Firstly, we propose an automatic step to refine the transfer function that improves the rendering of volume data by reducing overall occlusions with no previous assumptions of the data set. Furthermore, we propose two interactive methods that extend on the optimization technique in order to enhance specific intensity ranges within the data as identified by the user. The process is fast and intuitive and allows the user to provide customized views of the data to aid in visual exploration of the volume data set.

II. Related Work

Various strategies have been proposed to simplify transfer function specification [1]. Overlapping intensity intervals corresponding to different materials make boundary detection difficult. Classical approaches try to detect boundary information between tissues by introducing derived attributes such as first and second derivatives to isolate materials [3] [4] [5]. In this case, the transfer functions are extended to multidimensional feature spaces. As a result, the interaction of transfer functions becomes more complex and unintuitive as the dimensionality becomes higher. Even in the case of two-dimensional transfer functions, a considerable amount of user interaction is required in order to come up with meaningful results [2].

Rezk-Salama et al. [6] presented high-level semantics to abstract parametric models of transfer functions in order to automatically assign transfer function templates. Bruckner and Gröller introduced the concept of style transfer functions [7] which aim to produce more comprehensible images by using transfer functions that map input values to different non-photorealistic rendering styles. Wu and Qu [8] developed a method that uses editing operations and stochastic search of the transfer function parameters to maximize the similarity between volume-rendered images given by the user. Finding good transfer functions for time-varying volume data is more difficult than for static volume data, as data value ranges and distributions change over time. Jankun-Kelly and Ma [9] examined how to combine transfer functions for different time-steps to generate a coherent transfer function. Tikhonova et al. [10] presented an explorative approach based on a compact representation of each time step of the dataset in the form of ray attenuation functions. Ray attenuation functions are subsequently used for transfer function generation.

A number of approaches have been proposed to automate the design of transfer functions. [11]described a method to structure attribute space in order to guide users to regions of interest within the transfer function histogram. Chan et al. [12] developed a system to optimize transparency automatically in volume rendering based on Metelli's episcotister model to improve the perceptual quality of transparent structures. Correa and Ma [13] proposed the visibility histogram to guide the transfer function design. In a later work [14], they generalized the visibility histogram and proposed a semi-automatic method for generating transfer functions by maximizing the visibility of important structures based on the visibility histogram, which represents the contribution of voxels to the resulting image. Ruiz et al. [15] also used visibility as a main parameter for the transfer function specification. Their method obtains the opacity transfer function by minimizing the informational divergence between the visibility distribution captured by a set of viewpoints and a target distribution defined by the user. Later, Bramon et al. [16] extended this approach to deal with multi-modal information.

The approach in this paper is based on our previous work [17] on optimizing transfer functions by minimizing the variance of control point weighting, which is generated from the intensity distribution of the data set. User-selected regions are taken into account, in order to enhance priority intensity ranges of importance to the user in the resulting image, and the approach, in contrast to others discussed previously, is viewpoint-independent. In this paper, we extend this further by introducing an intensity-based method to facilitate interactive exploration of volume data sets. In addition, we provide a more generalized approach to the distribution of control points, which in the original paper was limited to simplified tent-shaped distrubtions. Finally, we describe how to propagate optimisation of the transfer functions through different time-steps of time-varying data volume sets.

III Background

A. Transfer Function Specification

In the specification of a 1D (intensity-based) transfer function, the user essentially assigns a color and/or opacity to a certain point in the histogram of scalar values in the data set. In practice, the user would be presented with an interface that allows them to set up several control points which corresponds to a certain

Table 1: Hounsfield units of some typical substances [18]

air	fat	soft tissue	bone (cancellous/dense)
1000	-100 to -50	+100 to +300	+700 to +3000

kind of material or structure. The user then defines a mapping from each control points to some visual property (e.g. color) resulting in voxels of the corresponding intensity to be rendered in that color. Fig. 1 displays four typical shapes used in transfer function design. If a volume data set contains complex structures, tent-like shapes are desirable in revealing isosurfaces of structures and seeing through inner structures. Otherwise, the ramp shape and other shapes can also reveal structures effectively. In order to design transfer functions effectively, it is commonly required that users have prior knowledge about which intensity ranges are relevant or which regions should be emphasized in the data. This is especially the case in medical visualization. For instance, in computed tomography (CT) data the intensity ranges are determined by the Hounsfield scale (*Table 1*). The user may expect the constituent's intensity of CT data to follow the Hounsfield scale and thus set up control points accordingly.

Another consideration is that interior structures are likely to comprise far fewer voxels and are often occluded by the surrounding material. Consider the transfer function in Fig. 2. The user finds three intensity intervals of interest and then sets up three sets of control points in order to visualize these intensity intervals. The opacity of the three peak control points are assigned equally as they are equally important. However, if the distribution of voxels follows $p(x)$ (the blue curve), the voxels of the leftmost intensity intervals may completely occlude voxels of the other two intensity intervals in the resulting image. The global optimization in our approach aims at reducing this kind of occlusion by modulating the opacity of the transfer function based on the distribution of scalar values of the volume data set.

B. Entropy of Volume Data

In computer graphics, information-theoretic measures, such as entropy and mutual information, have been applied to solve multiple problems in areas such as view selection [19] [20], flow visualization [21], multi-modal visualization [22], [16] and transfer function design [23] [24]. Information theory provides a theoretic framework to measure the information content (or uncertainty) of a random variable represented as a distribution [25]. Consider a discrete random variable X which has a set of possible values $\{a_0, a_1, \ldots, a_{\{n-1\}}\}$ with

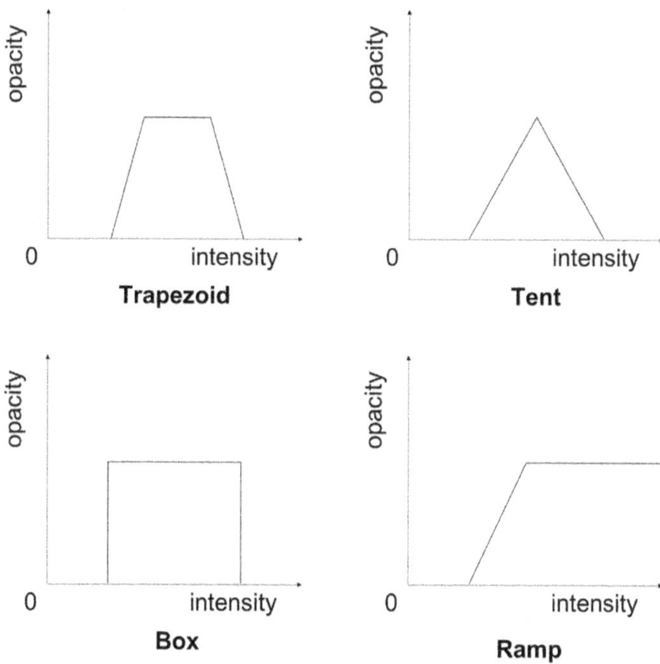

Fig. 1: Typical transfer function shapes [19]

probabilities of occurrence $\{p_0, p_1, \ldots, p_{\{n-1\}}\}$, we can measure the uncertainty of the outcome with the entropy $H(X)$, which is defined by

$$H(X) = -\sum_{x\in X} p(x)\log p(x)$$

where the summation is over the corresponding alphabet and the convention $0log0 = 0$ is taken. The term $-\log p(x)$ represents the information content associated with the result x. If the entire volume data set is treated as a random variable, $I(a_x) = -\log p(x)$ represents the information content of a voxel a_x with intensity x, and the entropy gives us the average amount of information of a volume data. The probability $p(x)$ is defined by $p(x) = \frac{n_x}{n}$, where n_x is the number of voxels with intensity x and n is the total number of voxels in the volume data.

Bordoloi and Shen [19] proposed a *noteworthiness factor* to denote the significance of a voxel to the visualization. The noteworthiness should be high for the

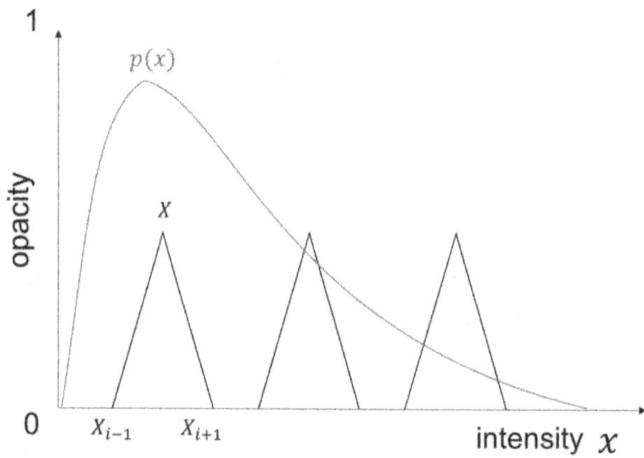

Fig. 2: A transfer function with tent-like shapes

voxels which are desired to be seen, and vice versa. The noteworthiness of voxel j is defined as $W_j = \alpha_j I_j = -\alpha_j \log f_j$, where α_j is the opacity of voxel j looked up from the transfer function, I_j is the information carried by voxel j, which can be derived from the frequency of its histogram bin f_j. $-\log f_j$ represents the amount of information associated with voxel j.

IV. Method

In this section, we present a transfer function refinement approach for modulating the opacity associated with the control points in a transfer function and combine it with user interaction to specify priority areas or intensity values of importance in the resulting image. In previous work [17], we proposed a weighting of control points of transfer functions and an optimization method to minimize the energy function based on this weighting. However, this approach was limited to refining transfer functions with tent-like shapes and processing static volume data sets. In this paper we extend this approach with a more general distribution of weightings and energy function, so that it can handle any form of one-dimensional transfer function based on control points. In addition, an interaction widget (as in Fig. 8) is introduced to allow users to explore the data sets by emphasizing certain intensity values and see the optimized output immediately.

In our approach, the user has control of the transfer functions by setting up control points as input for the optimization or tweaking the resulting transfer functions after the optimization. For example, the user can leave out less relevant data ranges by not covering the data ranges with shapes formed by control points (as in Fig. 1(. In the case of refining existing transfer functions, users also have the flexibility to refine the input transfer functions partially and keep certain control points constant during the optimization.

A. Weighting of Transfer Function Components

The goal of our transfer function refinement approach is to balance the opacity settings so that voxels of more significance contribute more, and voxels of less significance contribute less to the resulting images. Given the intensity of the control points $v_1,\ v_2,\ \ldots,\ v_n$ of the transfer function t are $x_1, x_2,\ \ldots,\ x_n$ and the corresponding opacity are $\alpha(x_1),\ \alpha(x_2),\ \ldots,\ \alpha(x_n)$. The intensity range of the transfer function is normalized to [0,1]. For the convenience of discussion, two control points v_0 and v_{n+1} are added to the lower bound and the upper bound respectively, and, $x_0 = 0, \alpha(x_0) = 0$, $x_{n+1} = 1$ and $\alpha(x_{n+1}) = 0$.

Similar to the noteworthiness factor by Bordoloi and Shen [19], opacity and probability (derived from the intensity histogram) are also used in our weighting. We define the significance factor of the intensity x as

$$s(x) = -\alpha(x)\, p(x) \log p(x),\ x \in [0, 1]$$

In the significance factor $s(x)$, $p(x)$ is computed from the histogram of the data set, and $\alpha(x)$ is the opacity function that we want to modulate. The significance factor should be high for the voxels which are desired to be seen, and vice versa. Then we define the weight of the i-th edge (the segment between v_i and v_{i+1}) as

$$\mathrm{e}(\mathrm{i}) = \int_{\mathrm{x} \in [\mathrm{x_i},\, \mathrm{x_{i+1}}]} s(x)\, dx$$

where $i \in [0, n]$ and $x \in [0, 1]$.

Hence the energy function with distance factors for the intensity x_0 of the transfer function can be defined as the variance of edge weights with the distance factors

$$E = \sum_{i=0}^{n} \left(e(i) - \overline{e(i)}\right)^2$$

where $\overline{e(i)}$ is the mean of edge weights, i.e.

$$\overline{e(i)} = \frac{\sum_{i=0}^{n} e(i)}{n}$$

Consequently, minimizing the energy function is equivalent to flattening the curve of the edge weights.

B. Optimizer

Constraints are introduced in the search of the parameter space. Control points would only be moved vertically in the transfer function space. In other words, only the opacity associated with control points would be changed. The intensity of control points remains the same. Also, those control points that are marked as constant would not be updated in the optimization process. These constraints are based on our assumption that the intensity intervals associated with control points are the user's intensity intervals of interest. The user has explored the volume data and set up the transfer function according to his/her needs. Our algorithm aims to help the user reduce occlusion while preserving the user's knowledge or judgements of the data set.

A greedy strategy is employed in our algorithm to minimize the energy function. In each iteration, two operations are performed:

- Find the edge with the highest weight in the transfer function and reduce the opacity of the control point at its upper end (the vertex with a larger significance factor in the edge's two adjacent vertices).
- Find the edge with the lowest weight in the transfer function and increase the opacity of the control point at its lower end (the vertex with a smaller significance factor in the edge's two adjacent vertices).

In our implementation, the optimization ends when it reaches a user-specified iteration count. The two step sizes in reducing opacity and increasing opacity can both be user-specified, or the first one is user-specified and the second one is computed based on the first one and the ratio of the significance factors of the two chosen control points. The ratio of the two step sizes affects the overall opacity of the resulting image, for instance, the image becomes more opaque or translucent.

C. Distance Factors for Prioritizing Intensity Ranges

The above described optimizer is an approach to balance the global opacity and thus reduce occlusions in the rendered images. In other words, we de-emphasize the most prevalent voxels, which are considered to have a high probability of occluding the rest of the scene and in particular interior structures of the data.

Although global optimization can help deliver images with better overall visibility, small details may be under-enhanced in the global optimization and certain structures in the image may have to be further enhanced for specific purposes. For instance, in an anatomical data set, the global optimization may guarantee that all structures of materials such as skin, bone and flesh are all visible however if the task of the user is specifically to study skin, this may be counter-productive. Thus, it is clear that a flexible method guided by user interactions is necessary to achieve various visualization goals.

In this section, we describe two alternative methods for prioritizing specific intensity ranges in the volume data. The first approach allows the user to interactively select a specific intensity range target that they are interested in and a hue-based distance factor is used to emphasize different intensity ranges based on the target identified by the user. Secondly, we describe a region-based optimization to provide an intuitive method of interaction by choosing regions of interest in the image that are then enhanced in the final rendering.

1) Distance Factors for User-Selected Intensity Values: We introduce distance factors for intensity values to prioritize the specific intensity ranges. In our implementation, the user can select an intensity value by clicking on a color palette, which uses the same color map in the transfer function. Assume each control point is assigned a unique intensity value, we can get the difference between a specific intensity value and the intensity value of a control point.

Given the selected intensity x_0, we define the distance factor of the i-th control point $x(i)$ as

$$D_h(i) = |x_0 - x(i)|, i \in [0, n+1]$$

Linear interpolation is used to obtain the distance factor $d_h(x)$ for the intensity $x \in [x_i,\ x_{i+1})$

$$d_h(x) = D_h(i) + (D_h(i+1) - D_h(i)) \frac{x - x_i}{x_{i+1} - x_i}$$

Therefore, we define the weight of the i-th edge (the segment between v_i and v_{i+1}) with the distance factors as

$$e_h(i) = \int_{x \in [x_i, x_{i+1}]} d_h(x)\, s(x)\, dx$$

where $s(x) = -\alpha(x)\, p(x) \log p(x), x \in [0, 1], i \in [0, n]$.

Hence the energy function with distance factors for the intensity of the transfer function can be defined as the variance of edge weights with the distance factors

$$E_h = \sum_{i=0}^{n} \left(e_h(i) - \overline{e_h(i)}\right)^2$$

where $\overline{e_h(i)}$ is the mean of edge weights $e_h(i)$.

To use the energy function E_h described in this section, we simply need to replace the original energy function with E_h in the previously described optimization algorithm.

2) Distance Factors for User-Selected Regions: In addition to selecting a specific intensity value, the distance factors can also be calculated using a region selected by the user. This region-based selection provides an intuitive method for the user to prioritize specific intensity ranges by selecting a region with the colors of interest in the resulting image.

We introduce region-based distance factors to prioritize the user's region of interest. Assume each control point is assigned a unique color, we can get the difference between the color of a pixel in the region (which is selected in image space) and the color of a control point (HSV color space is used in our implementation).

Given the color of the i-th control point is $c(i)$, the distance between the color of a pixel r in the region R and the color of the i-th control point is denoted by $d(r, c(i))$. In our approach, the sum of the distances D between each pixel in the region R and the i-th control point is used to measure the difference between the region and the control point.

$$D(R, i) = \sum_{r \in R} d(r, c(i)), i \in [0, n+1]$$

Given the selected region R, we define the distance factor of the i-th control point as

$$W_R(i) = \frac{D(R, i)}{\sum_{i=1}^{n} D(R, i)}, i \in [0, n+1]$$

Linear interpolation is used to obtain the distance factor $w_R(x)$ for the intensity $x \in [x, \ x_{i+1})$

$$w_R(x) = W_R(i) + (W_R(i+1) - W_R(i)) \frac{x - x_i}{x_{i+1} - x_i}$$

Therefore, we define the weight of the i-th edge (the segment between and) with the distance factors as

$$e_R(i) = \int_{x \in [x_i, x_{i+1}]} w_R(x) \, s(x) \, dx$$

where $s(x) = -\alpha(x) \, p(x) \log p(x), x \in [0, 1], i \in [0, n]$.

Hence the energy function with distance factors for the region of the transfer function can be defined as the variance of edge weights with the distance factors

$$E_R = \sum_{i=0}^{n} \left(e_R(i) - \overline{e_R(i)} \right)^2$$

where $\overline{e_R(i)}$ is the mean of edge weights $e_R(i)$.

Similarly, in order to use the energy function E_R, we need to replace the original energy function with E_R in the previously described optimization algorithm.

The distance factors described in this section measure the dissimilarity between a selected region and a control point. Therefore, the distance factor would be small if the region has an overall color similar to the color of the control point. Since we are minimizing the energy function, which is the variance of the edge weights, reducing the distance factors of those control points, which are related to the selected region, will result in their opacity values being increased. As a result, the features (in this case, the intensity intervals) in the selected regions will be enhanced and other features will be de-emphasized in the rendered image. Since the distance is measured in the color space, the choice of colors for the control points affects the distance measured and thus affects the weighting function.

D. Adaptive Transfer Functions for Time-Varying Data Sets

In time-varying data sets, the data value ranges and distributions change among time-steps. A single global transfer function may not be able to adequately catch the details of the data set. Therefore, we exploit the transfer function optimizer (as discussed in Section IV-B) to locally refine the transfer function for each time-step in the data set. In this case, the user specifies a transfer function for a single time-step of the time-varying data set and using either of the two interaction methods (as discussed in Section IV-C) to specify priority intensity ranges. The transfer function designed for this time-step is taken as an input transfer function and optimized again based on the histogram of the next time-step. Subsequently, the output of next time-step is taken as input of the time-step after it and so forth.

V. Results and Discussions

In this section, we present some results to demonstrate the effectiveness of our approach on the CT-knee ($379 \times 229 \times 305$) and VisMale head ($128 \times 256 \times 256$) datasets [26] and a time-varying data set of a simulated

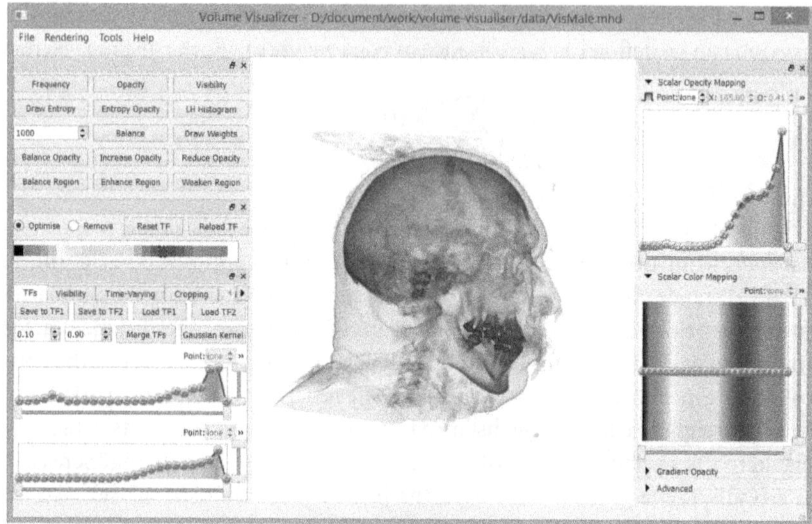

Fig. 3: A screenshot of our volume rendering system

turbulent vortex flow (128 × 128 × 128, 100 time-steps) [27]. Results were generated in our volume rendering system (Fig. 3) on a computer equipped with an Intel Core i5-2410M CPU, 8GB of RAM and a NVIDIA GeForce GT 540M graphics card.

Automatically generated transfer functions with ramps and tent-like shapes are provided as initial input to the optimizer. Fig. 4 displays a continuous transfer function. The ramps are formed by a series of control points with corresponding colors from the color map. Fig. 6 displays a transfer function with several evenly distributed tent-like shapes. Each tent-like shape consists of a peak control point and two bottom control points. The peak control points are movable while the bottom control points are marked as constant to maintain the tent-like shapes.

Note that the transfer functions consisting of equidistant and equal height tent-like shapes (as in Fig. 6) are just provided as naive examples of user designed transfer functions and used as input to the optimizer. In practice, users would design transfer functions (which consist of ramps, tent-like or other shapes) according to the characteristic of the data sets and the intensity ranges that they are interested in.

The initial opacity values of control points will affect the overall opacity level of the resulting image after optimization. Because there are omitted intensity ranges (the gaps) in transfer functions with tent-like shapes, the initial opacity values should be higher in transfer functions with tent-like shapes than in continuous transfer functions. In transfer functions with tent-like shapes (as discussed in Section III), the opacities of the top control points are set to 1/2 and the opacity of the bottom control points are set to 0. The bottom control points are fixed to 0 in order to keep the tent-like shapes in the transfer functions. By contrast, all control points in continuous transfer functions are movable vertically except that control points v_0 and v_{n+1} are fixed and serve as the boundary. The opacities of control points are set to 1/6 in continuous transfer functions. The color maps used in the transfer functions in our system are evenly sampled from a spectrum (with hue from 0° to 360° in HSV color space). In the optimization, the two step sizes for reducing opacity and increasing are both set to 1/256.

A. Automatic Transfer Function Refinement

Firstly, we demonstrate the global optimization with continuous transfer functions. In Fig. 4, the CT-Knee data set is rendered with a naive transfer function consisting of 6 tent-like shapes of various colors with equal opacity. Fig. 5 shows the resulting image rendered with the optimized transfer function. We tested this specific example as joints are popular regions of interest in medical visualization. The knee in particular is a commonly studied joint. In Fig. 4, only parts of the skeleton are visible. The rest is occluded by the surrounding material (such as the skin and muscles). After optimization (Fig. 5), the surrounding tissues become translucent, hence the skeleton is exposed and the knee joint is visible, while the overall context is preserved. We argue that in the absence of any previous assumptions on what the user is looking for, the global optimization provides a more balanced initial view before deeper exploration of the data.

Although the continuous transfer function is useful in automatic transfer function generation, in typical volume visualization programs, users often prefer much more simplified transfer functions, e.g. a small number of control points such as in transfer functions with tent-like shapes. Fig. 6 is the CT-Knee data set rendered with the transfer function. We show in Fig. 7 that our optimization can also benefit such simpler transfer functions.

In practice we observed that the energy function usually converges to a small but non-zero value. As the number of control points increases, it takes more iterations for the optimization to achieve a stable state - a number of tent-like shapes ranging from 4 to 16 was found to be the most effective. Our approach is relatively lightweight, e.g. we observed that the optimization finished within one second on the CT-Knee and VisMale data sets with transfer functions consisting of 24 control points and maximum iteration count of 1000.

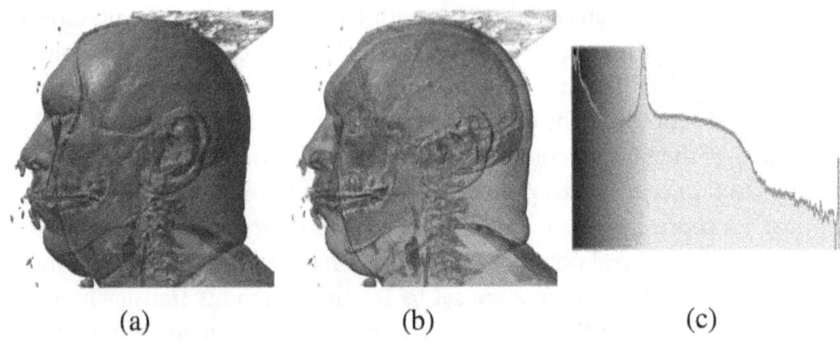

(a) (b) (c)

Fig. 4: Before optimization: CT-Knee with a continuous transfer function
(a) Preliminary view of data set (b) A continuous transfer function with a ramp
(c) Histogram of the dataset

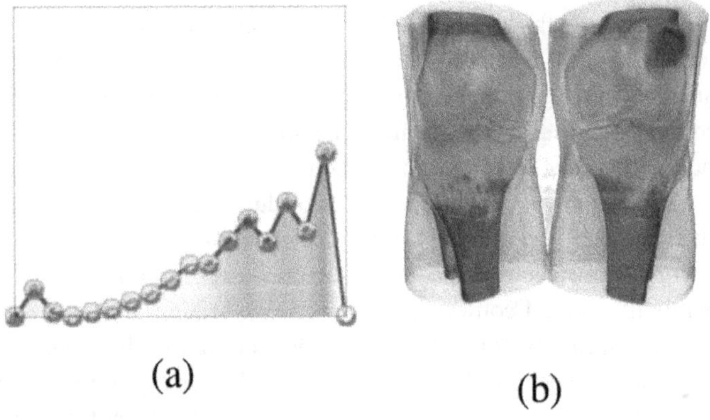

(a) (b)

Fig. 5: The transfer function from Fig. 4 after optimization: (a) Optimized transfer function) (b) Optimized output

B. Transfer Function Refinement with User-Selected Intensity Values

Fig. 9 shows three images of a CT-Knee data set while different colors are selected for optimization. Since the colors of the transfer function are generated in HSV color space by varying the hue component. The difference of intensity values are mapped to the difference of hue in HSV color space. By clicking on the color palette (Fig. 8), the transfer function is instantly optimized for the corresponding intensity values.

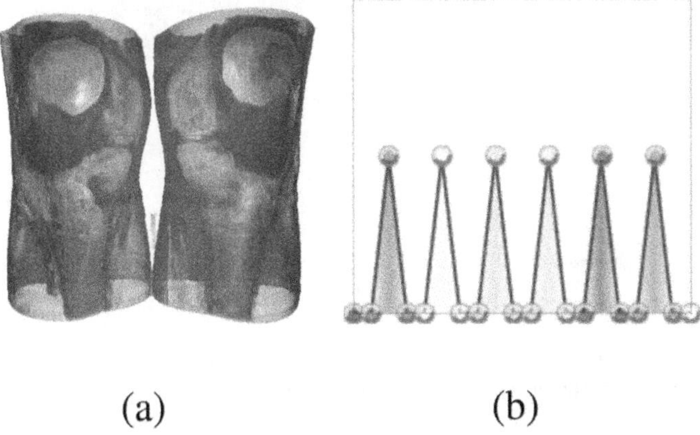

Fig. 6: Before optimization: CT-Knee rendered with a transfer function consisting of tent-like shapes (a) Preliminary view of data set (b) A transfer function with 6 tent-like shapes

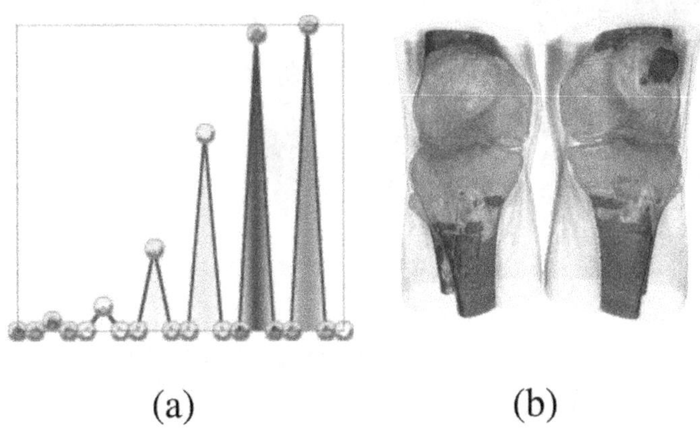

Fig. 7: The transfer function from Fig. 6 after optimization: (a) Optimized transfer function (b) Optimized output

C. Transfer Function Refinement with User-Selected Regions

Fig. 10a shows the VisMale data set with a generated transfer function of 4 tent-like shapes. Fig. 10c shows the intensity histogram of the data set. After the optimization (Fig. 10b), the outside of the head is less opaque so the inner structures are revealed to the user. However, the intermediate material (i.e. the skull) also

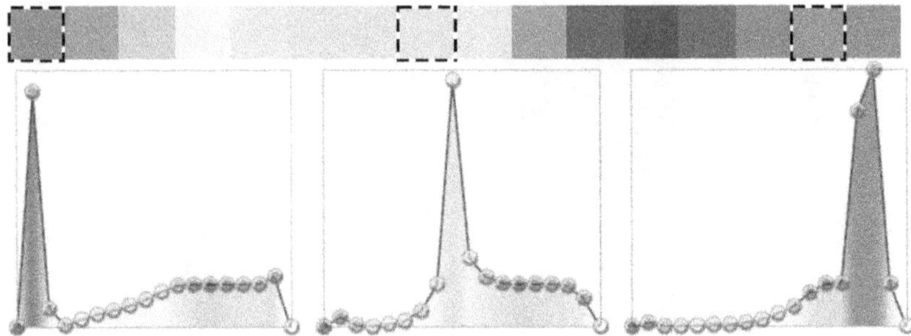

Fig. 8: The 3 chosen colors (corresponding to different intensity values) and the transfer functions after optimization for each intensity value. The same transfer function as shown in Fig. 4 is used as input to the optimizer. Note how the transfer functions are enhanced for the specific intensity ranges as compared to the result of global optimization in Fig. 5.

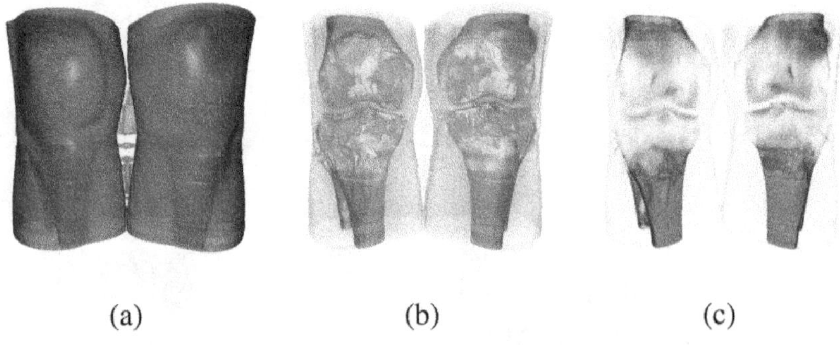

Fig. 9: The CT-Knee data with transfer functions optimized for the 3 colors in Fig. 8. (a) The materials with intensity values mapped to red are enhanced. Similarly, the materials in green and magenta are enhanced respectively in (b) and (c).

becomes less clear. If the goal is to make the skull more visible, the user could select a region consisting of parts of the skull to generate a weighting and perform further optimization of the transfer function. If the material of interest is occluded by surrounding materials, the user could use an axis-aligned clipping plane in order to accurately select voxels of the skull whilst minimizing the accidental tagging of the surrounding material (Fig. 10d). As shown in Fig. 10e, the skull becomes more clear after the region-based optimization.

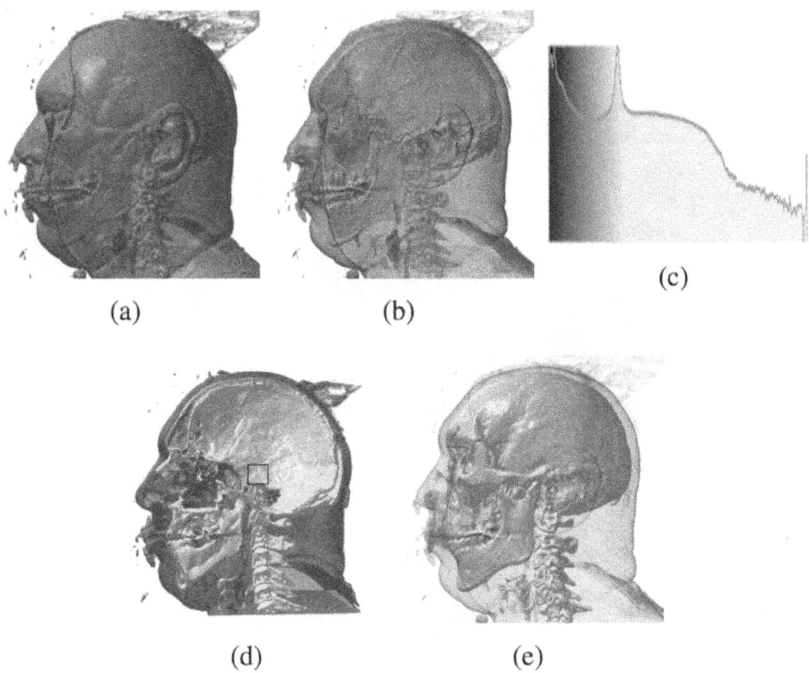

(a) (b) (c)

(d) (e)

Fig. 10: (a) Preliminary view of the VisMale data set. (b) Optimized output. (c) Histogram of the data set. (d) The user selects a region on the skull under a clipping plane. (e) Optimized output: skull is enhanced, and the outer layer is de-emphasized.

D. Adaptive Transfer Functions for Time-Varying Data Sets

We demonstrate our approach on a turbulent vortex data set [28], which consists of 100 time-steps. Our optimizer propagates transfer functions for the time-varying data set in an adaptive way. Specifically, the transfer function of the previous time-step is taken as input to generate the transfer function for the next time-step. Therefore, the transfer functions are locally optimized for each time-step of the time-varying data set, but as the input transfer function comes from the immediate preceding timestep, the resulting transfer functions also exhibit reasonable temporal coherency assuming the dataset is also coherent. As the difference of intensity histograms among consecutive time-steps is hard to notice, only the images of the first time-step and a time-step in the middle of the data set are displayed here. Fig. 12 shows three images of the vortex data set at time-step 0 while optimized for the three colors chosen in the color palette in Fig. 11.

Fig. 11: The 3 chosen colors (corresponding to different intensity values) for optimization. Note how different parts of the data set are enhanced respectively later in Fig. 12 and Fig. 13.

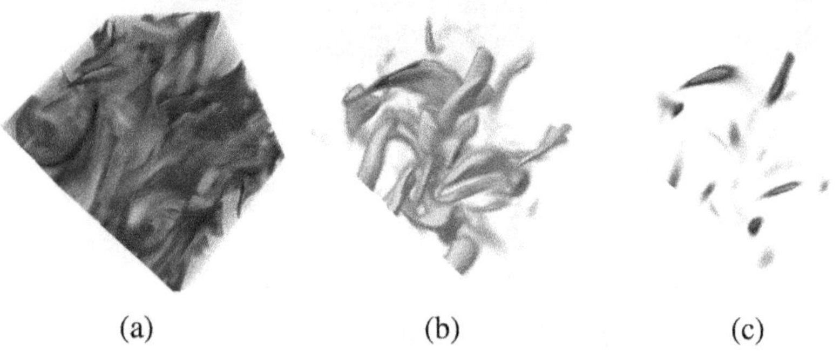

Fig. 12: The vortex data at time-step 0. (a) The materials with intensity values mapped to red are enhanced. Similarly, the materials in green and magenta are enhanced respectively in (b) and (c).

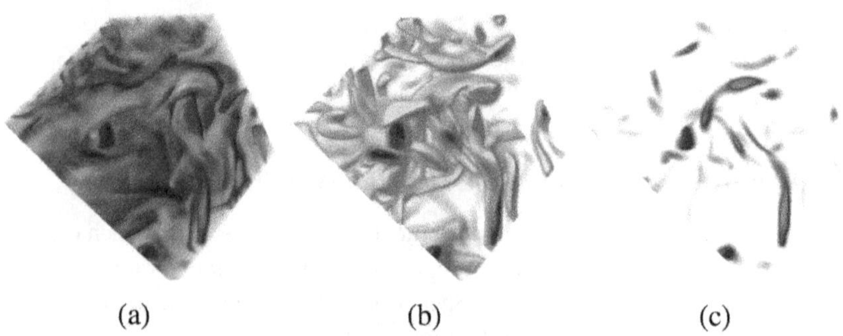

Fig. 13: The vortex data at time-step 50. These images show similar results as those for time-step 0, because there are only limited changes among the histograms of different time-steps.

Fig. 13 displays the vortex data set at time-step 50 with the transfer functions optimized for the three chosen colors. Fig. 14 shows the intensity histograms of the two time-steps discussed above and the corresponding optimized transfer functions for the colors selected in Fig. 11.

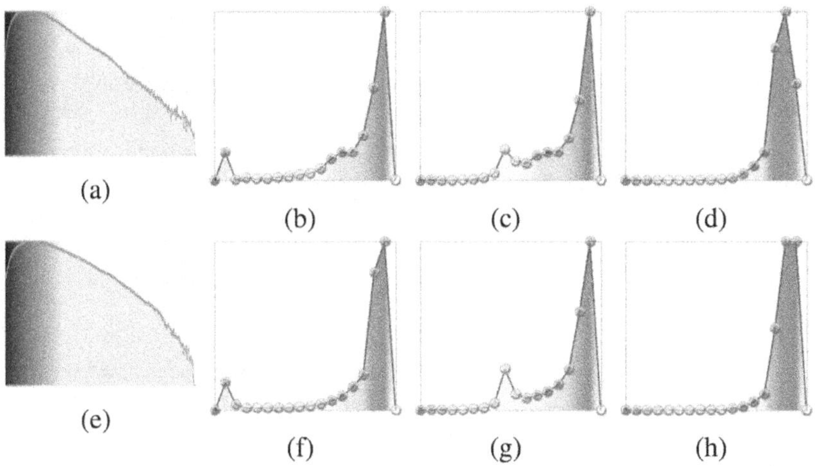

Fig. 14: (a) Histogram of time-step 0. (b) Transfer function (TF) for Fig. 12a. (c) TF for Fig. 12b. (d) TF for Fig. 12c. (e) Histogram of time-step 50. (f) TF for Fig. 13a. (g) TF for Fig. 13b. (h) TF for Fig. 13c.

VI. Conclusion

In this paper, the global optimization aims to alleviate excessive occlusion problems in volume rendering. However, instead of computing the view-dependent visibility of each voxel as is necessitated in other similar approaches [14] [15], we achieve this by balancing the opacity of voxels based on the distribution of intensity values. Our view-independent approach is relatively lightweight and should have better performance in contrast to other techniques. In addition, we propose two interactive methods that extend on the optimization technique in order to enhance specific intensity ranges within the data as identified by the user. This mechanism provides the ability for users to specify priority intensity ranges in an intuitive way, thus facilitating the exploration of both static and time-varying volume data sets.

The main limitation of the approach is that variations to the transfer function are limited to opacity values, which limits, to some degree, the resulting variations in output. The initial choice of intensity ranges, number of control points and color mapping across the histograms can affect the quality of the final output and some prior knowledge of the data sets may be of benefit for optimal results. On the other hand, the simple and straightforward techniques presented in this paper should be fully compatible with independent mechanisms for choosing optimal combinations of other visual parameters or indeed if the user wishes to combine these with more manual choices of parameters such as the color map. In addition, the transfer functions in our proposed system are always editable through the user interface. Users may benefit from the flexibility of further tweaking the intensity or opacity of the control points after the application of the automatic optimizations discussed in this paper.

This work is part-funded by Science Foundation Ireland Project 13-IA-1895 and Trinity College Dublin.

References

[1] H. Pfister, B. Lorensen, C. Bajaj, G. Kindlmann, W. Schroeder, L. S. Avila, K. M. Raghu, R. Machiraju and J. Lee, "The transfer function bake-off," *IEEE Computer Graphics and Applications*, vol. 21, pp. 16–22, 2001.

[2] S. Arens and G. Domik, "A survey of transfer functions suitable for volume rendering," in *Proceedings of the 8th IEEE/EG international conference on Volume Graphics*, Aire-la-Ville, Switzerland, Switzerland, 2010.

[3] G. Kindlmann and J. W. Durkin, "Semi-automatic generation of transfer functions for direct volume rendering," in *IEEE Symposium on Volume Visualization, 1998*, 1998.

[4] J. Kniss, G. Kindlmann and C. Hansen, "Multidimensional Transfer Functions for Interactive Volume Rendering," *IEEE Transactions on Visualization and Computer Graphics*, vol. 8, pp. 270–285, 7 2002.

[5] G. Kindlmann, "Transfer functions in direct volume rendering: Design, interface, interaction," *Course notes of ACM SIGGRAPH*, 2002.

[6] C. Rezk-Salama, P. Hastreiter, J. Scherer and G. Greiner, "Automatic Adjustment of Transfer Functions for 3D Volume Visualization," in *In Proc. Workshop Vision, Modeling, and Visualization (VMV*, 2000.

[7] S. Bruckner and M. E. Gröller, "Style Transfer Functions for Illustrative Volume Rendering," *Computer Graphics Forum*, vol. 26, pp. 715–724, 2007.

[8] Y. Wu and H. Qu, "Interactive Transfer Function Design Based on Editing Direct Volume Rendered Images," *IEEE Transactions on Visualization and Computer Graphics*, vol. 13, pp. 1027–1040, 2007.

[9] T. J. Jankun-Kelly and K.-L. Ma, "A study of transfer function generation for time-varying volume data," in *Proceedings of the 2001 Eurographics conference on Volume Graphics*, Aire-la-Ville, Switzerland, Switzerland, 2001.

[10] A. Tikhonova, C. D. Correa and K.-L. Ma, "An Exploratory Technique for Coherent Visualization of Time-varying Volume Data," *Computer Graphics Forum*, vol. 29, pp. 783–792, 2010.

[11] R. Maciejewski, I. Woo, W. Chen and D. S. Ebert, "Structuring Feature Space: A Non-Parametric Method for Volumetric Transfer Function Generation," *IEEE Transactions on Visualization and Computer Graphics*, vol. 15, pp. 1473–1480, 2009.

[12] M.-Y. Chan, Y. Wu, W.-H. Mak, W. Chen and H. Qu, "Perception-Based Transparency Optimization for Direct Volume Rendering," *IEEE Transactions on Visualization and Computer Graphics*, vol. 15, pp. 1283–1290, 12 2009.

[13] C. Correa and K.-L. Ma, "Visibility-driven transfer functions," in *Visualization Symposium, 2009. PacificVis '09. IEEE Pacific*, 2009.

[14] C. D. Correa and K.-L. Ma, "Visibility Histograms and Visibility-Driven Transfer Functions," *IEEE Transactions on Visualization and Computer Graphics*, vol. 17, pp. 192–204, 2011.

[15] M. Ruiz, A. Bardera, I. Boada, I. Viola, M. Feixas and M. Sbert, "Automatic Transfer Functions Based on Informational Divergence," *IEEE Transactions on Visualization and Computer Graphics*, vol. 17, pp. 1932–1941, 2011.

[16] R. Bramon, M. Ruiz, A. Bardera, I. Boada, M. Feixas and M. Sbert, "Information Theory-Based Automatic Multimodal Transfer Function Design," *IEEE Journal of Biomedical and Health Informatics*, vol. 17, pp. 870–880, 2013.

[17] S. Luo and J. Dingliana, "Information-Guided Transfer Function Refinement," in *Eurographics (Short Papers)*, Strasbourg, 2014.

[18] T. G. Feeman, The Mathematics of Medical Imaging: A Beginner's Guide, Springer, 2009.

[19] A. König and M. Gröller, "Mastering transfer function specification by using VolumePro technology," Technical Report. Institute of Computer Graphics and Algorithms, Vienna University of Technology, Vienna, Austria, 2000.

[20] U. D. Bordoloi and H.-W. Shen, "View selection for volume rendering," in *IEEE Visualization*, 2005.

[21] R. Bramon, M. Ruiz, A. Bardera, I. Boada, M. Feixas and M. Sbert, "An Information-Theoretic Observation Channel for Volume Visualization," *Computer Graphics Forum*, vol. 32, pp. 411–420, 2013.

[22] L. Xu, T.-Y. Lee and H.-W. Shen, "An Information-Theoretic Framework for Flow Visualization," *IEEE Transactions on Visualization and Computer Graphics*, vol. 16, pp. 1216–1224, 12 2010.

[23] M. Haidacher, S. Bruckner, A. Kanitsar and M. E. Gröller, "Information-based transfer functions for multimodal visualization," in *Proceedings of the First Eurographics conference on Visual Computing for Biomedicine*, Aire-la-Ville, Switzerland, Switzerland, 2008.

[24] S. Bruckner and T. Möller, "Isosurface Similarity Maps," *Computer Graphics Forum*, vol. 29, pp. 773–782, 2010.

[25] C. Y. Ip, A. Varshney and J. Jaja, "Hierarchical Exploration of Volumes Using Multilevel Segmentation of the Intensity-Gradient Histograms," *IEEE Transactions on Visualization and Computer Graphics*, vol. 18, pp. 2355–2363, 2012.

[26] C. Wang and H.-W. Shen, "Information Theory in Scientific Visualization," *Entropy*, vol. 13, pp. 254–273, 1 2011.

[27] S. Roettger, *The Volume Library*, 2006.

[28] K.-L. Ma and D. M. Camp, "High Performance Visualization of Time-Varying Volume Data over a Wide-Area Network," in *Supercomputing, ACM/IEEE 2000 Conference*, 2000.

[29] K.-L. Ma, *Time-Varying Data Repository*, 2003.

22. Simple Vertical Human Climbing Control with End Effector State Machines

Zümra Kavafoğlu
Dept. of Mathematics Hacettepe UniversityAnkara, TURKEY zdemir@hacettepe.edu.tr

Hacer İlhan
Dept. of Mathematics Hacettepe University Ankara, TURKEY hacerilhan@hacettepe.edu.tr

Ersan Kavafoğlu
Dept. of Comp. Graphics Hacettepe University Ankara, TURKEY ersan@avamstudios.com

Haşmet Gürçay
Dept. of Computer Eng. Hacettepe University Ankara, TURKEY gurcay@hacettepe.edu.tr

Tolga Çapın
Dept. of Computer Eng. TED University Ankara, TURKEY tolga.capin@tedu.edu.tr

Abstract— We present a framework for physics based vertical climbing control for human like articulated figures. In our work motion of end effectors of the figure are categorized under four states: Holding, Swinging, Reached and Gave-Up. Each end effector owns a finite state machine for managing the transitions between these states. Different target poses are determined for the limbs for swinging and holding states and they are tracked with PD Controller and gravity compensation. At Reached and Gave-Up states the aim of the end effector is to touch the closest point on the wall. Physics based movement of end effectors for touching the wall is simulated by using Jacobian Transpose Control. Our framework also includes a strategy manager which enables building different climbing strategies.

Keywords— physics based animation; climbing; character animation

I. Introduction

Climbing motion is generated by carrying body upward by using both lower and upper body limbs. In mechanical point of view, the aim of climbing is to increase the potential energy of the body with muscular work [1]. Approaches on synthesizing climbing motion can be categorized under two in terms of contact planning and motion planning: Contact-before-motion and motion-before-contact [2, 3, 4]. In the former one, contact points are determined first and then necessary motion of body parts are synthesized which will drive end effectors to these contact points. In the latter one, motion of the body parts is synthesized first and then contact positions are determined according to the final position of end effectors. Our system follows motion-before-contact strategy. Desired pose for each body part is pre-determined during each phase of climbing motion. Body parts are driven to desired poses by using physics based controllers. When desired pose is reached or climber decides to give up, contact point on the wall is determined and end effector is driven to that point again with physics based controllers.

We introduce a physics based controller approach which enables creating a wide variety of climbing motions. Each end effector owns a cyclic behaviour and the coordination, timing and ordering between end effector behaviours are organized resulting in different motions. We don't use any motion capture data, but our give-up strategy enables creating realistic, non-monotonic climbing motions.

Climbing is achieved by coordinated continuous motion of hands and feet and their associated limbs. The behaviour of each end effector can be summarised as follows: End effector and its associated limbs try to reach a target pose by swinging. If they don't reach the target pose in a reasonable amount of time, the climber decides to give up. If the climber gives up, the aim is to reach the nearest point on the wall. When the end effector touches the wall as a consequence of reaching desired pose or giving up, the holding phase begins. While end effector is holding the wall, its associated limbs can either stay fixed or can be moved. After some time, end effector should begin swinging again to make climber keep on climbing. This cyclic behaviour of each end effector transitioning between specific states is very convenient to be represented by a finite state machine. We used finite state machine since it's fairly enough to represent the behaviour of an end-effector and simple to implement.

Coordinated movement of end effectors determine the overall appearance and strategy of climbing motion. While end effectors behave in a cyclic nature on their own, timing and ordering of each effector according to others should also be determined. For this purpose, we included a simple strategy manager that is similar to Subsumption Architecture [5] which is mostly used in behavior based robotics. This manager enables the programmer to easily arrange ordering and timing of end effector behaviors according to each other.

There are also several physics based character animation approaches on synthesizing climbing motion [6, 7, 8, 9]. In [8] complex motions including contact behaviors are synthesized with contact invariant optimization. In [6], an interactive control strategy is proposed which can be used for several motions including climbing motions. They use state machines for controlling actions but the climbing motion generated by their state machine is much less detailed. In [7], a physics based controller is proposed for several motions in a dynamically varied environment. They demonstrate their optimization-based method with a climbing motion of type contact-before-motion. They also use state machine for action planning but state machine is built for defining overall motion not for planning action of each end-effector. Our method enables generating a wide variety of physics based 3D climbing motions easily. Our simulation runs in real time and can be implemented by using common physics engines. It doesn't need to access equations of motion or doesn't need any inverse dynamics process, so it can be easily integrated into video games.

Several approaches have been described on the analysis of climbing activities. A technique of tracking the entire body movement of non-professional climbers and their center of mass is presented using an optoelectronic system to describe the effects of the body center of mass in rock climbing [10]. The use of body attached sensors to show the difference between the movements of several climbers with different qualifications is an example of movement analysis in rock climbers [11]. To make a design of safer and more suitable rock climbing equipment for children, data are collected using a rock-climbing wall equipped with embedded sensors and a model for the prediction of climbing behavior is created [12]. In another work, computer vision technology is utilized to give augmented real-time and delayed feedback for analyzing body movement for climbers [13].

The climbing problem has also been examined in the domain of motion planning and reconstruction. A tool for designing, testing and visualizing climbing route on a virtual climbing wall using a simulated climber are provided to create possible interesting and challenging routes [14]. A kinematic, 2D and grasp-based motion planning algorithm for virtual characters is proposed to solve a variety of problems such as determining and synthesizing the required transitions between varied modes of locomotion like walking, crawling, climbing, and swinging using pose heuristics when a constrained environment with specified possible contact points is given [15]. To reconstruct static poses of a climber by optimization, contact forces are captured by using an instrumented bouldering wall and several data are recorded by motion capture technology simultaneously [16]. A real-time method to create automated contact configurations between a virtual creature and the environment for various motions tasks such as getting

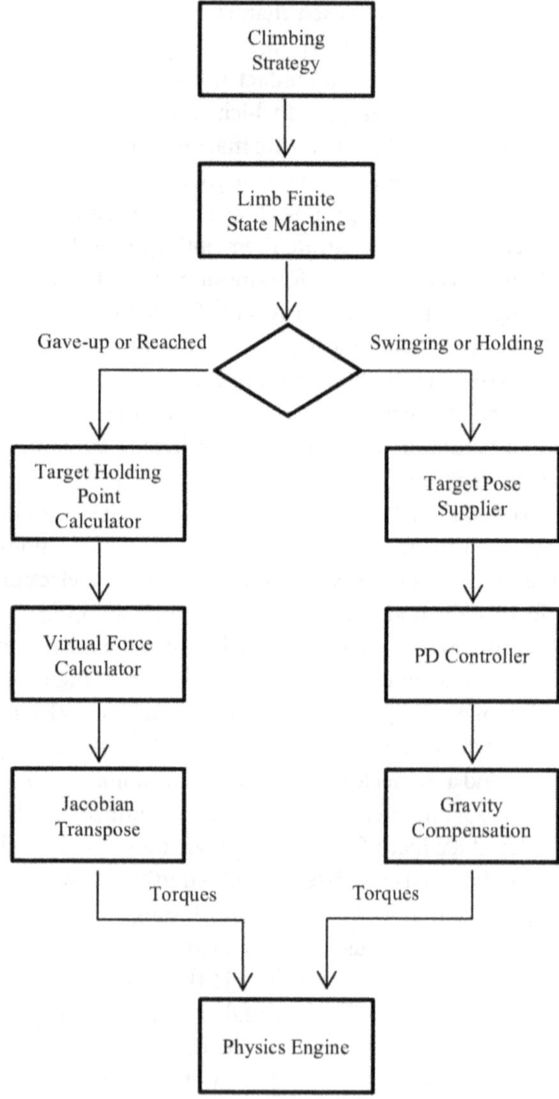

Fig. 1. System overview for each end effector.

up, pushing or pulling objects, climbing and insect locomotion is addressed as a means of combining a random sampling approach with a heuristic related to the force transmission ratio [17]. Our framework doesn't consist of planning contact positions of end effectors except in the state of reaching and giving up. Instead, target poses are defined for limbs and physics based tracking is used to drive the limbs to target poses. When target pose is reached or climber decides to give up, nearest point on the wall is determined as contact point and end effector is driven to that point by using Jacobian Transpose Control.

Several approaches have also been proprosed in robotics area covering contact planning, motion planning and motion controlling together [2, 3, 4, 9, 18, 19, 20, 21, 22, 23, 24, 25,]. A wall climbing robot equipped with suction pads on each foot is developed and an optimal gait called "wall gait" is introduced to provide high motion stability and maximum locomotion speed. This is achieved by analyzing different kinds of gait styles and specifying an ideal order and phases of supporting and swing legs and also determining leg positions [18]. Robots with four, six and eight legs are described to climb in complicated formed pipes by pushing against the walls using a hierarchical control architecture which is similar to subsumption architecture [9]. A ladder climbing robot motion is planned with a fixed strategy and limbs are controlled by non-linear feedback controller and joint space controller with motion equations [21]. A genetic algorithm is proposed for building a climbing strategy of a three limbed robot that is compliant to physical constraints [20]. For the same three limbed robot a simplified Cartesian computed torque control algorithm is built which includes Jacobian transpose control and PD Controller [19]. Our work deals with simpler climbing actions. Only two desired poses are determined for each end effector: One for holding state and the other one for swinging state. We propose a flexible climbing strategy that can be built easily. For limb control we use PD Controller and Jacobian Transpose Control.

Since climbing is done by coordinated movement of all of four limbs, it can be thought that it resembles quadruped locomotion. A large body of work exists for quadruped animation of ground locomotion [26, 27, 28, 29, 30, 31, 32]. A physics based character control of quadruped locomotion is developed to obtain realistic set of gaits such as walk, trot, pace, canter, transverse gallop with desired styles for dog characters using a dual leg frame model connected by a flexible spine. It also uses gait graphs from tracking of video data, a Jacobian transpose and proportional derivative (PD) controller system with virtual forces [26]. An optimization technique is presented to generate convincing gaits and morphologies of a variety of virtual creatures with two, four and five legged creatures [27]. Also, an extensive survey on quadruped animation techniques is given by comparing varied methods including video based systems, physics based models, inverse kinematics, or some combination of these [28]. However none of these work include controlling quadruped climbing motion.

II. Overview

In our framework, a state machine runs for each end effector, which manages the motion of the end effector during climbing. Also a strategy manager is included for managing the ordering and timing of each end effector state according to other end effectors' states. End effector states and strategy manager are described in Section 4. In each state there are one or more tasks for the associated limb chain to achieve, like driving the end effector to a desired position, or keeping limbs fixed. To achieve these tasks limbs are controlled by using Proportional Derivative Controller and Jacobian Transpose Control. Details of limb control are explained in Section 5. System overview of each end effector is illustrated in Fig.1.

A. Human Body Model

Our articulated human body model consists of 15 body parts and 14 joints with 26 degrees of freedom in total. Right hand, right foot, left hand and left foot are the end-effectors of the body. Fig. 2 illustrates our human body model with joint degrees of freedom and end effectors. Each end effector has an associated limb chain. Fig. 3 illustrates limb-chains.

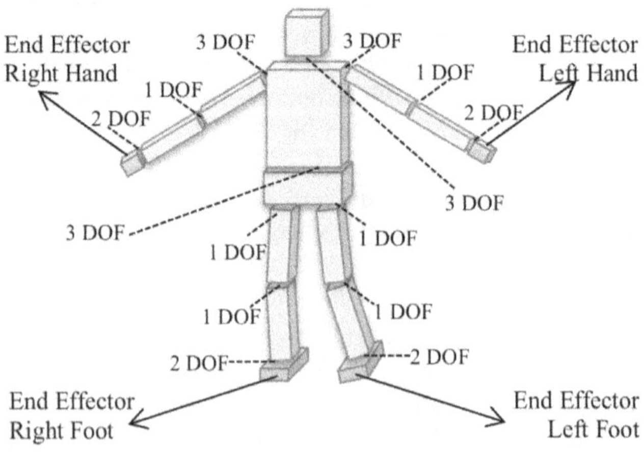

Character degrees of freedom and end effectors.

B. End Effector State Machine

Movement of an end effector during a vertical wall climbing motion can be categorized as follows: Holding the wall, swinging limb to reach the wall, reaching the wall within a desired duration or giving up to swing by holding the nearest point on the wall. During climbing, each end effector transitions between these movements and each of them have different tasks, starting and ending conditions. In our system, each end effector has its own finite state machine for managing the transitions between these movements. The states are: Holding State, Swinging State, Reached State and Gave-up State. States and transitions of end effector state machine are illustrated in Fig. 4.

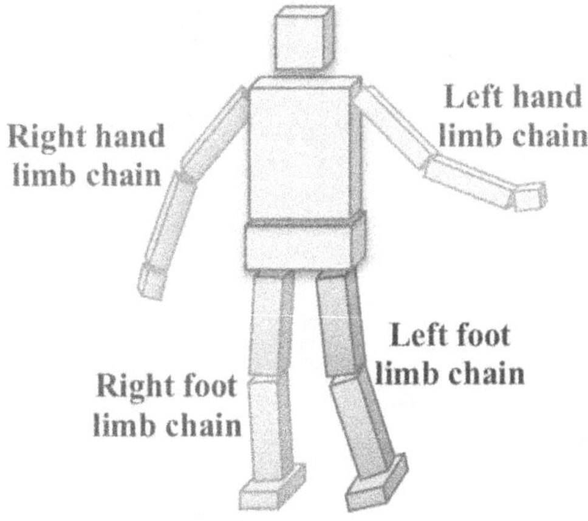

Each end effector has an associated limb chain used for desired pose tracking.

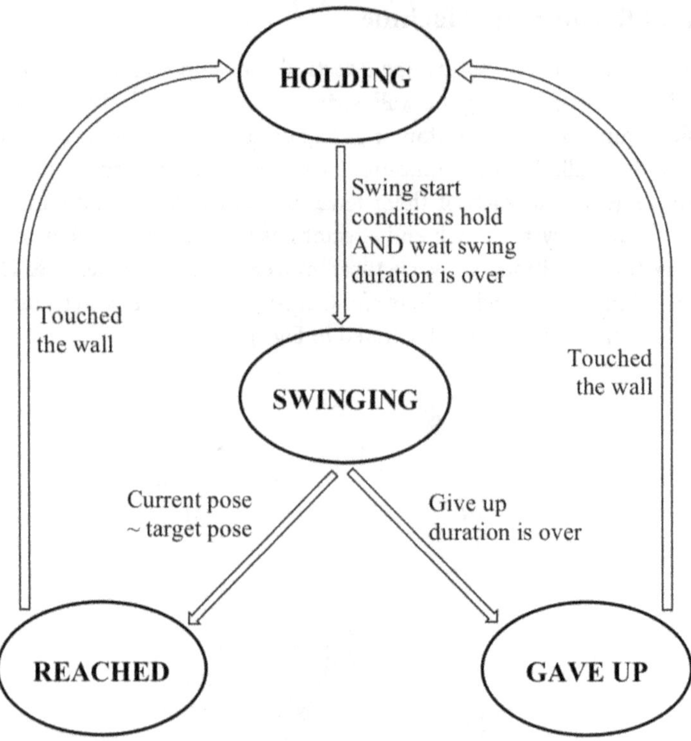

End effector state machine.

Holding State

Holding state of the state machine stands for the holding movement of an end-effector. A climber can try to keep his/her associated limb-chain in a fixed pose while holding the wall or can prefer to move the limb-chain towards a desired pose. Both of these options are modeled in our system within the holding state as illustrated in Fig. 5. Conditions for starting the transition from fixed pose to movement can be defined by the user within the climb strategy. When the pose of moving limbs get similar to the desired pose or a pre-determined time passes, limb poses get fixed again. Fig. 6 illustrates the desired poses for holding and swinging states.

Swinging State

Swinging state manages the swinging movement of an end effector. While swinging, the limb-chain associated with the end effector tries to reach a given desired pose within a given duration. For this purpose, swinging state consists of two phases: one for physics based tracking of the desired pose with PD(Proportional Derivative) Controller and the other one for controlling whether the desired pose is reached. The reaching control is done with a simple orientation metric that measures the difference between two poses. Until desired pose is reached or give up time is over, swinging limb continues to track the desired pose.

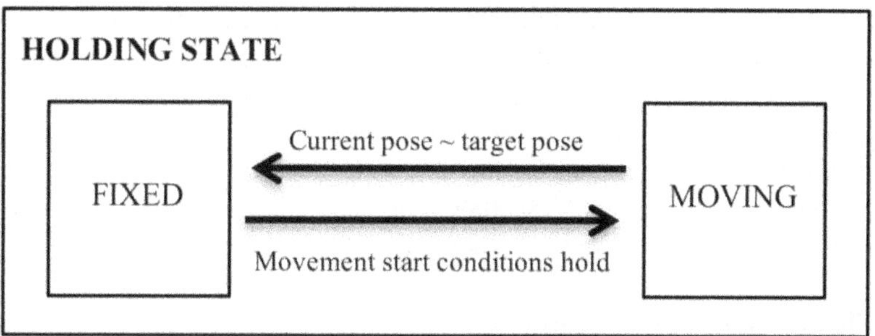

Holding state consists of two states: Fixed and moving. In fixed state, limb chain associated with the end effector remains in a fixed pose, while in moving state limb chain tries to reach a given target pose.

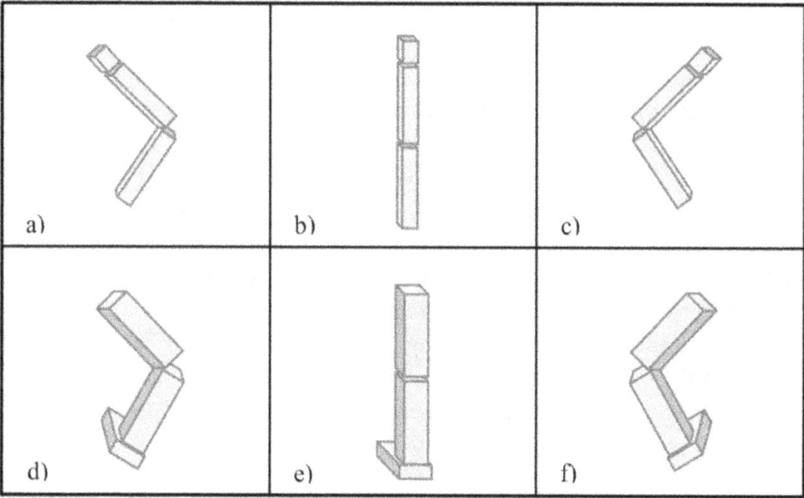

Desired poses for end effector limb chains. a) Right hand limb chain desired pose in holding state. b) Right and left hand limb chain desired pose in swining state (just straight arm). c) Left hand limb chain desired pose in holding state. d) Right foot limb chain desired pose in swinging state. e) Right and left foot limb chain desired pose in holding state (just straight leg). f) Left foot limb chain desired pose in swinging state.

Reached State and Gave-up State

Swinging motion of an end-effector can conclude in one positive and one negative situation. Positive one is to reach the desired position in a reasonable time. But negatively, climber can spend too much time on trying to reach and eventually decides to give up. For handling this situation, a give-up duration is determined by the user for each end effector. In both situations climber aims to touch the wall at its nearest point. In Section 5, the controller designed for the motion of trying to touch the wall is expressed in detail.

Building a climbing strategy

There doesn't exist only one way for vertically climbing a wall. Climbers usually decide to move some limbs before others, or some limbs in a coordinated manner, according to the grips or obstacles on the wall. Since the main aim of vertical climbing is to carry the body upwards, it's important for a climber to move right limbs in the right time for carrying the center of mass upwards. Our system doesn't aim to make a climbing plan according to the grips or obstacles, but includes a strategy manager which enables the user to build a desired climbing strategy. Strategy manager enables to determine the relative starting time of an end effector's state according to other end effectors' states as well as the ordering of different end effectors' states. For this purpose, required conditions are determined for an end effector to start the swinging state. As an example, the required condition for right hand to start swinging can be determined as follows: 5 seconds passed after left hand and right foot started holding state. Moreover, the timing of passing from fixed limb state to moving limbs within holding state can be determined as a part of the strategy. Determining the correct strategy plays an important role for carrying center of mass of the body upwards. In Fig. 7, an example strategy is shown, which enables coordinated movement of crosswise end effectors.

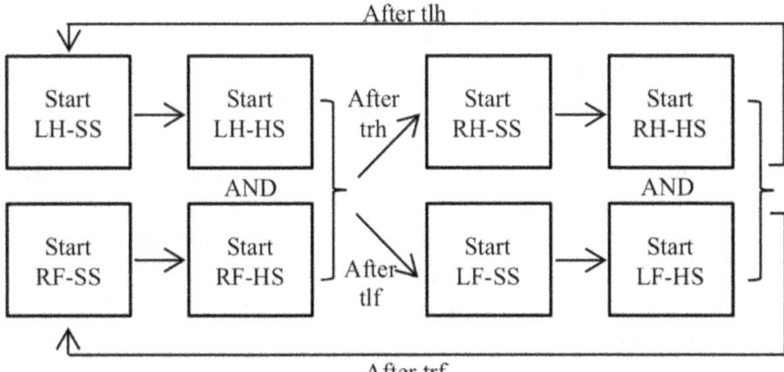

SS: Swinging state
HS: Holding state
LH: Left hand
RH: Right hand
LF: Left foot
RF: Right foot
trh: Right hand swing wait duration
tlh: Left hand swing wait duration
trf: Right foot swing wait duration
tlf: Left foot swing wait duration

A climbing strategy example: trh time after left hand's and right foot's holding states start, right hand starts to swing. And tlf time after left hand's and right foot's holding states start, left foot starts to swing. Similar holds for left hand's and right foot's swing states start. If trh is equal to tlf and tlh is equal to trf then a synchronized strategy is obtained in which left hand and right foot are swinging together while right hand and left are holding and vice versa.

C. Limb Control

Our climbing controller system is based on tracking desired limb poses which are determined according to the state of the associated end effector. If an end effector is in swinging state, then associated limbs try to reach the desired pose by using PD Controller and gravity compensation. If an end effector is in reached or gave-up state, then torques for associated limbs are calculated by Jacobian Transpose Control which will make the end effector touch the closest point on the wall. In this section, fundamental concepts like Proportional Derivative Controller, Jacobian Transpose Control and gravity compensation are mentioned. Also the controller designed for making the end effector try to touch the wall is explained in detail.

Proportional Derivative Controller

Proportional derivative controller is one of the most widely used controllers, which has a simple formulation [33]. PD Controller calculates the torque to move a joint from its current state (orientation and angular velocity) to the desired one in the next time step. This calculation is performed as the weighted sum of differences between current and desired states. In an explicit manner, let q_c and $\dot{q}_c$ be the current orientation and current angular velocity of a body segment respectively and q_d and $\dot{q}_d$ be the desired orientation and desired angular velocity in the next frame. Then the appropriate torque τ to take the segment from current state to the desired one is calculated with PD-Controller as follows

$$\tau = k_p (q_d - q_c) + k_d (\dot{q}_d - \dot{q}_c) \tag{1}$$

where k_p and k_d are proportional and derivative gains respectively. It's clear that, calculated torque will change as proportional and derivative gains change, and appropriate gains should be found for each character and motion. We find the appropriate gains by manual tuning which will result in non-stiff and also not loosely character motion.

Jacobian Transpose Control

To simulate a desired motion, applying a force directly to a body part of the character results in an unnatural motion, since it contradicts with the joint based structure of character. To simulate the effect of this linear force, appropriate joint torques should be computed. For calculating these torques, we've used Jacobian Transpose Control method [34, 35]. We first determine a chain of limbs on whose joints the emulating torques are applied. Fig. 8 shows the Jacobian Transpose chains for emulating a force applied at end effectors.

Let p be the global coordinates of the point that the linear force will be applied and let F denote the linear force. Here F is not actually applied at the point but only its effect is emulated. Let q_i denote the position of the ith joint in the limb chain. Then the torque τ_i that must be exerted on this joint is calculated as below

$$\tau_i = F \times (p - q_i) \tag{2}$$

Torque for each joint in the chain is calculated in the same way and the effect of linear force at point p is emulated by applying these torques to the joints.

Jacobian Transpose Control is used in gravity compensation and "try to touch wall" control parts of our system.

Gravity Compensation

Gravity force creates an undesired effect on body parts which impedes them from reaching a desired pose with PD Controller. This is caused by the simple formulation of PD Controller which discards gravity. To overcome this effect gravity compensation torques are calculated, which compensate the effect of gravity [36]. Gravity effects a body part by leading to the force mg applied on the center of mass of the body part. Here m is the mass of the body part and g is the gravity force. To compensate this effect, the effect of a counter force should be emulated by internal joint torques. By using Jacobian Transpose control, gravity compensation torques are calculated and applied to the joints in the limb chain.

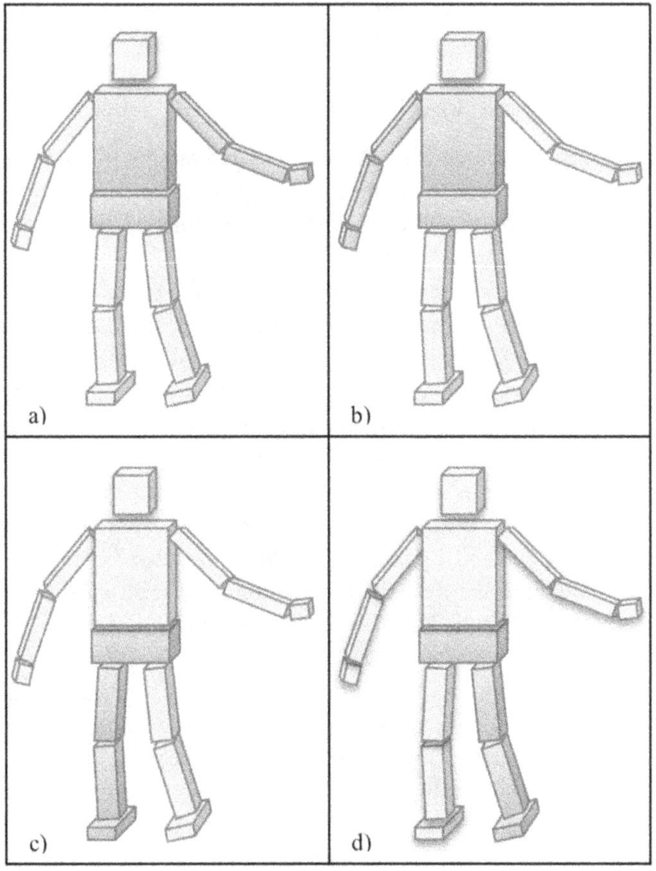

Jacobian transpose chain for a) left hand b) right hand c) right foot d) left foot.

"Try to touch wall" control

In the states of reached or gave-up, the aim of an end effector is to touch the wall at its closest point. For this purpose, first the projection of the current hand position (p_c) on the wall is taken. Let p_w denote this projected point. A virtual force is calculated which will drive the end effector to p_w with a simple PD Controller formula as below.

$$F_V = k_p (p_w - p_c) - k_d \dot{p}_c \tag{3}$$

Here F_V denotes the virtual force and $\dot{p}_c$ denotes the end effector velocity. We cannot apply F_V directly to the end effector for sake of reality, so we find appropriate torques of joints of Jacobian transpose chain by using Jacobian transpose control. Applying these torques drives the end effector directly to the wall in a realistic manner.

III. Implementation and Results

We have presented a framework that enables interactive physics based synthesis of vertical wall climbing motions. We tested our framework with different climbing strategies. Our experiments show that coordinated movement of limbs is of great importance for carrying body upwards.

We have made our experiments on an Intel Core i7 2.4 GHz CPU. Our system simulates vertical climbing motion in real time. Our simulation time step is 0.0001. Bullet physics engine is used for dynamic simulations.

Proportional and derivative gains are set manually for body joints with a reference-scaling method. We determined the shoulder joint as reference and we set proportional and derivative gains of it manually. These are called reference gains. To find gains of another joint, we scale the reference gains with the ratio of the joint's mass to reference joint's mass. Our reference joint proportional and derivative gains are 200 and 20 respectively. Proportional and derivative gains of virtual force PD Controller described in Section 5-d are also set manually. We don't determine a fixed value for these gains. They can be tuned according to the desired speed of the end effector towards wall.

Modeling hand or foot interactions with wall or grips on the wall is beyond the scope of this work. When an end-effector makes transition to holding state, we freeze it at its current position to emulate a scene like it's holding a grip on the wall. Associated limbs of holding end effector keep their current orientations by using PD controller and gravity compensation.

We've tested our controller with several climbing strategies. Although we use the same desired target poses (Fig. 6) the overall motion can have completely different characteristics. This situation arises by means of our timer and condition based strategy. Different climbing strategies and their results are illustrated in Fig. 9 -13.

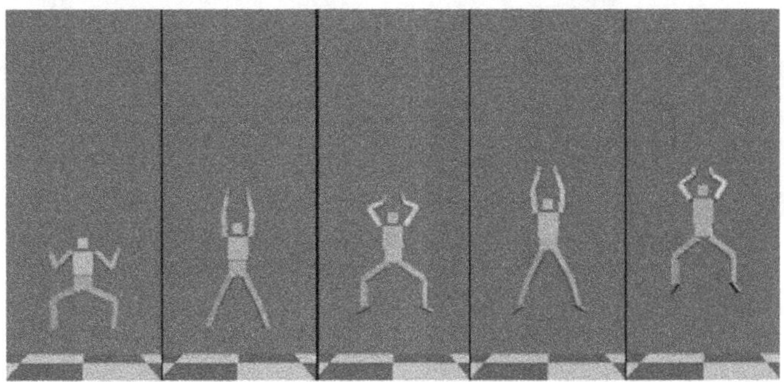

Frog like synchronized climbing strategy. For this motion arms swing cooperatively while legs are moving in hold state. And legs swing cooperatively while arms are moving in hold state.

Crosswise synchronized climbing strategy. Cross arm-leg pairs move cooperatively. This is the result of the strategy explained in Fig. 7.

Three to one climbing strategy. Right arm moves seperately while other three move cooperatively.

Asynchronized frog like climbing strategy. Asynchronized version of frog like strategy in Fig. 9. Asynchronization is achieved by altering swing wait durations. Swing wait duration of left arm and left leg is 0.5 and swing wait duration of right arm and right leg is 0.05.

Leg waiting frog like strategy. This is also a synchronized frog like strategy but legs start moving lately in the holding state.

IV. Future Work

Climbing motion is created by coordinated movement of nearly all body parts and it can be realized in very different ways. With our framework different climbing strategies can be defined by the user. But we constrained body parts to some predefined key poses during climbing. Thus our results are rich in movement strategy but very simple in point of the motion of body parts. Following Contact Before Motion planning instead of Motion Before Contact planning can be a future direction to overcome this situation. Another possible future direction can be to utilize motion capture in our current system for achieving a more human like style.

Acknowledgements

This work is supported by the Scientific and Technological Research Council of Turkey (TUBITAK, project number 112E105).

References

1. A.E. Chapman, "Biomechanical Analysis of Fundamental Human Movements", Human Kinetics, 2008.
2. T. Bretl, J. Latombe, and S. Rock, "Toward autonomous free-climbing robots," in Int. Symp. Rob. Res., Siena, Italy, 2003.
3. T. Bretl, S. Lall, J. Latombe, and S. Rock, "Multi-step motion planning for free-climbing robots," in WAFR, 2004.
4. K. Hauser, T. Bretl, and J.C. Latombe, "Non-gaited humanoid locomotion planning. In Humanoids", Tsukuba, Japan, 2005.
5. R. Brooks, "A robust layered control system for a mobile robot", IEEE Journal of Robotics and Automation, RA-2(1):14–23, 1986.
6. J. Laszlo, M. van de Panne, and E. Fiume. "Interactive control for physically-based animation." Proceedings of SIGGRAPH 2000, pages 201–208, July 2000. ISBN 1-58113-208-5.
7. S. Jain, Y. Ye, and C. K. Liu, "Optimization-based interactive motion synthesis," *ACM Trans. Graph.*, vol. 28, no. 1, pp. 1–12, Jan. 2009.
8. I. Mordatch, E. Todorov, and Z. Popović, "Discovery of complex behaviors through contact-invariant optimization," *ACM Trans. Graph.*, vol. 31, no. 4, pp. 1–8, Jul. 2012.
9. W. Neubauer, "A Spider-like Robot That Climbs Vertically in Ducts or Pipes," *Proceedings of IEEE/RSJ International Conference on Intelligent Robots and Systems (IROS'94)* 2, 1178–1185, 1994.
10. F. Sibella, I. Frosio, F. Schena, and N.A. Borghese, "3D analysis of the body center of mass in rock climbing," Human Movement Science, 26, 6 (Dec. 2007), 841–852.

11. T. Schmid, R. Shea, J. Friedman, M.B. Srivastava, "Movement Analysis in Rock-Climbers," *2007 6th International Symposium on Information Processing in Sensor Networks* (April 2007): 567–568.
12. H. Ouchi, Y. Nishida, I. Kim, Yoichi Motomura, Hiroshi Mizoguchi, "Detecting and Modeling Play Behavior Using Sensor-Embedded Rock-Climbing Equipment," *Proceedings of the 9th International Conference on Interaction Design and Children - IDC '10* (2010): 118.
13. R. Kajastila, P. Hämäläinen "Augmented Climbing : Interacting With Projected Graphics on a Climbing Wall," *CHI '14 Extended Abstracts on Human Factors in Computing System* (2014): 1279–1284.
14. J. Pfeil, J. Mitani, and T. Igarashi, "Interactive Climbing Route Design Using a Simulated Virtual Climber," *SIGGRAPH Asia 2011 Sketches on - SA '11* (2011): 1.
15. M. Kalisiak and M. van de Panne, "A Grasp-Based Motion Planning Algorithm for Character Animation," *The Journal of Visualization and Computer Animation* 12, no. 3 (July 2001): 117–129, doi:10.1002/vis.250.
16. R. Aladdin and P.G. Kry, "Static Pose Reconstruction with an Instrumented Bouldering Wall," *Proceedings of the 18th ACM Symposium on Virtual Reality Software and Technology - VRST '12* (2012): 177.
17. S. Tonneau, J Pettré, and F Multon, "Task Efficient Contact Configurations for Arbitrary Virtual Creatures," *Proceedings of the 2014 Graphics ...* (2014): 9–16.
18. A. Nagakubo and Shigeo Hirose, "Walking and Running of the Quadruped Wall-Climbing Robot," *Robotics and Automation, 1994. ...* (1994): 1005–1012.
19. D. Bevly, Steven Dubowsky, and C. Mavroidis, "A Simplified Cartesian-Computed Torque Controller for Highly Geared Systems and Its Application to an Experimental" 122, no. March (2000): 27–32.
20. D. M. Bevly, S. Farritor and S. Dubowsky, "Action Module Planning and Its Application to an Experimental Climbing Robot" no. April (2000).
21. H. Amano, K. Osuka, and T.J. Tarn, "Development of Vertically Moving Robot with Gripping Handrails for Fire Fighting," *Proceedings 2001 IEEE/RSJ International Conference on Intelligent Robots and Systems. Expanding the Societal Role of Robotics in the the Next Millennium (Cat. No.01CH37180)* 2 (2001): 661–667.
22. T. Bretl, S. Rock, and J.C. Latombe, "Motion Planning for a Three-Limbed Climbing Robot in Vertical Natural Terrain," *Robotics and Automation, 2003. ...* (2003): 2946–2953.

23. T. Bretl, S. Rock, J.C. Latombe, B. Kennedy and H. Aghazarian., "Free-Climbing with a Multi-Use Robot", In Int. Symp. Exp, (2004): 1–10.
24. T. Bretl, "Motion Planning of Multi-Limbed Robots Subject to Equilibrium Constraints : The Free-Climbing Robot Problem" (2006): 1–35.
25. K. Hauser, T. Bretl, K, Harada and J.C. Latombe, "Using Motion Primitives in Probabilistic Sample-Based Planning for Humanoid Robots" (2008): 1–16.
26. S. Coros, A. Karpathy, B. Jones, L. Reveret, M. van de Panne "Locomotion Skills for Simulated Quadrupeds," *ACM SIGGRAPH 2011 Papers on - SIGGRAPH '11* 1, no. 212, 2011.
27. K. Wampler and Z. Popović, "Optimal Gait and Form for Animal Locomotion," *ACM Transactions on Graphics* 28, no. 3, 2009.
28. L. Skrba, L. Reveret, F. Hétroy, M.P. Cani, and C. O'Sullivan, "Quadruped animation", In Eurographics State-of-the-Art Report, 2008.
29. D.W. Marhefka, D.E. Orin, J.P. Schmiedeler, K.J. Waldron "Intelligent Control of Quadruped Gallops," *IEEE/ASME Transactions on Mechatronics* 8, no. 4, 2003.
30. N. Torkos, M. van de Panne: Footprint–based quadruped motion synthesis. In Graphics Interface, pp. 151–160, 1998.
31. Z. Bhatti, A. Shah and F. Shahidi, "Procedural Model of Horse Simulation", VRCAI '13. 139–146, 2013.
32. S. Levine and J. Popovi, "Physically Plausible Simulation for Character Animation", SCA '12, 221–230, 2012.
33. J.K. Hodgins, W.L. Wooten, D.C. Brogan, and J.F. O'Brien. "Animating human athletics", Proceedings of the 22nd annual conference on Computer graphics and interactive techniques, Pp. 71–78, 1995.
34. C. Sunada, D. Argaez, S. Dubowsky, and C. Mavroidis, "A coordinated jacobian transpose control for mobile multi-limbed robotic systems", In Proc. IEEE Int'l Conf. on Robotics and Automation, 1994, 1910–1915.
35. J. Pratt, C. Chew, A. Torres, P. Dilworth, and G. Pratt, "Virtual model control: An intuitive approach for bipedal locomotion", 2001, Int'l J. Robotics Research 20, 2, 129.
36. S. Coros, P. Beaudoin, M. van de Panne, "Generalized biped walking control", ACM Transactions on Graphics 29 (2010).

23. Interacting with Boids in an Incompressible Fluid Environment

Aytek Aman
Department of Computer Engineering Bilkent University aytek.aman@cs.bilkent.edu.tr

Ateş Akaydın
Department of Computer Engineering Bilkent University akaydin@cs.bilkent.edu.tr

Uğur Güdükbay
Department of Computer Engineering Bilkent University gudukbay@cs.bilkent.edu.tr

Abstract – With the increasing power of computing hardware, computer simulations are being realized in real-time. It is now possible to simulate real-life phenomena on standard personal computers. In this study, we present a school of fish and fluid interaction environment where the school of fish in the environment depict flocking behaviors affected by the flow of an incompressible fluid that can be manipulated interactively by a depth sensor controller. The proposed approach can be used in virtual reality applications and video games to provide immersive underwater scenarios for schools of fish. We developed our application on a commercially available game engine that provides real-time experience for the users.

Keywords: interaction; simulation; crowd simulation

I. Introduction

Crowds are collections of a number of individuals who come together in an unorganized way. With the increasing application of virtual, augmented, and mixed reality technologies in military simulations, safety in architectural design, such as emergency evacuation simulations, entertainment industry, including game development and film production, the simulation of crowd behavior gained significant attention in the research community. The realistic simulation of crowds, including human crowds, schools of fish, flocks of birds, and herds of sheep, is especially important for the entertainment industry, including games and film production. For example, in the popular game Assassin's Creed Unity, the crowd is treated as the main character in the game 0. The crowds can be modeled and simulated at the micro-scale where the individuals in the crowd are modeled (agent-based) and at the macro-scale where the crowd behavior as a whole (flow-based).

The simulation of animal crowds is important to increase the realism of three-dimensional scenes, including the behavior of schools of fish in underwater simulations, flocks of birds, such as starlings, murmurating while moving from one location to another, and herds of cows moving wildly in the jungle. The realistic simulation of herd behavior is not a trivial task because both psychological and physical factors affect the nature of the herd behavior. One way to simulate the behaviors of the groups of individuals in an organized way is to define some simple behaviors and let the individuals interact with each other with these behaviors. This approach was first proposed by Reynolds 0 and it is called the *Boids Model*.

In this study, we focus on the behavior of a school of fish while fish are interacting with each other and their environment. We incorporate velocity fields into the Boids model to simulate the interaction of boids with the environment. This allows us to experiment with the behavior of a school of fish under the influence of external physical forces like wind and water current. The approach we take allows the real-time simulation of thousands of boids, namely fish. We also provide an interactive method to manipulate the fluid's velocity field using a Microsoft Kinect device 0. To manipulate the velocity field of the fluid, we use hand gestures that are recognized by the Kinect device. Hand gestures allow the user to change the direction of the school of fish while moving in water.

II. Related Work

Craig W. Reynolds developed a computer model describing the coordinated motion of animals such as flocks of birds, or schools of fish. He named this model the *Boids Model* 0. Every individual animal in this model is called a boid. In the *Boids Model*, there are three main force components that affect the motion of a boid. These are the *separation*, *alignment*, and *cohesion* forces. Separation describes the force that keeps a boid away from its crowding neighbors so that they do not collide. The alignment force orients the boid towards the average direction of the neighboring boids. Finally, the cohesion force steers the boid towards the average position of the neighboring boids. For each boid in the system, these three driving forces are calculated separately and then a final velocity is calculated. In this way, a completely deterministic yet naturally looking crowd simulation is produced. Reynolds's approach has a number of extensions which incorporate other dynamic effects. For instance, Delgado-Mata et al. 0

study the effects of fear on flock behavior by extending the Reynolds model. Hemerlijk and Hildenbrandt 0 study flocks of birds. They propose a physically accurate model by incorporating the effects of fixed-wing aerodynamics into the basic Reynolds model.

Other than the simulation of herd behaviors, this work also focusses on computational fluid simulation. Almost all of the fluid simulation methods are based on a set of nonlinear differential equations called the Navier-Stokes equations. The analytical solution of this set of equations is available only for several, greatly simplified and constrained (i.e., incompressible fluids) cases. Numerical approximations are more common in the field of Computer Graphics. A number of simplifications are performed on the Navier-Stokes equations to approximate the fluid behavior in real-time. Stam propose a stable animation model for fluid-like objects 0. Fedkiw et al. propose a method for visual simulation of smoke 0. Later, Stam extend his previous model to cover real-time applications such as video games 0. Foster and Fedkiw solve the Navier-Stokes equations efficiently using an adaptation of a semi-Lagrangian method and produce a high-quality implicit surface obtained from the velocity field of the liquid to represent the fluid surface 0. For a comprehensive survey of common fluid simulation techniques for Computer Graphics, the reader is referred to the book that by Bridson 0.

III. The Proposed Approach

Our framework consists of three different components. The first component provides the interface between the user and the system. The second component is a fast and stable fluid solver based on the approach developed by Stam 0. The third component is the flock simulator and it is responsible for driving the autonomous agents (fish in particular) around. In order to create flocking behavior in a fluid environment, we combine these three components together, as it is seen in Figure 1.

To represent fluid velocity in the environment, we keep a three dimensional grid, where each grid cell has an associated fluid velocity. We call this grid as the cell-wise velocity field and denote it with the function $\vec{v}\left(\vec{i},\ t\right)$. $\vec{i}$ is a vector that specifies the grid index in three dimensions and t is the simulation time. For each cell, the associated velocity represents the overall movement of the fluid at the center. The cell-wise velocity field is common to all three components and can be influenced by adding (or injecting) additional velocities. To compute $\vec{v}\left(\vec{i},\ t\right)$, we use a three-dimensional version of the Stam's approach 0. The extension of the Stam's approach to three dimensions is trivial.

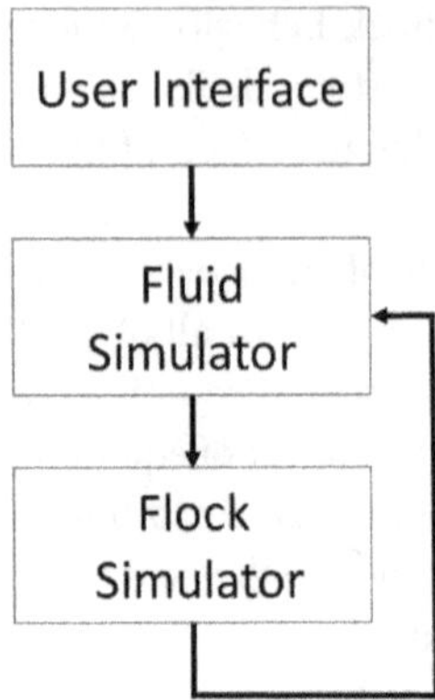

Fig. 1. The flow chart demonstrating the system components.

The flow chart given in Figure 1 demonstrates the main dependencies among the three components. For each frame, user input is gathered from the Kinect controller 0, and then appropriate movement information is fed to the fluid simulator. With the user input, fluid simulator updates its state. Then, agents in the simulation environment are simulated with respect to flocking rules and fluid motion. The following sections explain these three components in detail.

A. *User Interface*

The user interaction in our framework involves the localization of the user's hand position within the simulation environment. For simplicity, we represent the hand as a semi-transparent spherical object centered at the hand's detected position. This object is also called the *Velocity Generator.* We determine the position of the *Velocity Generator* using the functions provided by the Microsoft Kinect API 0. The *Velocity Generator* tracks its own velocity across successive frames. At each frame, we sample random points inside the boundary of the *Velocity Generator.* Then, we use these points to inject velocity into the cell-wise velocity field. To do so, we determine the grid cells that contain these sampled points. We then add velocities to these corresponding grid cells. After we inject velocities to the fluid, we advance the simulation time by a predetermined time step. Then, the fluid simulator (see Section 3.2) updates the velocity field for the new time step.

B. *Fluid Simulator*

To compute the fluid velocity at an arbitrary position, we keep a continuous velocity field that is separate from the cell-wise velocity field. We denote the continuous velocity field with the vector function $\vec{v}_f(\vec{x}, t)$, where t is the simulation time and $\vec{x}$ is an arbitrary position vector within the field. To compute $\vec{v}_f(\vec{x}, t)$ at some $\vec{x}$, we first determine the cell that contains $\vec{x}$. Then, the index vectors $(\vec{i})$ of the eight local neighbors of this cell are computed separately. For each of the eight neighbors, we retrieve the velocities from the cell-wise velocity field. Finally, we use trilinear interpolation to compute fluid velocity at $\vec{x}$. Having computed the cell-wise velocity field, the control is passed to the Flock Simulator to simulate the flocking behavior.

C. *Flock Simulator*

The flock simulator uses flocking rules proposed by Reynolds 0. In addition to basic alignment, cohesion and separation forces, we also have boundary forces that make the fish stay in the simulation environment (please note that fish and boid are used interchangeably in the following parts). In order to achieve real-time performance, we keep a secondary grid structure (different from the cell-wise velocity field) to cache the positions of the boids in the simulation space. This grid structure significantly speeds up the neighborhood search calculations. Figure 2 depicts the secondary grid structure that we use for neighborhood search. Additionally, threading is used for the movement of boids. Each boid is assigned to a particular thread for processing.

At the beginning, the flock simulator computes the accelerations of the boids with respect to the Reynold's rules. We use a simplified model to compute boid acceleration which is given in Equation 1. For simplicity, we assume that the boids' mass can be omitted. Hence, we drop the mass out of the equation for motion. In Equation 1, the vectors $\vec{f}_a(m)$, $\vec{f}_c(m)$, and $\vec{f}_s(m)$ correspond to the alignment, cohesion and separation forces, respectively. These forces are computed separately for each boid (with index m) using the common formulation proposed by Reynolds [1]. The coefficients ω_a, ω_c, and ω_s control the magnitudes of the three components, respectively. We tune these coefficients empirically to obtain a natural looking simulation.

$$\vec{a}(m, t) = \omega_a \vec{f}_a(m) + \omega_c \vec{f}_c(m) + \omega_s \vec{f}_s(m) + \vec{f}_o(m)$$

The only additional force we introduce is the obstacle force $(\vec{f}_o(m))$, which keeps the boids away from the aquarium boundaries and the terrain. The obstacle response force is computed using Equation 2.

$$\vec{f}_o(m) = \frac{sgn(\delta_x(m))\,\vec{e}_x}{|\delta_x(m)| + h} + \frac{sgn(\delta_y(m))\,\vec{e}_y}{|\delta_y(m)| + h} + \frac{sgn(\delta_z(m))\,\vec{e}_z}{|\delta_z(m)| + h}$$

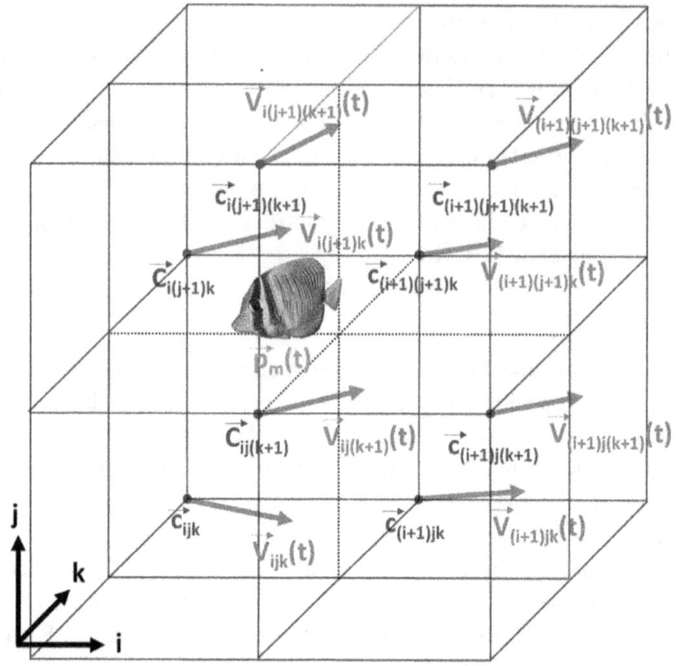

Fig. 2. The grid structure used for neighborhood search.

In Equation 2, $\vec{e}_x$, $\vec{e}_y$, and $\vec{e}_z$ represent the basis vectors pointing at right (x), up (y) and forward (z) directions, respectively. *sgn* represents the sign function. A response force is activated only when a boid moves close to any of the six boundaries along the three directions. $\delta_x(m)$, $\delta_y(m)$, and $\delta_z(m)$ are scalars that store the difference between the boid's position and the closest boundary for the respective direction. For any of the six boundaries along the three directions, if the boid's distance to the boundary is greater than a threshold, then the respective direction component is dropped from the equation. Finally, h is the parameter used to set an upper bound for the obstacle avoidance force. The magnitude of the obstacle avoidance force along any direction is never higher than $1/h$.

For simulation environments where we use a terrain at the bottom of the aquarium, we slightly modify the obstacle response force along the up direction. We represent the terrain as a height map. In this case, the upward difference $\delta_y(m)$ represents the difference between the position of the boid and the height map (at the boid's position) along with the up direction.

We then compute an intermediate boid velocity at the current simulation time by summing up the boid's current velocity and acceleration (Equation 3). If the speed of boid exceeds the maximum permissible speed, we clamp the velocity such that its magnitude becomes the maximum speed.

$$\overrightarrow{vI}\,(m,\ t) = \vec{v}\,(m,\ t) + \vec{a}\,(m,\ t)$$

Each boid is affected by the fluid around it. For simplicity, we directly displace the boid with some portion of the fluid velocity vector at its position. The fluid velocity at the boid's position is determined from the continuous velocity field ($\vec{v}_f\,(\vec{x},\ t)$). In our test runs, we observed that, just displacing the boids with respect to the fluid velocity around it creates rather synthetic movement. To overcome this problem, we add some other portion of the fluid velocity to the boid itself. Therefore, boid movement under the fluid force can be formulated using Equation 4.

$$\vec{p}\,(m,\ t+1) = \vec{p}\,(m,\ t) + \overrightarrow{vI}\,(m,\ t) + (1-\alpha)\,\vec{v}_f\,(\vec{p}\,(m,\ t)\,,\ t)$$

$$\vec{v}\,(m,\ t+1) = \overrightarrow{vI}\,(m,\ t) + \ \alpha\vec{v}_f\,(\vec{p}\,(m,\ t)\,,\ t)$$

In Equation 4, α is an experimental weight coefficient. Increasing α creates the illusion of stronger fluid behavior where boids are dragged along the velocity field of the fluid. The displacement does not affect the direction of the boids, thus, the dragging effect is more pronounced as it should be. We use spherical linear interpolation 0 to slowly update the directions of the boids in such a way that they are aligned with their velocity vectors.

At any instance, the user can interact with the system by relocating the velocity generator. When such an interrupt occurs, the system schedules a user interface update for the next time step. At the beginning of the next time step, the control is delivered to the user interface component and the new velocities are added to the cell-wise velocity field, as described in Section 3.1.

IV. Results

Using the methods described above, we developed a multi-threaded sample application using the Unity 3D Framework 0. Users can interact with the system via either mouse or the Kinect Controller 0. Both controllers are used to localize the velocity generator. Several snapshots from the application are presented in Figure 3. Each of the snapshots are taken during run-time and they depict different flocking behaviour, including group formation and vortex movement.

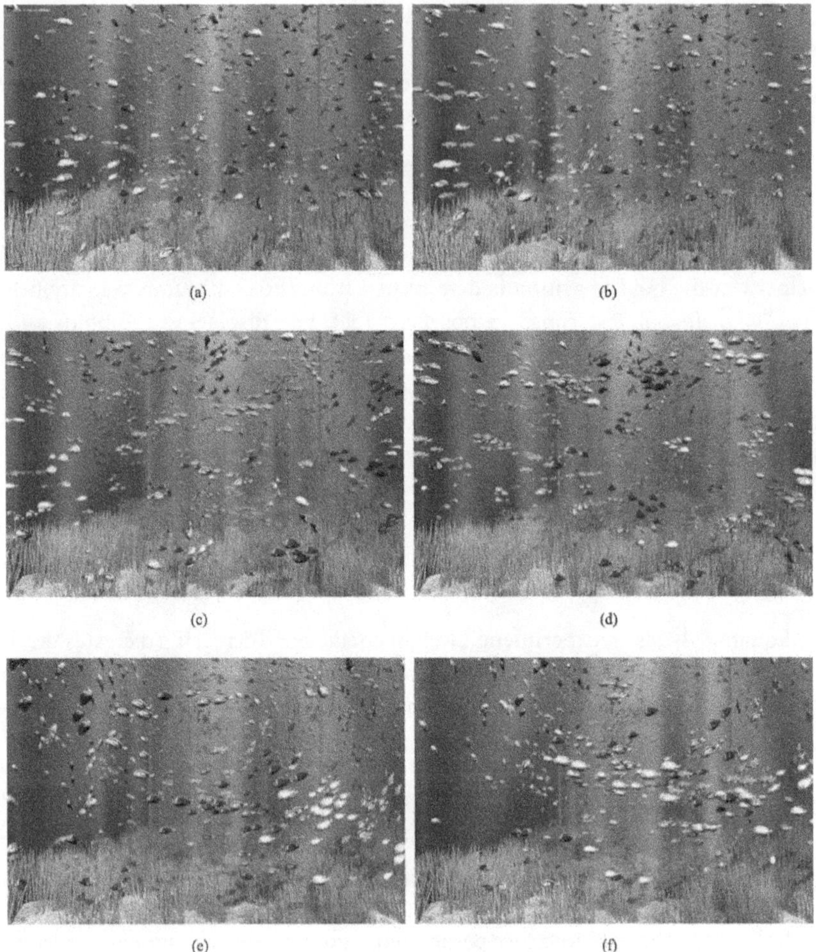

Fig. 3. Snapshots taken during run-time from the application. (a) and (b): The first two still frames show the school of fish where the groups are not formed yet and the flocking behaviour is hardly visible. This correspond to low cohesion (ω_c) and alignment (ω_a) and high separation (ω_s) coefficients. As the simulation proceeds, groups are formed and fish start to demonstrate flocking behaviour, which corresponds to high cohesion (ω_c) and alignment (ω_a) and low separation (ω_s) coefficients. The alignment and cohesion forces between fish in different subgroups (shown in different colors) are not taken into account. In this way, separation behavior will be dominant between fish in different subgroups and fish in the same subgroups form their own schools. (e) and (f): still frames demonstrating the influence of fluid motion on the flock. The user created a vortex by doing circular hand motion around the center of the aquarium. The vortex motion also directs the boids accordingly.

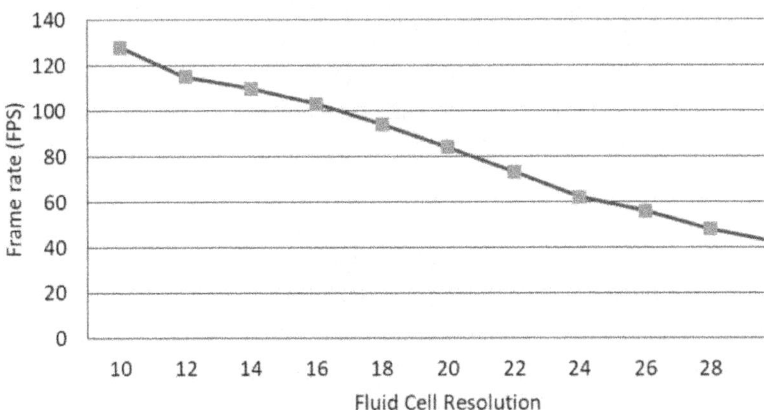

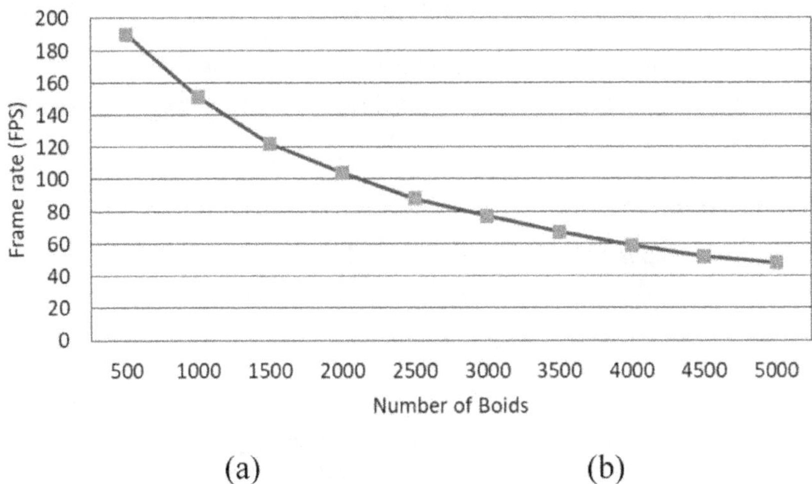

(a) (b)

Figure 4. The computational performance of the application in terms of frame rates (frames per second-FPS) with respect to different parameters. (a) The frame rates with respect to the resolution of the fluid velocity field. (b) The frame rates with respect to the number of boids in the system.

The performance of the system is bound to two parameters. The number of boids in the scene and the number of grid cells of the fluid solver. We tested our application on a workstation with the following hardware configuration: 8 cores, 2.8 GHz, 16 GB of RAM and Quadro K3000 GPU. The graphs given in Figure 4 demonstrate the system performance with respect to the fluid cell resolution and the number of boids in the system.

V. Conclusion

In this study, we demonstrate a framework where boids, specifically fish, and the fluid interact with each other. We have primarily focused on visually appealing school of fish simulation. The physical correctness of the simulation was a secondary concern. Our application runs at real-time frame rates for thousands of fish. The approach we have taken is simple to implement and it is suitable to be used in modern video games or other computer graphics applications. The application provides immersive user experience for the simulations of schools of fish in water where fish are interacting with each other and with the environment. We also provide a hand-gesture based interface to interact with the fluid. The user can manipulate the velocity field of the fluid with the help of a depth camera controller. In this way, hand gestures manipulate the school of fish by changing the directions and velocities of fish in water.

Acknowledgements

This work is supported by The Scientific and Technological Research Council of Turkey (TÜBİTAK) under Grant No. 112E110.

References

[1] Jamin Warren, "In Assassin's Creed Unity, the crowd is the main character", *Kill Screen*, Available at https://killscreen.com/articles/assassins-creed-unity-making-crowd-main-character/, Access date: 24 May 2019.

[2] Craig. W. Reynolds, "Flocks, herds and schools: A distributed behavioral model," *ACM Computer Graphics (*Proceedings of SIGGRAPH'87), vol. 21, no. 4, pp. 25–34, 1987.

[3] Microsoft, Inc., "Xbox 360 + Kinect, XBOX," http://www.xbox.com/en-US/kinect, 2014.

[4] Carlos Delgado-Mata, Jesus Ibanez Martinez, Simon Bee, Rocio Ruiz-Rodarte, and Ruth Aylett, "On the use of virtual animals with artificial fear in virtual environments," *New Generation Computing*, vol. 25, no. 2, pp. 145–169, 2007.

[5] Charlotte K. Hemelrijk and Hanno Hildenbrandt, "Some causes of the variable shape of flocks of birds," *PloS One*, vol. 6, no. 8, article no. e22479, 13 pages, 2011.

[6] Jos Stam, "Stable fluids," in *Proceedings of the 26th Annual Conference on Computer Graphics and Interactive Techniques*, ser. SIGGRAPH '99. New York, NY, USA: ACM Press/Addison-Wesley Publishing Co., 1999, pp. 121–128.

[7] Ronald Fedkiw, Jos Stam, and Henrik Wann Jensen, "Visual simulation of smoke," in *Proceedings of the 28th Annual Conference on Computer Graphics and Interactive Techniques*, ser. SIGGRAPH '01. New York, NY, USA: ACM, 2001, pp. 15–22.

[8] Jos Stam, "Real-time fluid dynamics for games," in *Proceedings of the Game Developer Conference*, vol. 18, 2003, p. 25.

[9] Nick Foster and Ronald Fedkiw, "Practical animation of liquids," in *Proceedings of the 28th Annual Conference on Computer Graphics and Interactive Techniques*", ser. SIGGRAPH '01. New York, NY, USA: ACM, 2001, pp. 22–30.

[10] Robert Bridson, *Fluid Simulation for Computer Graphics*. A. K. Peters Ltd., 2008.

[11] Ken Shoemake, "Animating Rotation with Quaternion Curves," *Proceedings of the 12th Annual Conference on Computer Graphics and Interactive Techniques*", ser. SIGGRAPH '85. New York, NY, USA: ACM, 1985, pp. 245–254.

[12] Unity Technologies, "Unity 3d," http://unity3d.com, 2014.

www.ingramcontent.com/pod-product-compliance
Lightning Source LLC
Chambersburg PA
CBHW020909210326
41598CB00018B/1817